ATLAS
A-Z

 Penguin
Random
House

FOR THE SIXTH EDITION
Senior Cartographic Editor Simon Mumford **Designer** Nimbus Design
Editors Cambridge International Reference on Current Affairs (CIRCA)
3–D Globes Planetary Visions Ltd., London
Senior Producer Mandy Inness **Producer, Pre-Production** Nikoleta Parasaki

Publisher Andrew Macintyre **Publishing Director** Jonathan Metcalf
Associate Publishing Director Liz Wheeler **Art Director** Philip Ormerod

FOR PREVIOUS EDITIONS
Cartographic Director Andrew Heritage
Cartography Roger Bullen, Rob Stokes, Iorwerth Watkins
Project Editor Sam Atkinson **Art Editor** Karen Gregory

First American Edition, 2001. This revised Edition, 2015
Published in the United States by DK Publishing,
345 Hudson Street, New York, New York 10014

Published in Great Britain by Dorling Kindersley Limited.

A catalog record for this book is available from the Library of Congress.
ISBN 978-1-4654-2985-8

DK books are available at special discounts when purchased in bulk for sales promotions,
premiums, fund-raising, or educational use. For details, contact: DK Publishing Special Markets,
345 Hudson Street, New York, New York 10014 or SpecialSales@dk.com.

Printed and bound in Hong Kong

A WORLD OF IDEAS:
SEE ALL THERE IS TO KNOW

www.dk.com

Key to map symbols

ELEVATION

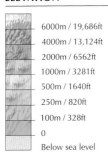

6000m / 19,686ft
4000m / 13,124ft
2000m / 6562ft
1000m / 3281ft
500m / 1640ft
250m / 820ft
100m / 328ft
0
Below sea level

▲ Mountain

• Depression

BORDERS

Full international

Disputed *de facto*

Territorial claim

Cease-fire line

Undefined

State/Province

DRAINAGE FEATURES

River

Seasonal river

Canal

Lake

Seasonal lake

SETTLEMENTS

● Capital city

◎ Major town

○ Minor town

● Major port

COMMUNICATIONS

Major road

Rail

✈ International airport

◆ Insight; facts, figures, and amazing information from around the world

4

Atlas contents

North & Central America 16–17

South America 38–39

Africa 50–51

Europe 62–63

Atlas contents

Factfile contents

The Political World

KEY TO NUMBERS
1. Germany
2. Liechtenstein
3. Czech Republic
4. Austria
5. Slovakia
6. Hungary
7. Slovenia
8. Croatia
9. Bosnia & Herzegovina
10. Serbia
11. Montenegro
12. Kosovo *(disputed)*
13. San Marino
14. Vatican City

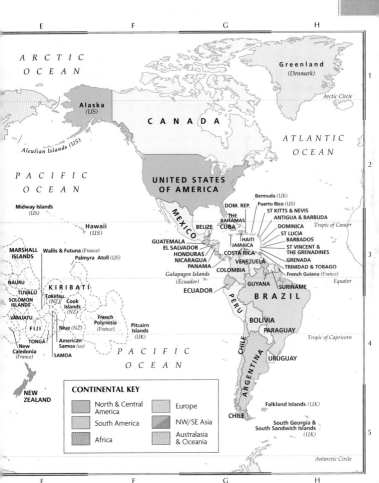

ARCTIC OCEAN

Greenland *(Denmark)*

Arctic Circle

Alaska *(US)*

CANADA

Aleutian Islands *(US)*

ATLANTIC OCEAN

PACIFIC OCEAN

UNITED STATES OF AMERICA

Midway Islands *(US)*

Bermuda *(UK)*
Puerto Rico *(US)*
ST KITTS & NEVIS
ANTIGUA & BARBUDA

Hawaii *(US)*

MEXICO
DOM. REP.
THE BAHAMAS
BELIZE   CUBA
DOMINICA
ST LUCIA
BARBADOS

Tropic of Cancer

MARSHALL ISLANDS
Wallis & Futuna *(France)*
Palmyra Atoll *(US)*

GUATEMALA   HAITI
EL SALVADOR   JAMAICA
HONDURAS
NICARAGUA   COSTA RICA
PANAMA   VENEZUELA
ST VINCENT & THE GRENADINES
GRENADA
TRINIDAD & TOBAGO
French Guiana *(France)*

NAURU
TUVALU
SOLOMON ISLANDS
KIRIBATI
Tokelau *(NZ)*
Cook Islands *(NZ)*
Galapagos Islands *(Ecuador)*
COLOMBIA

VANUATU
French Polynesia *(France)*
ECUADOR
GUYANA
SURINAME

FIJI
Niue *(NZ)*
Pitcairn Islands *(UK)*
PERU
BRAZIL

Equator

TONGA
American Samoa *(us)*
New Caledonia *(France)*
SAMOA
BOLIVIA
PARAGUAY

Tropic of Capricorn

PACIFIC OCEAN

CHILE
ARGENTINA
URUGUAY

NEW ZEALAND

CONTINENTAL KEY

North & Central America

South America

Africa

Europe

NW/SE Asia

Australasia & Oceania

Falkland Islands *(UK)*

CHILE

South Georgia & South Sandwich Islands *(UK)*

Antarctic Circle

E   F   G   H

The Physical World

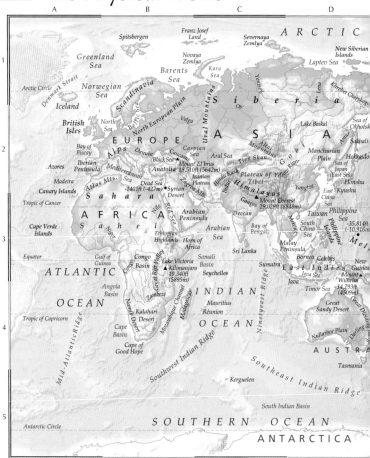

Spitsbergen
Franz Josef Land
Severnaya Zemlya
ARCTIC
New Siberian Islands

Greenland Sea
Novaya Zemlya
Laptev Sea

Norwegian Sea
Barents Sea
Kara Sea
Khrebet Cherskogo

Arctic Circle
Denmark Strait
Yenisey
Lena
Sea of Okhotsk

Iceland
Scandinavia
Baltic Sea
Siberia
Amur
Sakhali

North Sea
Ob'
Lake Baikal

British Isles
EUROPE
Ural Mountains
ASIA
Manchurian Plain
Hokkaido

Bay of Biscay
Alps
Danube
Volga
Caspian Sea
Altai Mountains
Gobi
Sea of Japan (East Sea)
Honshu

Azores
Iberian Peninsula
Mediterranean Sea
Caucasus
Black Sea
Aral Sea
Tien Shan
Yellow River
East China Sea
Kyushu

Madeira
Mount El'brus 18,510ft (5642m)
Iranian Plateau
Hindu Kush
Plateau of Tibet
Yangtze

Canary Islands
Atlas Mts.
Dead Sea -1401ft (-427m)
Zagros Mts.
Syrian Desert
Himalayas
Mount Everest 29,029ft (8848m)
Taiwan
Philippine Islands
-35,814ft (-10,916m)

Tropic of Cancer
Sahara
Anatolia
Ganges
Deccan
South China Sea
Philippine Sea

Cape Verde Islands
AFRICA
Nile
Red Sea
Arabian Peninsula
Bay of Bengal
Mel

Sahel
Ethiopian Highlands
Horn of Africa
Arabian Sea
Sri Lanka
Malay Peninsula
Borneo
New Guinea

Equator
Gulf of Guinea
Congo
Congo Basin
Great Rift Valley
Lake Victoria
Somali Basin
Seychelles
Sumatra
Celebes
East Indies
Mount Wilhelm 14,793ft (4509m)

ATLANTIC
Angola Basin
Zambezi
Kilimanjaro 19,340ft (5895m)
INDIAN
Java Sea
Java
Timor Sea
Great Sandy Desert

OCEAN
Namib Desert
Kalahari Desert
Mozambique Channel
Madagascar
Mauritius
Réunion
OCEAN
Ninetyeast Ridge
Nullarbor Plain
AUSTRA
Darling

Tropic of Capricorn
Cape Basin
Cape of Good Hope
Southwest Indian Ridge
Southeast Indian Ridge
Tasmania

Mid-Atlantic Ridge
Kerguelen
South Indian Basin

Antarctic Circle
SOUTHERN
OCEAN

ANTARCTICA

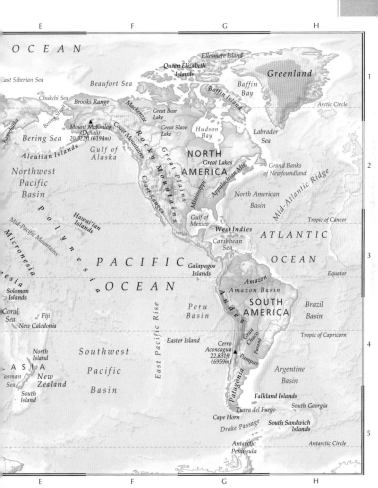

E F G H

OCEAN

1

Ellesmere Island

Queen Elizabeth
Islands

Greenland

East Siberian Sea

Beaufort Sea

Baffin
Bay

Baffin Island

Chukchi Sea

Bering Strait

Brooks Range

Mackenzie

Great Bear
Lake

Arctic Circle

Kamchatka

Mount McKinley
(Denali)
20,322ft (6194m)

Coast Mountains

Rocky Mountains

Great Slave
Lake

Hudson
Bay

Labrador
Sea

2

Bering Sea

Aleutian Islands

Gulf of
Alaska

Great Plains

NORTH
AMERICA

Great Lakes

Appalachian Mts.

Grand Banks
of Newfoundland

Northwest
Pacific
Basin

Coast Ranges

Mississippi

North American
Basin

Mid-Atlantic Ridge

Polynesia

Hawaiian
Islands

Gulf of
Mexico

West Indies

Tropic of Cancer

Mid-Pacific Mountains

Caribbean
Sea

ATLANTIC

Micronesia

3

PACIFIC

Galapagos
Islands

OCEAN

esia

OCEAN

Amazon

Equator

Andes

Amazon Basin

SOUTH
AMERICA

Solomon
Islands

Peru
Basin

Brazil
Basin

Coral
Sea

Fiji

New Caledonia

East Pacific Rise

Gran
Chaco

4

North
Island

Southwest

Easter Island

Cerro
Aconcagua
22,831ft
(6959m)

Pampas

Tropic of Capricorn

ASIA

New
Zealand

Pacific

Argentine
Basin

Tasman
Sea

South
Island

Basin

Patagonia

Falkland Islands

South Georgia

Tierra del Fuego

Cape Horn

South Sandwich
Islands

Drake Passage

5

Antarctic
Peninsula

Antarctic Circle

E F G H

Standard Time Zones

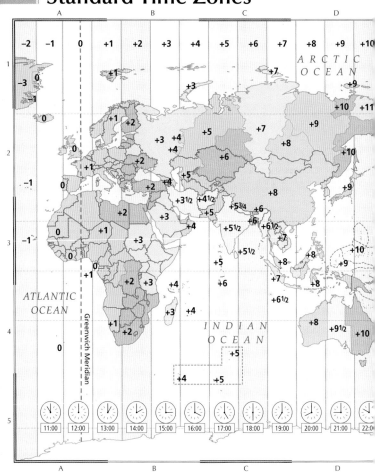

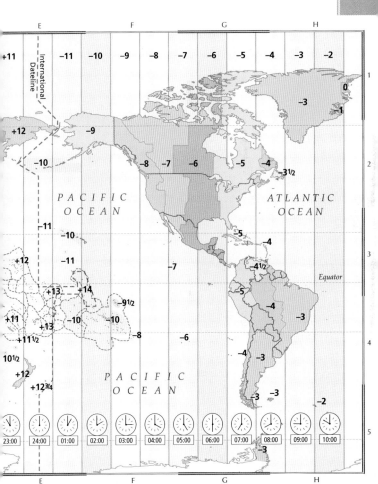

The world's regions

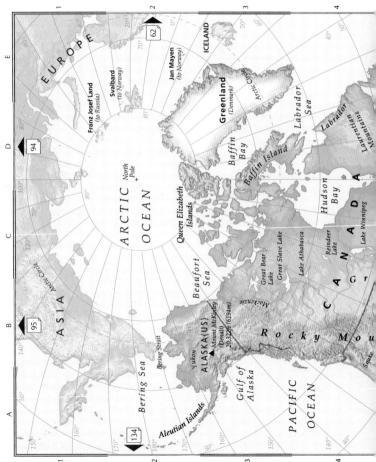

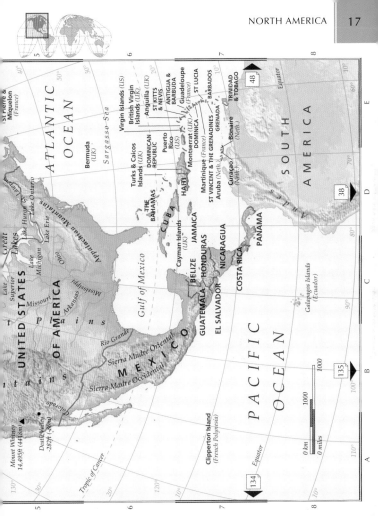

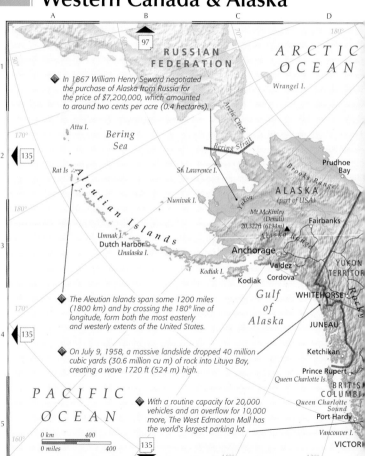

In 1867 William Henry Seward negotiated the purchase of Alaska from Russia for the price of $7,200,000, which amounted to around two cents per acre (0.4 hectares).

The Aleutian Islands span some 1200 miles (1800 km) and by crossing the 180° line of longitude, form both the most easterly and westerly extents of the United States.

On July 9, 1958, a massive landslide dropped 40 million cubic yards (30.6 million cu m) of rock into Lituya Bay, creating a wave 1720 ft (524 m) high.

With a routine capacity for 20,000 vehicles and an overflow for 10,000 more, The West Edmonton Mall has the world's largest parking lot.

RUSSIAN FEDERATION

ARCTIC OCEAN

Wrangel I.

Bering Strait

Arctic Circle

Attu I.

Bering Sea

Rat Is

Aleutian Islands

St. Lawrence I.

Nunivak I.

Yukon

Prudhoe Bay

Brooks Range

ALASKA (part of USA)

Mt McKinley (Denali) 20,322ft (6194m)

Alaska Range

Fairbanks

Umnak I.
Dutch Harbor
Unalaska I.

Kodiak I.

Kodiak

Anchorage

Valdez
Cordova

Gulf of Alaska

YUKON TERRITORY

WHITEHORSE

Rocky

JUNEAU

Ketchikan

Prince Rupert
Queen Charlotte Is.

BRITISH COLUMBIA

Queen Charlotte Sound

Port Hardy

PACIFIC OCEAN

0 km   400
0 miles   400

Vancouver I.

VICTOR

97

135

135

135

E   160°   140°   120°   100°   80°   60°   80°   60°

◆ Sought by explorers for centuries as a trade route between Europe and Asia, the famous Northwest Passage is now often navigable during the summer months without the need for an icebreaker because of reduced volumes of sea ice.

64

Greenland
(Danish external territory)

◆ Despite an area of 787,155 sq miles (2,038,722 sq km), the northerly province of Nunavut has just 530 miles (850 km) of roads with only around 4000 vehicles registered in the entire territory.

Ellesmere Island

Axel Heiberg Island

Queen Elizabeth Islands

Bathurst I.

Melville Island

Devon Island

Resolute
(Qausuittuq)

Lancaster Sound

Baffin Bay

Davis Strait

64

Viscount Melville Sound

Somerset Island

Banks Island

Prince of Wales I.

Baffin Island

Beaufort Sea

Amundsen Gulf

Victoria Island

King William I.

Arctic Circle

IQALUIT
(Frobisher Bay)

Inuvik

Kugluktuk
(Coppermine)

N U N A V U T

Hudson Strait

Great Bear Lake

NORTHWEST

Southampton I.

TERRITORIES

YELLOWKNIFE

Mackenzie

Great Slave Lake

Dubawnt

Rankin Inlet

Hudson Bay

QUÉBEC

Hay River

Fort Smith

Churchill

20

ALBERTA

Fort McMurray

Lake Athabasca

SASKATCHEWAN

MANITOBA

Fort St. John

C A N A D A

rince George

Grande Prairie

Flin Flon

Thompson

ONTARIO

EDMONTON

Saskatchewan

Leduc

Red Deer

Prince Albert

Saskatoon

Yorkton

Lake Winnipeg

WINNIPEG

◆ Only just over 1% of Canada's 3.5 million sq miles (9.1 million sq km) land area is devoted to grain production, yet this yields around 25 million tons (tonnes) of wheat every year.

amloops

Kelowna

Vancouver

Calgary

REGINA

Brandon

Estevan

25

Lethbridge

U S A

E   F   F   100°   G   H   80°

4000

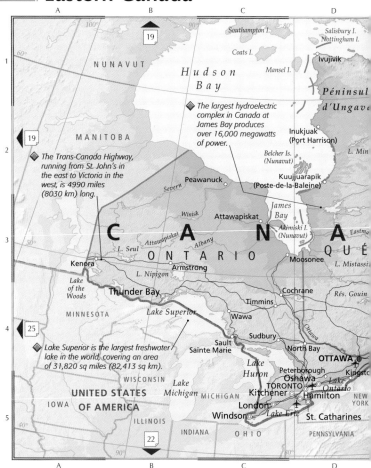

The largest hydroelectric complex in Canada at James Bay produces over 16,000 megawatts of power.

The Trans-Canada Highway, running from St. John's in the east to Victoria in the west, is 4990 miles (8030 km) long.

Lake Superior is the largest freshwater lake in the world, covering an area of 31,820 sq miles (82,413 sq km).

Southampton I.
Coats I.
Mansel I.
Salisbury I.
Nottingham I.
Ivujivik
Hudson Bay
Péninsul d'Ungave
NUNAVUT
MANITOBA
Inukjuak (Port Harrison)
L. Min
Belcher Is. (Nunavut)
Peawanuck
Kuujjuarapik (Poste-de-la-Baleine)
Severn
James Bay
C A N A
Winisk
Attawapiskat
Akimiski I. (Nunavut)
Eastma
QUÉ
Attawápiskat
L. Seul
Albany
O N T A R I O
Moosonee
L. Mistassi
Kenora
Armstrong
L. Nipigon
Rés. Gouin
Lake of the Woods
Thunder Bay
Cochrane
MINNESOTA
Lake Superior
Timmins
Wawa
Ottawa
Sudbury
Sault Sainte Marie
North Bay
OTTAWA
WISCONSIN
Lake Huron
Peterborough
Oshawa
Kingsto
IOWA
Lake Michigan
MICHIGAN
TORONTO
Lake Ontario
UNITED STATES OF AMERICA
Kitchener
Hamilton
London
NEW YORK
ILLINOIS
Windsor
Lake Erie
St. Catharines
INDIANA
OHIO
PENNSYLVANIA

19
19
25
22

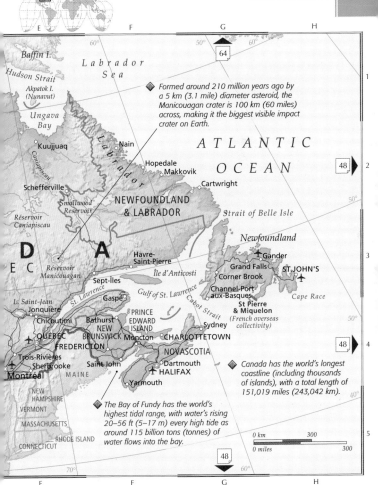

Formed around 210 million years ago by a 5 km (3.1 mile) diameter asteroid, the Manicouagan crater is 100 km (60 miles) across, making it the biggest visible impact crater on Earth.

Canada has the world's longest coastline (including thousands of islands), with a total length of 151,019 miles (243,042 km).

The Bay of Fundy has the world's highest tidal range, with water's rising 20–56 ft (5–17 m) every high tide as around 115 billion tons (tonnes) of water flows into the bay.

NORTH AMERICA
USA: The Northeast

MINNESOTA

95° 90° 85° 80°

20

1

Lake Superior

C A

ONTARIO

Superior

Ironwood

Marquette

Sault Ste Marie

45°

25

2

Ladysmith

Iron Mountain

Cheboygan

Lake Huron

WISCONSIN

MICHIGAN

Eau Claire

Green Bay

Traverse City

La Crosse

Lake Michigan

Bay City

Oshkosh

IOWA

MADISON

Grand Rapids

Saginaw

Flint

Milwaukee

LANSING

Waukegan

Detroit

Lake Erie

Erie

◆ The Chicago River
originally flowed into
Lake Michigan, but was
reversed in 1900 by
the completion of
a canal.

Rockford

Chicago

Ann Arbor

Aurora

Joliet

South Bend

Toledo

Cleveland

Rock Island

Gary

Youngstown

Galesburg

Peoria

Fort Wayne

Akron

Mansfield

Canton

Wheeling

ILLINOIS

INDIANA

OHIO

40°

Champaign

Muncie

SPRINGFIELD

INDIANAPOLIS

Dayton

COLUMBUS

25

4

Decatur

Cincinnati

MISSOURI

Effingham

Terre Haute

Huntington

East St Louis

Bloomington

CHARLESTON

◆ Many US freight trains are
over 2 miles (3.2 km) long,
made up of almost 200 cars,
and can take around 5 minutes
to pass a level-crossing.

Mt. Vernon

Louisville

FRANKFORT

Lexington

WEST VIRGINIA

Evansville

Richmond

Carbondale

Owensboro

London

KENTUCKY

5

Paducah

Hopkinsville

Bowling Green

ARKANSAS

30

A B C D

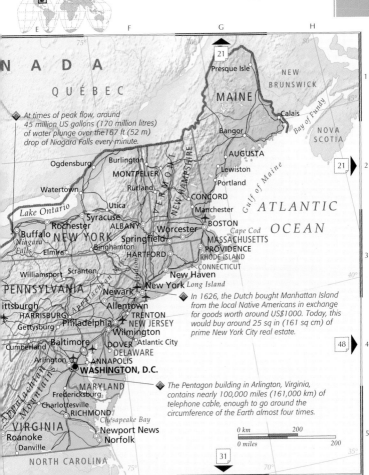

E   75°   F   70°   G   80°   H

C A N A D A

QUÉBEC

NEW BRUNSWICK

Presque Isle

MAINE

◆ At times of peak flow, around
45 million US gallons (170 million litres)
of water plunge over the 167 ft (52 m)
drop of Niagara Falls every minute.

Calais

Bay of Fundy

NOVA SCOTIA

Ogdensburg   Burlington   Bangor

AUGUSTA

Watertown   MONTPELIER   Lewiston   Gulf of Maine

Rutland   Portland

Lake Ontario   Utica   CONCORD   ATLANTIC

Syracuse   Manchester   OCEAN

Buffalo   Rochester   ALBANY   Worcester   BOSTON   Cape Cod

Niagara   NEW YORK   Springfield   MASSACHUSETTS

Falls   Elmira   Binghamton   HARTFORD   PROVIDENCE

Williamsport   Scranton   RHODE ISLAND

CONNECTICUT

PENNSYLVANIA   Newark   New Haven

New York   Long Island

ittsburgh   Allentown   ◆ In 1626, the Dutch bought Manhattan Island

HARRISBURG   TRENTON   from the local Native Americans in exchange
for goods worth around US$1000. Today, this

Gettysburg   Philadelphia   NEW JERSEY   would buy around 25 sq in (161 sq cm) of
prime New York City real estate.

Baltimore   Wilmington

Cumberland   DOVER   Atlantic City

Arlington   DELAWARE

ANNAPOLIS

WASHINGTON, D.C.   ◆ The Pentagon building in Arlington, Virginia,
contains nearly 100,000 miles (161,000 km) of

MARYLAND   telephone cable, enough to go around the
circumference of the Earth almost four times.

Fredericksburg

Charlottesville   0 km   200

RICHMOND   0 miles   200

VIRGINIA   Chesapeake Bay

Roanoke   Newport News

Danville   Norfolk

NORTH CAROLINA   75°   70°   35°

E   F   G   H

21

21

48

31

USA: Central States

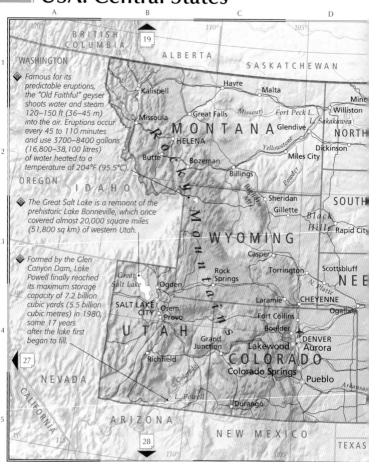

◆ Famous for its predictable eruptions, the "Old Faithful" geyser shoots water and steam 120–150 ft (36–45 m) into the air. Eruptions occur every 45 to 110 minutes and use 3700–8400 gallons (16,800–38,100 litres) of water heated to a temperature of 204°F (95.5°C).

◆ The Great Salt Lake is a remnant of the prehistoric Lake Bonneville, which once covered almost 20,000 square miles (51,800 sq km) of western Utah.

◆ Formed by the Glen Canyon Dam, Lake Powell finally reached its maximum storage capacity of 7.2 billion cubic yards (5.5 billion cubic metres) in 1980, some 17 years after the lake first began to fill.

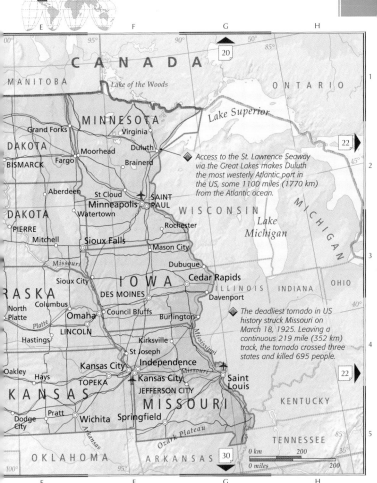

C A N A D A

MANITOBA

Lake of the Woods

ONTARIO

Lake Superior

MINNESOTA

Grand Forks

Virginia

Duluth

◆ Access to the St. Lawrence Seaway via the Great Lakes makes Duluth the most westerly Atlantic port in the US, some 1100 miles (1770 km) from the Atlantic ocean.

DAKOTA

Moorhead

Fargo

Brainerd

BISMARCK

St Cloud

SAINT

PAUL

Aberdeen

Minneapolis

WISCONSIN

DAKOTA

Watertown

Lake Michigan

PIERRE

Rochester

MICHIGAN

Mitchell

Sioux Falls

Mason City

Missouri

Dubuque

IOWA

Cedar Rapids

Sioux City

DES MOINES

ILLINOIS **INDIANA** **OHIO**

RASKA

Columbus

Council Bluffs

Davenport

North Platte

Omaha

Burlington

◆ The deadliest tornado in US history struck Missouri on March 18, 1925. Leaving a continuous 219 mile (352 km) track, the tornado crossed three states and killed 695 people.

Platte

LINCOLN

Hastings

Kirksville

St Joseph

Oakley

Kansas City

Independence

Mississippi

Hays

Kansas City

Missouri

Saint Louis

KANSAS

TOPEKA

JEFFERSON CITY

KENTUCKY

Pratt

MISSOURI

Dodge City

Wichita

Springfield

Arkansas

Ozark Plateau

TENNESSEE

OKLAHOMA **ARKANSAS**

0 km 200

0 miles 200

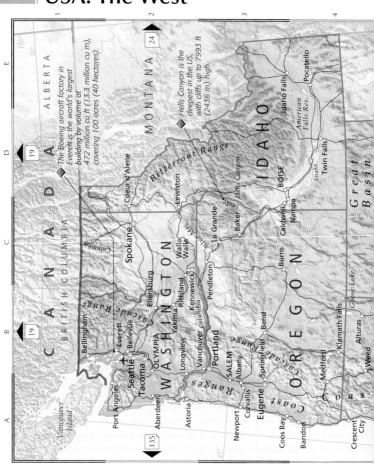

The Boeing aircraft factory in Everett is the world's largest building by volume at 472 million cu ft (13.3 million cu m), covering 100 acres (40 hectares)

Hells Canyon is the deepest in the US, with cliffs up to 7993 ft (2436 m) high.

CANADA

ALBERTA

BRITISH COLUMBIA

MONTANA

IDAHO

OREGON

WASHINGTON

Great Basin

Vancouver Island

Port Angeles
Bellingham
Everett
Bellevue
Seattle
Tacoma
OLYMPIA
Aberdeen
Astoria
Newport
Corvallis
Coos Bay
Bandon
Crescent City

Cascade Range
Coeur d'Alene
Spokane
Columbia
Ellensburg
Yakima
Longview
Vancouver
Portland
SALEM
Albany
Springfield
Eugene
Coast Ranges
Cascade Range

Lewiston
Walla Walla
Richland
Kennewick
Pendleton
La Grande
Baker
Bend
Burns
Medford
Klamath Falls
Alturas
Weed

Bitterroot Range
Salmon
Hells Canyon
Snake River

Idaho Falls
Pocatello
American Falls Res.
Twin Falls
BOISE
Caldwell
Nampa

Goose Lake

24

19

19

19

135

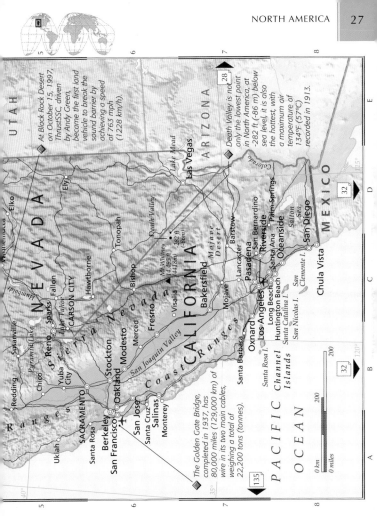

At Black Rock Desert on October 15, 1997, ThrustSSC, driven by Andy Green, became the first land vehicle to break the sound barrier by achieving a speed of 763 mph (1228 km/h).

Death Valley is not only the lowest point in North America, at -282 ft (-86 m) below sea level, it is also the hottest, with a maximum air temperature of 134°F (57°C) recorded in 1913.

The Golden Gate Bridge, completed in 1937, has 80,000 miles (129,000 km) of wire in its two main cables, weighing a total of 22,200 tons (tonnes).

UTAH

NEVADA

ARIZONA

CALIFORNIA

MEXICO

Sierra Nevada

Coast Ranges

Ranges

San Joaquin Valley

Mojave Desert

Death Valley

PACIFIC OCEAN

Channel Islands

Winnemucca
Elko
Eureka
Ely
Lake Mead
Colorado
Las Vegas
Tonopah
Hawthorne
Bishop
Mt. Whitney
14,495 ft
(4418m)
-282 ft
(-86m)
Barstow
Visalia
Bakersfield
Mojave
Lancaster
Palm Springs
San Bernardino
Pasadena
Riverside
Salton Sea
Santa Ana
Oceanside
San Diego
Chula Vista
Los Angeles
Long Beach
Huntington Beach
Oxnard
Santa Barbara
Santa Rosa I.
Santa Cruz I.
Santa Catalina I.
San Nicolas I.
San Clemente I.
Fresno
Merced
Modesto
Stockton
Oakland
Berkeley
San Francisco
San Jose
Santa Cruz
Salinas
Monterey
SACRAMENTO
Santa Rosa
Ukiah
Redding
Chico
Yuba City
Susanville
Sparks
Reno
Fallon
CARSON CITY
Lake Tahoe
Pyramid Lake
Humboldt

0 km   200
0 miles   200

28

32

32

135

32

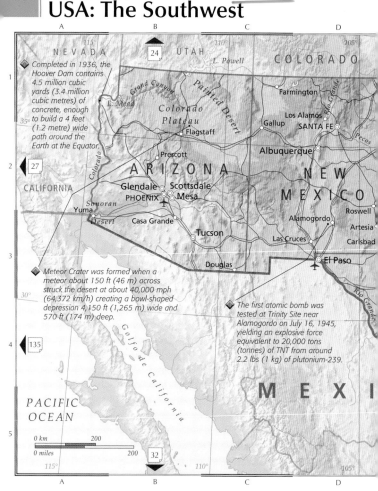

◆ Completed in 1936, the Hoover Dam contains 4.5 million cubic yards (3.4 million cubic metres) of concrete, enough to build a 4 feet (1.2 metre) wide path around the Earth at the Equator.

◆ Meteor Crater was formed when a meteor about 150 ft (46 m) across struck the desert at about 40,000 mph (64,372 km/h) creating a bowl-shaped depression 4,150 ft (1,265 m) wide and 570 ft (174 m) deep.

◆ The first atomic bomb was tested at Trinity Site near Alamogordo on July 16, 1945, yielding an explosive force equivalent to 20,000 tons (tonnes) of TNT from around 2.2 lbs (1 kg) of plutonium-239.

NEVADA
UTAH
COLORADO

L. Powell

Grand Canyon
Painted Desert
L. Mead
Colorado Plateau
Farmington
Los Alamos
Gallup
SANTA FE
Flagstaff
Albuquerque
Prescott

A R I Z O N A
N E W
Glendale Scottsdale
PHOENIX Mesa
M E X I C O
CALIFORNIA
Sonoran
Desert
Yuma
Roswell
Casa Grande
Alamogordo
Artesia
Tucson
Carlsbad
Las Cruces
Douglas
El Paso

Rio Grande
Pecos
Rio Grande

Golfo de California

PACIFIC
OCEAN

M E X I

0 km   200
0 miles   200

24
27
135
32

KANSAS

OKLAHOMA

Ponca City
Enid
Tulsa
Broken Arrow
OKLAHOMA CITY
Borger
Pampa
Norman
Shawnee
Amarillo
Canadian
Clovis
Lawton
Red River
Red River
Paris
Vernon
Wichita Falls
Lubbock
Denton
Brownfield
Fort Worth
Arlington
Longview
Hobbs
Abilene
Dallas
Tyler
Sweetwater
Jacksonville
Big Spring
Brazos
Toledo Bend Res.
Odessa
Midland
Waco
Neches
Pecos
San Angelo
Colorado
T E X A S
LOUISIANA
Bryan
Beaumont
Edwards
L. Travis
Port Arthur
Plateau
AUSTIN
Houston
Pasadena
San Antonio
Texas City
Galveston
Del Rio
Victoria
Freeport
Eagle Pass
San Antonio
Corpus Christi
Laredo
Kingsville
C O
Rio Grande
Padre Island
Brownsville

Gulf

of

Mexico

◆ The world's first parking
meter was installed in
Oklahoma City on
July 16, 1935.

◆ On January 10, 1901,
the Lucas Gusher blew
oil 150 ft (46 m) into the
air, flowing at 100,000
barrels a day until it was
eventually capped nine
days later.

◆ With winds estimated at over
145 mph (233 km/h), the 1900
Galveston Hurricane claimed over
8000 lives, making it the deadliest
natural disaster in US history.

ARKANSAS

25

30

30

33

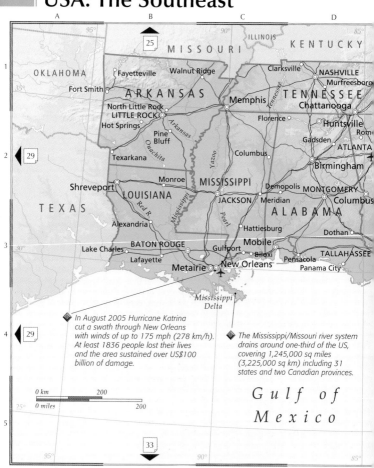

In August 2005 Hurricane Katrina cut a swath through New Orleans with winds of up to 175 mph (278 km/h). At least 1836 people lost their lives and the area sustained over US$100 billion of damage.

The Mississippi/Missouri river system drains around one-third of the US, covering 1,245,000 sq miles (3,225,000 sq km) including 31 states and two Canadian provinces.

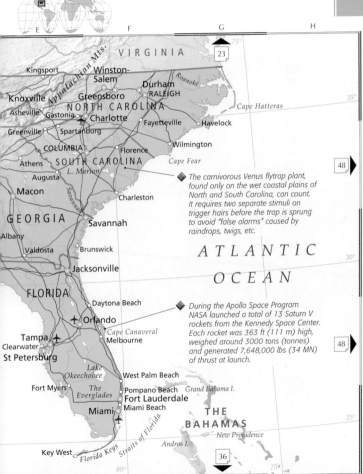

VIRGINIA

Kingsport   Winston-Salem

Knoxville   Greensboro   Durham   RALEIGH

Roanoke

Asheville   Gastonia   Charlotte   Fayetteville   Havelock

Greenville   Spartanburg

COLUMBIA   Florence   Wilmington

Athens   SOUTH CAROLINA   Cape Fear

Augusta   L. Marion

Macon   Charleston

GEORGIA   Savannah

Albany   Brunswick

Valdosta   Jacksonville

FLORIDA   Daytona Beach

Orlando   Cape Canaveral   Melbourne

Tampa   Clearwater   Lake Okeechobee

St Petersburg

Fort Myers   West Palm Beach

The Everglades   Pompano Beach   Grand Bahama I.

Miami   Fort Lauderdale   Miami Beach

Key West   Florida Keys   Straits of Florida   Andros I.   New Providence

Kingsport   Knoxville   Asheville   Greenville

NORTH CAROLINA

Cape Hatteras

ATLANTIC

OCEAN

THE BAHAMAS

The carnivorous Venus flytrap plant, found only on the wet coastal plains of North and South Carolina, can count. It requires two separate stimuli on trigger hairs before the trap is sprung to avoid "false alarms" caused by raindrops, twigs, etc.

During the Apollo Space Program NASA launched a total of 13 Saturn V rockets from the Kennedy Space Center. Each rocket was 363 ft (111 m) high, weighed around 3000 tons (tonnes) and generated 7,648,000 lbs (34 MN) of thrust at launch.

Appalachian Mts.

Savannah

23

48

48

36

E   F   G   H

Mexico

115°

110°

105°

Mexicali

Tijuana

Desierto de Altar

28

NEW MEXICO

ARIZONA

Ensenada

Ciudad Juárez

Río Grande

UNITED

30°

I. Ángel de la Guarda

Hermosillo

Chihuahua

Sierra

Sierra Madre

135

I. Cedros

Golfo de California

Conchos

M

Monclov

Baja California

Ciudad Obregón

Gómez Palacio

Torreón

Large examples of the Saguaro cactus, found in the Altar Desert, can take nearly 150 years to grow to their full height of around 45 ft (14 m), and can hold several tons (tonnes) of water.

25°

Los Mochis

Culiacán

E

Durango

La Paz

Fresnillo

Zacatecas

Tropic of Cancer

Mazatlán

Gray whales have one of the longest migrations of any mammal, traveling some 12,500 miles (20,000 km) every year from the Arctic Ocean to their winter breeding grounds in the Golfo de California.

Islas Marías

Tepic

Aguascaliente

20°

Guadalajara

Puerto Vallarta

L. de Chapala

P A C I F I C

O C E A N

Islas Revillagigedo (part of Mexico)

0 km   200

0 miles   200

135

The cliff divers of Acapulco must time their dive from the 148 ft (45 m) cliff at La Quebrada to coincide with the incoming swells to avoid being dashed on the rocks in the shallow inlet.

115°   110°   105°

A   B   C   D

STATES OF AMERICA

LOUISIANA

TEXAS

Rio Grande

Nuevo Laredo

In spring 2001, the Rio Grande stopped flowing into the Gulf of Mexico for the first time in recorded history, allowing illegal immigrants to simply walk into the US.

Reynosa

Matamoros

Monterrey

Saltillo

G u l f   o f

M e x i c o

Tropic of Cancer

Oriental

Ciudad Victoria

It is thought that "The Ballgame," a ritual sport played by Maya and Aztec civilizations, and a forerunner of volleyball, often ended with members of the losing team being sacrificed.

Cancún

San Luis Potosí

Tampico

Mérida

Isla Cozumel

Río Verde

Ciudad Valles

Yucatan Peninsula

Dolores Hidalgo

León

Poza Rica

Campeche

Querétaro

Tulancingo

Bahía de Campeche

Pachuca

Xalapa

MEXICO CITY

Veracruz

Morelia

Coatzacoalcos

Villahermosa

Cuernavaca

Puebla

ruapan

Minatitlán

BELIZE

Tehuacán

O

Balsas

GUATEMALA

Oaxaca

Tuxtla

Sierra Madre del Sur

Acapulco

Golfo de Tehuantepec

Tapachula

HONDURAS

EL SALVADOR

Central America

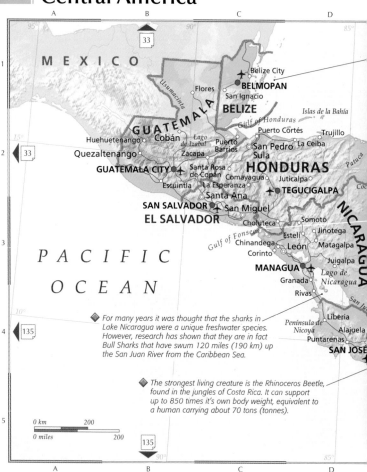

MEXICO

GUATEMALA

BELIZE

HONDURAS

EL SALVADOR

NICARAGUA

PACIFIC OCEAN

Belize City
BELMOPAN
Flores
San Ignacio
Huehuetenango
Cobán
Lago de Izabal
Puerto Barrios
GUATEMALA CITY
Quezaltenango
Escuintla
Santa Ana
SAN SALVADOR
Santa Rosa de Copán
La Esperanza
Comayagua
San Miguel
Islas de la Bahía
Puerto Cortés
Trujillo
La Ceiba
San Pedro Sula
TEGUCIGALPA
Juticalpa
Zacapa
Gulf of Honduras
Patuca
Choluteca
Somoto
Estelí
Jinotega
Chinandega
León
Matagalpa
Corinto
Juigalpa
Gulf of Fonseca
MANAGUA
Granada
Lago de Nicaragua
Rivas
Liberia
Península de Nicoya
Alajuela
Puntarenas
SAN JOSÉ

For many years it was thought that the sharks in Lake Nicaragua were a unique freshwater species. However, research has shown that they are in fact Bull Sharks that have swum 120 miles (190 km) up the San Juan River from the Caribbean Sea.

The strongest living creature is the Rhinoceros Beetle, found in the jungles of Costa Rica. It can support up to 850 times it's own body weight, equivalent to a human carrying about 70 tons (tonnes).

0 km   200
0 miles   200

The Great Blue Hole in Lighthouse Reef, a submerged cave some 1000 ft (303 m) in diameter and 400 ft (120 m) deep, was originally explored by Jacques Cousteau, co-inventor of the aqualung.

HAITI

Greater Antilles

JAMAICA

Islas Santanilla (part of Honduras)

Bajo Nuevo (part of Colombia)

Cayos Miskitos

Mosquito Coast

C a r i b b e a n

S e a

I. de Providencia (part of Colombia)

I. de San Andrés (part of Colombia)

Islas del Maíz

Bluefields

Each chamber at Gatun Locks on the Panama Canal is 110 ft (33 m) wide and 1000 ft (303 m) long. The locks took four years to build and required 2 million cubic yards (1.5 million cu m) of concrete.

COSTA RICA

Limón

Cartago

Cordillera de Talamanca

PANAMA

David   Penonomé

Colón

✛ **PANAMA CITY**

Panama Canal

Gulf of Darien

Golfo de Chiriquí

Santiago

Chitré

Las Tablas

Isla del Rey

Golfo de Panamá

COLOMBIA

36

36

40

40

The Caribbean

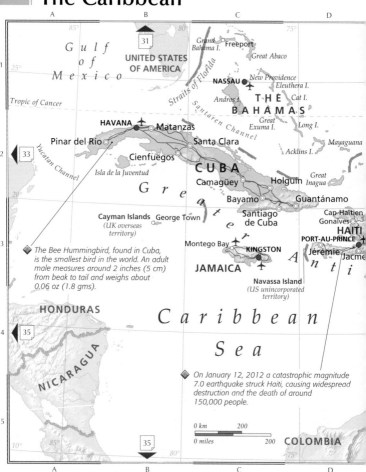

Gulf of Mexico

UNITED STATES OF AMERICA

31

85° 80° 75°

25°

Tropic of Cancer

Straits of Florida

Grand Bahama I. Freeport

Great Abaco

NASSAU *New Providence*
Eleuthera I.

Andros I.

THE BAHAMAS

Cat I.

Great Exuma I.

Long I.

Mayaguana

Acklins I.

Santaren Channel

HAVANA ○ Matanzas

Pinar del Rio ○

Cienfuegos ○ ○ Santa Clara

Isla de la Juventud

CUBA

G r e

Camagüey ○

Bayamo ○ ○ Holguín

Great Inagua

○ Guantánamo

Cayman Islands George Town
(UK overseas territory)

Santiago de Cuba ○

Yucatan Channel

20°

33

The Bee Hummingbird, found in Cuba, is the smallest bird in the world. An adult male measures around 2 inches (5 cm) from beak to tail and weighs about 0.06 oz (1.8 gms).

Montego Bay

KINGSTON

JAMAICA

Navassa Island
(US unincorporated territory)

Cap-Haïtien
Gonaïves

HAITI
PORT-AU-PRINCE

Jérémie Jacme

t i

HONDURAS

C a r i b b e a n

S e a

35

NICARAGUA

On January 12, 2012 a catastrophic magnitude 7.0 earthquake struck Haiti, causing widespread destruction and the death of around 150,000 people.

0 km   200
0 miles   200

10°

85° 80°

35

75°

COLOMBIA

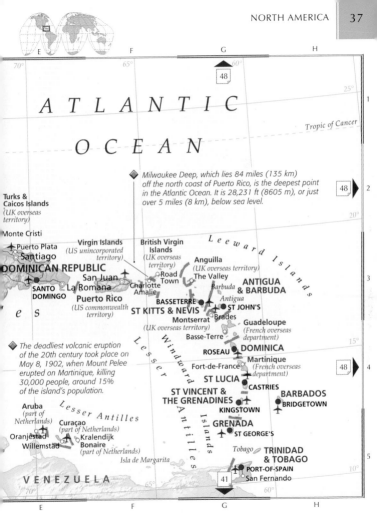

ATLANTIC

OCEAN

Tropic of Cancer

◆ *Milwaukee Deep, which lies 84 miles (135 km) off the north coast of Puerto Rico, is the deepest point in the Atlantic Ocean. It is 28,231 ft (8605 m), or just over 5 miles (8 km), below sea level.*

Turks & Caicos Islands *(UK overseas territory)*

Monte Cristi
♦Puerto Plata
Santiago
DOMINICAN REPUBLIC
San Juan
SANTO DOMINGO La Romana
Puerto Rico *(US commonwealth territory)*

Virgin Islands *(US unincorporated territory)*
Road Town
Charlotte Amalie
British Virgin Islands *(UK overseas territory)*

Leeward Islands

Anguilla *(UK overseas territory)*
The Valley
Barbuda
ANTIGUA & BARBUDA
Antigua
BASSETERRE
ST KITTS & NEVIS
ST JOHN'S
Montserrat *(UK overseas territory)*
Brades
Guadeloupe *(French overseas department)*
Basse-Terre
ROSEAU **DOMINICA**

◆ *The deadliest volcanic eruption of the 20th century took place on May 8, 1902, when Mount Pelée erupted on Martinique, killing 30,000 people, around 15% of the island's population.*

Lesser

Fort-de-France
Martinique *(French overseas department)*
ST LUCIA
CASTRIES
ST VINCENT & THE GRENADINES
KINGSTOWN
BARBADOS
BRIDGETOWN

Aruba *(part of Netherlands)*
Oranjestad
Curaçao *(part of Netherlands)*
Willemstad
Kralendijk
Bonaire *(part of Netherlands)*
Isla de Margarita

Lesser Antilles

Antilles

Windward Islands

GRENADA
ST GEORGE'S
Tobago
TRINIDAD & TOBAGO
PORT-OF-SPAIN
San Fernando

VENEZUELA

es

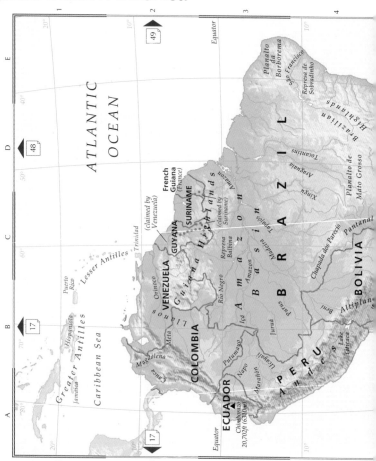

ATLANTIC

OCEAN

49

48

17

17

Equator

Equator

Puerto Rico

Hispaniola

Jamaica

Greater Antilles

Lesser Antilles

Trinidad

Caribbean Sea

COLOMBIA

VENEZUELA

G u i a n a   H i g h l a n d s

(claimed by
Venezuela)

GUYANA

SURINAME

(claimed by
Suriname)

French
Guiana
(France)

L l a n o s

Orinoco

Meta

Magdalena

Cauca

ECUADOR

Chimborazo
20,702ft (6310m)

Napo

Putumayo

Içá

Marañón

Ucayali

Juruá

Purús

Rio Negro

A   m   a   z   o   n

Amazon

Madeira

Represa
Balbina

Japurá

Tapajós

B   a   s   i   n

Xingu

Tocantins

Araguaia

B   R   A   Z   I   L

Planalto
da
Borborema

São Francisco

Represa de
Sobradinho

B r a z i l i a n   H i g h l a n d s

Planalto de
Mato Grosso

Chapada dos Pareis

Pantanal

BOLIVIA

Beni

Altiplano

Lake
Titicaca

PERU

A n d e s

20°

10°

10°

40°

50°

60°

70°

20°

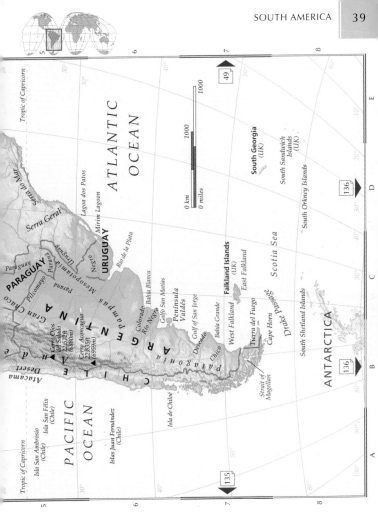

ATLANTIC OCEAN

PACIFIC OCEAN

Tropic of Capricorn

Serra do Mar

Serra Geral

Lagoa dos Patos

Mirim Lagoon

URUGUAY

PARAGUAY

Paraguay

Pilcomayo

Paraná

Uruguay

Río de la Plata

Mesopotamia

A R G E N T I N A

Gran Chaco

A N D E S

C H I L E

Cerro Ojos del Salado 22,572ft (6880m)

Cerro Aconcagua 22,835ft (6959m)

Atacama Desert

Colorado

Río Negro

Bahía Blanca

Golfo San Matías

Península Valdés

Gulf of San Jorge

Deseado

Chico

Bahía Grande

P A T A G O N I A

Strait of Magellan

Isla de Chiloé

Isla San Ambrosio (Chile)

Isla San Félix (Chile)

Islas Juan Fernández (Chile)

Tierra del Fuego

Cape Horn

Drake Passage

West Falkland

East Falkland

Falkland Islands (UK)

Scotia Sea

South Shetland Islands

South Orkney Islands

South Georgia (UK)

South Sandwich Islands (UK)

ANTARCTICA

0 km 1000

0 miles 1000

49

136

136

135

Caribbean Sea

PANAMA

PACIFIC OCEAN

Santa Marta
Ríohacha
Gulf of Venezuela
Maicao
Coro
CARACAS
Maracay
Maracaibo
Barranquilla
Cartagena
Valledupar
Cabimas
Ciudad Ojeda
Valencia
Sincelejo
Lago de Maracaibo
Barquisimeto
Acarigua
Montería
Valera
Mérida
Guanare
San Juan de los Morros
Cúcuta
Barinas
San Cristóbal
San Fernando
Bello
Bucaramanga
Barrancabermeja
Arauca
Puerto Carreño
Medellín
Itagüí
Quibdó
Tunja
Yopal
Manizales
Pereira
BOGOTÁ
Armenia
Ibagué
Buenaventura
Villavicencio
Cali
Popayán
Neiva
San José del Guaviare
Mitú
Pasto
Mocoa
Florencia
Esmeraldas
Tulcán
Ibarra
QUITO
Santo Domingo de los Colorados
Ambato
Manta
Portoviejo
Riobamba
Guayaquil
Milagro
Golfo de Guayaquil
Cuenca
Machala
Loja

COLOMBIA
VENE
ECUADOR
PERU

Caribbean Sea
Magdalena
Cauca
Arauca
Apure
Guanare
Meta
Guaviare
Negro
Caquetá
Putumayo

Colombia has the highest number of species by area in the world. There are over 1700 endemic bird species; more than all of Europe and North America combined.

The first coffee seedlings were brought to Colombia in 1804 by Jesuit missionaries; today, Colombia produces over 700,000 tons (tonnes) of coffee beans every year.

Lesser

36
35
135
42

GRENADA

Antilles

TRINIDAD & TOBAGO

Isla de Margarita

Carúpano

Cumaná

Barcelona

Maturín

Tucupita

The Serpent's Mouth

El Tigre

Ciudad Bolívar

Ciudad Guayana

Orinoco

n o s

ZUELA

Embalse de Gurí

Caura

Caroní

Paragua

Orinoco

Salto Ángel

Guiana Highlands

Cuyuni

Bartica

Rockstone

Linden

GEORGETOWN

New Amsterdam

GUYANA

PARAMARIBO

Nieuw Amsterdam

St.-Laurent-du-Maroni

Sinnamary

Kourou

CAYENNE

W.J. van Blommesteinmeer

SURINAME

French Guiana

(French overseas department)

Essequibo

Courantyne

Maroni

Acarai Mts.

Amazon Basin

B R A Z I L

A T L A N T I C O C E A N

◆ The Guiana Shield is one of the Earth's oldest surfaces, formed around 2 billion years ago.

(claimed by Venezuela)

(claimed by Suriname)

◆ Angel Falls (Salto Ángel) plunge a total of 3212 ft (979 m) to form the world's highest waterfall.

(claimed by Suriname)

◆ The European Space Agency launch facility at Kourou takes advantage of the Earth's spin near the equator to gain 10 percent more payload than an equivalent launch at Cape Canaveral in the US.

Equator

◆ 2.47 acres (one hectare) of Amazon rain forest can contain more than 750 types of trees and 1500 plant species, amounting to around 900 tons (tonnes) of living plant material.

37

49

43

43

0 km 200

0 miles 200

65° 60° 55° 50°

10°

5°

5°

E F G H

Peru, Bolivia & North Brazil

Lake Titicaca is the largest lake in South America at 3220 sq miles (8340 sq km). With an altitude of 12,500 ft (3810 m) it is also the world's highest navigable lake.

BOLIVIA'S TWO CAPITALS

La Paz - seat of government

Sucre - legal capital

Map labels:

COLOMBIA

VENEZUELA

GUYANA

Guiana Highlands

ECUADOR

Boa Vista

Equator

Rio Negro

Represa Balbina

Amazon

Putumayo

Napo

Iquitos

Amazon

Manaus

Marañón

Moyobamba

Amazon Basin

Juruá

Madeira

Piura

Tarapoto

Ucayali

B R A Z I L

Chiclayo

Purus

Saña

Pucallpa

Trujillo

Porto Velho

Chimbote

Huaraz

Rio Branco

Huacho

Huánuco

Madre de Dios

Riberalta

La Orpya

Puerto Maldonado

Guaporé

Callao

LIMA

Huancayo

Beni

PACIFIC OCEAN

Ayacucho

Cusco

Trinidad

Pisco

Ica

Puno

BOLIVIA

Nazca

Lago Titicaca

LA PAZ

Cochabamba

Montero

Arequipa

Oruro

Santa Cruz

Tacna

SUCRE

Puerto Suárez

Lago Poopó

Potosí

PARAGUAY

Uyuni

Tupiza

Tarija

CHILE

ARGENTINA

40

135

135

46

Scale:

0 km 400

0 miles 400

80°

70°

60°

10°

20°

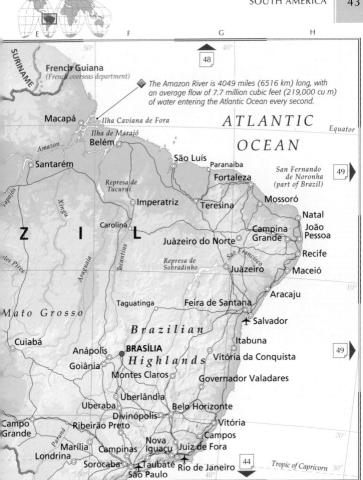

SURINAME

French Guiana
(French overseas department)

The Amazon River is 4049 miles (6516 km) long, with an average flow of 7.7 million cubic feet (219,000 cu m) of water entering the Atlantic Ocean every second.

48

Macapá · *Ilha Caviana de Fora*

ATLANTIC

Amazon

Ilha de Marajó

Belém

OCEAN

Equator

Santarém

São Luís

Paranaiba

San Fernando de Noronha *(part of Brazil)*

49

Represa de Tucuruí

Fortaleza

Mossoró

Imperatriz

Teresina

Natal

Z I L

Carolina

Juàzeiro do Norte

Campina Grande

João Pessoa

Tapajós

Xingu

Represa de Sobradinho

Recife

Juàzeiro

Maceió

10°

Araguaia

Tocantins

São Francisco

Taguatinga

Feira de Santana

Aracaju

Mato Grosso

Pires

Brazilian

Salvador

Cuiabá

Anápolis

Highlands

Itabuna

49

Goiânia

Vitória da Conquista

Montes Claros

Governador Valadares

BRASÍLIA

Uberaba

Uberlândia

Campo Grande

Divinópolis

Belo Horizonte

Ribeirão Preto

Vitória

20°

Marília

Paraná

Campos

Londrina

Campinas

Nova Iguaçu

Juiz de Fora

Sorocaba

Taubaté

Rio de Janeiro

São Paulo

Tropic of Capricorn

30°

44

50°

40°

Paraguay, Uruguay & South Brazil

◆ Formed by river deposits washed down from the Andes and Brazilian Shield, the Gran Chaco is virtually free of stones. It is composed of sand and silt sediments that are up to 10,000 ft (3050 m) thick.

◆ With a maximum height of 269 ft (82 m) and a total width of 1.7 miles (2.7 km) Iguaçu Falls has a peak flow rate of 452,000 cu ft/s (12,799 cu m/s) which would fill five Olympic size swimming pools every second.

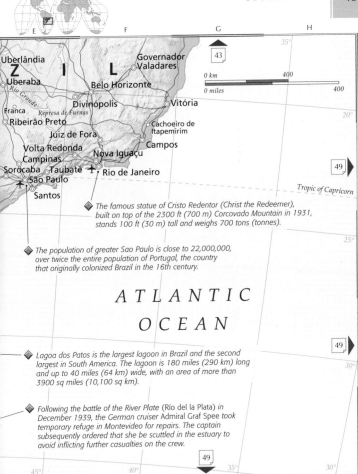

ZIL L

Uberlândia
Governador
Valadares
Uberaba
Belo Horizonte
Rio Grande
Divinópolis
Vitória
Franca
Represa de Furnas
Ribeirão Preto
Cachoeiro de
Itapemirim
Juiz de Fora
Campos
Volta Redonda
Nova Iguaçu
Campinas
Sorocaba Taubaté
Rio de Janeiro
São Paulo
Santos

Tropic of Capricorn

0 km 400
0 miles 400

◆ The famous statue of Cristo Redentor (Christ the Redeemer),
built on top of the 2300 ft (700 m) Corcovado Mountain in 1931,
stands 100 ft (30 m) tall and weighs 700 tons (tonnes).

◆ The population of greater Sao Paulo is close to 22,000,000,
over twice the entire population of Portugal, the country
that originally colonized Brazil in the 16th century.

ATLANTIC

OCEAN

◆ Lagoa dos Patos is the largest lagoon in Brazil and the second
largest in South America. The lagoon is 180 miles (290 km) long
and up to 40 miles (64 km) wide, with an area of more than
3900 sq miles (10,100 sq km).

◆ Following the battle of the River Plate (Río del la Plata) in
December 1939, the German cruiser Admiral Graf Spee took
temporary refuge in Montevideo for repairs. The captain
subsequently ordered that she be scuttled in the estuary to
avoid inflicting further casualties on the crew.

Southern South America

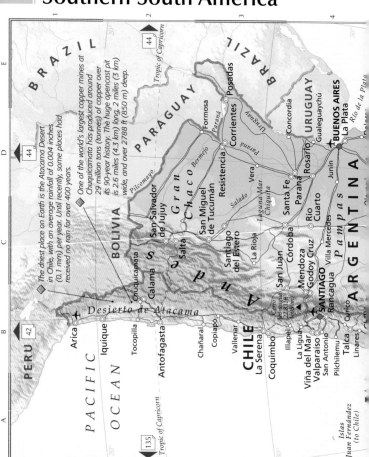

44

The driest place on Earth is the Atacama Desert in Chile, with an average rainfall of 0.004 inches (0.1 mm) per year. Until recently, some places had received no rain for over 400 years.

One of the world's largest copper mines at Chuquicamata has produced around 29 million tons (tonnes) of copper over its 90-year history. The huge opencast pit is 2.6 miles (4.3 km) long, 2 miles (3 km) wide, and over 2788 ft (850 m) deep.

42

135

44

PERU

BRAZIL

BRAZIL

PARAGUAY

BOLIVIA

URUGUAY

ARGENTINA

CHILE

PACIFIC

OCEAN

Tropic of Capricorn

Tropic of Capricorn

Gran Chaco

Pampas

Desierto de Atacama

Puna

Posadas
Formosa
Corrientes
Concordia
BUENOS AIRES
Gualeguaychú
La Plata
Rio de la Plata
Resistencia
Vera
Santa Fe
Paraná
Rosario
Junín
San Salvador
de Jujuy
Salta
San Miguel
de Tucumán
Santiago
del Estero
Río
Cuarto
Villa Mercedes
La Rioja
Córdoba
Chuquicamata
Calama
San Juan
Mendoza
Godoy Cruz
SANTIAGO
Rancagua
Arica
Iquique
Tocopilla
Antofagasta
Chañaral
Copiapó
Vallenar
La Serena
Coquimbo
Illapel
La Ligua
Viña del Mar
Valparaíso
San Antonio
Pichilemu
Curicó
Talca
Linares

Pilcomayo
Bermejo
Paraná
Paraguay
Uruguay
Salado
Laguna Mar Chiquita

Cerro
Aconcagua
22,831 ft
(6959 m)

*Islas
Juan Fernández
(to Chile)*

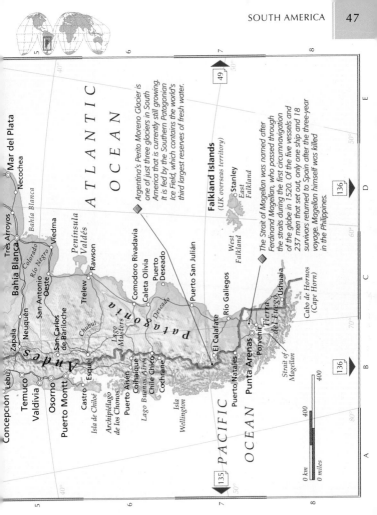

ATLANTIC OCEAN

Argentina's Perito Moreno Glacier is one of just three glaciers in South America that is currently still growing. It is fed by the Southern Patagonian Ice Field, which contains the world's third largest reserves of fresh water.

Falkland Islands
(UK overseas territory)

Stanley
East Falkland

West Falkland

The Strait of Magellan was named after Ferdinand Magellan, who passed through the straits during the first circumnavigation of the globe in 1520. Of the five vessels and 237 men that set out, only one ship and 18 survivors returned to Spain after the three-year voyage. Magellan himself was killed in the Philippines.

Mar del Plata
Necochea
Tres Arroyos
Bahía Blanca
Bahía Blanca
Viedma
Colorado
Río Negro
San Antonio
Oeste
Rawson
Península
Valdés
Neuquén
Zapala
Trelew
Chubut
Comodoro Rivadavia
Caleta Olivia
Puerto
Deseado
Deseado
San Carlos
de Bariloche
Lago
Musters
Puerto San Julián
Esquel
Lago Buenos Aires
Chile Chico
Río Gallegos
Coihaique
Puerto Aisén
Cochrane
El Calafate
Puerto Natales
Punta Arenas
Porvenir
Ushuaia
Tierra del Fuego
Cabo de Hornos
(Cape Horn)
Strait of Magellan

Concepción (Lebu)
Temuco
Valdivia
Osorno
Puerto Montt
Castro
Isla de Chiloé
Archipiélago
de los Chonos
Isla
Wellington

Andes
Patagonia

ATLANTIC OCEAN

PACIFIC OCEAN

0 km   400
0 miles 400

The Atlantic Ocean

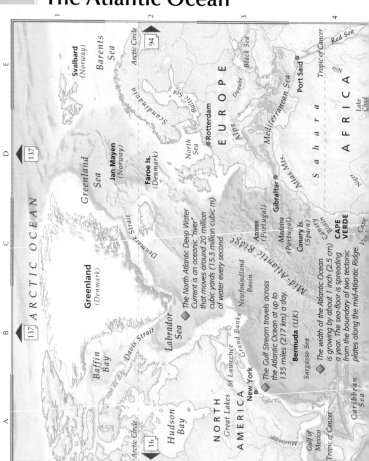

ARCTIC OCEAN

Svalbard (Norway)

Barents Sea

Arctic Circle

Scandinavia

Baltic Sea

E U R O P E

Danube

Black Sea

Red Sea

Tropic of Cancer

Port Said

Nile

Rotterdam

North Sea

Alps

A F R I C A

S a h a r a

Lake Chad

Mediterranean Sea

Greenland Sea

Jan Mayen (Norway)

Faroe Is. (Denmark)

Atlas Mts.

Gibraltar

Azores (Portugal)

Madeira (Portugal)

Canary Is. (Spain)

Canary Current

CAPE VERDE

Cape Verde Is.

Greenland (Denmark)

Denmark Strait

The North Atlantic Deep Water Current is an oceanic "river" that moves around 20 million cubic yards (15.3 million cubic m) of water every second.

Labrador Sea

Newfoundland Basin

Mid-Atlantic Ridge

Baffin Bay

Davis Strait

Grand Banks

Bermuda (UK)

Sargasso Sea

The Gulf Stream travels across the Atlantic Ocean at up to 135 miles (217 km) a day.

The width of the Atlantic Ocean is growing by about 1 inch (2.5 cm) a year. The sea-floor is spreading from the boundary of two tectonic plates along the mid-Atlantic Ridge.

New York

N O R T H A M E R I C A

St Lawrence

Great Lakes

Hudson Bay

Mississippi

Gulf of Mexico

Caribbean Sea

Arctic Circle

Tropic of Cancer

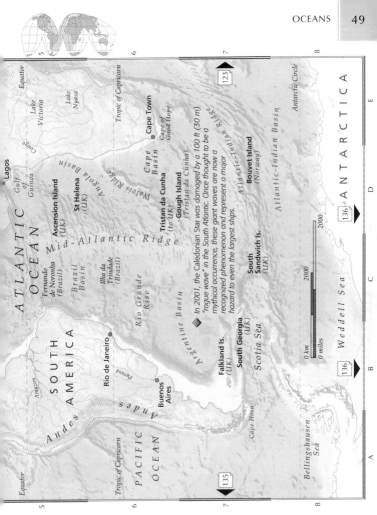

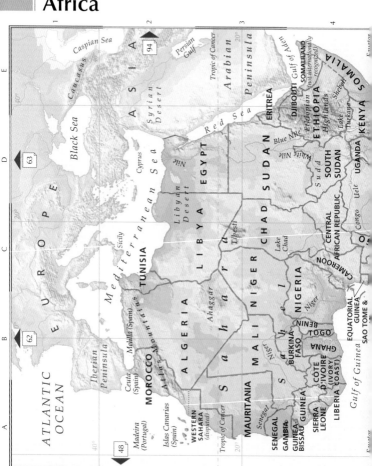

ATLANTIC OCEAN

Caspian Sea

Black Sea

Caucasus

Mediterranean Sea

Sicily

Cyprus

Iberian Peninsula

Madeira (Portugal)

Islas Canarias (Spain)

Ceuta (Spain)

Melilla (Spain)

Atlas Mountains

Persian Gulf

Tropic of Cancer

Arabian Peninsula

Gulf of Aden

Red Sea

Syrian Desert

Nile

Libyan Desert

Ahaggar

Sahara

S a h e l

Tibesti

Lake Chad

Blue Nile

White Nile

Sudd

Congo

Uele

Niger

Niger

Lake Turkana

Gulf of Guinea

Tropic of Cancer

A S I A

E U R O P E

SOMALILAND (not internationally recognized)

SOMALIA

ERITREA

DJIBOUTI

ETHIOPIA

Ethiopian Highlands

Shebeli

KENYA

UGANDA

SOUTH SUDAN

SUDAN

EGYPT

LIBYA

CHAD

CENTRAL AFRICAN REPUBLIC

CAMEROON

NIGER

NIGERIA

MALI

ALGERIA

TUNISIA

MOROCCO

WESTERN SAHARA (disputed)

MAURITANIA

SENEGAL

GAMBIA

GUINEA-BISSAU

GUINEA

SIERRA LEONE

LIBERIA

CÔTE D'IVOIRE (IVORY COAST)

GHANA

TOGO

BENIN

BURKINA FASO

EQUATORIAL GUINEA

SAO TOME &

Senegal

94

63

62

48

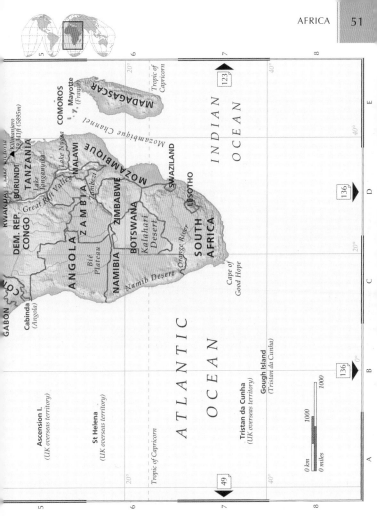

Northwest Africa

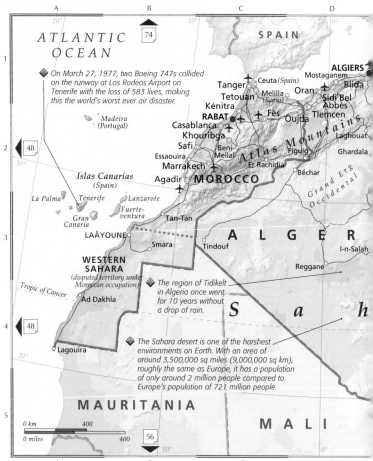

ATLANTIC
OCEAN

SPAIN

74

◆ On March 27, 1977, two Boeing 747s collided
on the runway at Los Rodeos Airport on
Tenerife with the loss of 583 lives, making
this the world's worst ever air disaster.

Madeira
(Portugal)

ALGIERS
Mostaganem
Tanger   Ceuta (Spain)   Oran   Blida
Tetouan   Melilla   Sidi Bel
          (Spain)   Abbès
Kénitra   Fès   Oujda   Tlemcen
RABAT   Laghouat
Casablanca   Figuig
Khouribga   Er Rachidia   Ghardaïa
Safi   Beni-
        Mellal
Essaouira,   Atlas Mountains
Marrakech
Agadir   **MOROCCO**   Béchar

Islas Canarias
(Spain)

La Palma   Tenerife   Lanzarote
                      Fuerte-
Gran   ventura
Canaria
                Tan-Tan

LAÂYOUNE   A L G E R I
          Smara   Tindouf   I-n-Salah

**WESTERN
SAHARA**
(disputed territory under
Moroccan occupation)   Reggane

Tropic of Cancer
          Ad Dakhla

◆ The region of Tidikelt   S a h
in Algeria once went
for 10 years without
a drop of rain.

◆ The Sahara desert is one of the harshest
environments on Earth. With an area of
around 3,500,000 sq miles (9,000,000 sq km),
roughly the same as Europe, it has a population
of only around 2 million people compared to
Europe's population of 721 million people.

Lagouira

M A U R I T A N I A

48

48

M A L I

0 km   400
0 miles   400

56

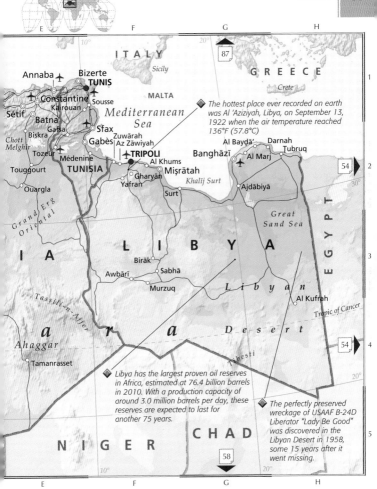

ITALY

Sicily

MALTA

GREECE

Crete

87

Annaba

Bizerte

TUNIS

Constantine

Sousse

Kairouan

Sétif

Batna

Gafsa

Mediterranean

Sea

Biskra

Sfax

Zuwārah

Tozeur

Gabès

Az Zāwiyah

Médenine

TRIPOLI

Al Khums

Al Baydā'

Darnah

Banghāzī

Al Marj

Tubruq

Chott
Melghir

Touggourt

Ouargla

TUNISIA

Gharyān

Miṣrātah

Yafran

Khalīj Surt

Ajdābiyā

Surt

Grand Erg

Oriental

L I B Y A

EGYPT

Great

Sand

Sea

I A

Birāk

Awbārī

Sabhā

Libyan

Murzuq

Al Kufrah

Tassili-n-Ajjer

Ahaggar

a

r

a

D e s e r t

Tropic of Cancer

Tamanrasset

54

54

◆ The hottest place ever recorded on earth
was Al 'Azīzīyah, Libya, on September 13,
1922 when the air temperature reached
136°F (57.8°C)

◆ Libya has the largest proven oil reserves
in Africa, estimated at 76.4 billion barrels
in 2010. With a production capacity of
around 3.0 million barrels per day, these
reserves are expected to last for
another 75 years.

◆ The perfectly preserved
wreckage of USAAF B-24D
Liberator "Lady Be Good"
was discovered in the
Libyan Desert in 1958,
some 15 years after it
went missing.

N I G E R

C H A D

58

E

F

G

H

Northeast Africa

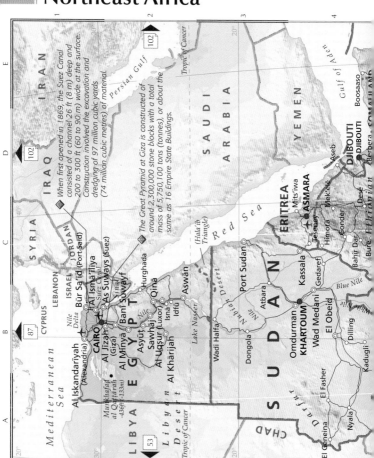

When first opened in 1869, the Suez Canal consisted of a channel 26 ft (8 m) deep and 200 to 300 ft (60 to 90 m) wide at the surface. Construction involved the excavation and dredging of 97 million cubic yards (74 million cubic metres) of material.

The Great Pyramid at Giza is constructed of around 2,300,000 stone blocks with a total mass of 5,750,100 tons (tonnes), or about the same as 16 Empire State Buildings.

IRAN

IRAQ

SYRIA

LEBANON

ISRAEL JORDAN

CYPRUS

Mediterranean Sea

Al Iskandarīyah (Alexandria)

Nile Delta

CAIRO

Būr Saʿīd (Port Said)

As Suways (Suez)

Suez Canal

Al Ismāʿīlīya

Sinai

Banī Suwayf

Bur Safājah

Al Jīzah (Giza)

Al Minyā

Asyūţ

Sawhāj

Al Uqşur (Luxor)

Isnā

Idfū

At-Ţūr

Qinā

Hurghada

Aswān

Nile

E G Y P T

Al Khārijah

Munkhafad al Qaţţārah
–436ft (–133m)

L i b y a n   D e s e r t

Libyan Desert

LIBYA

Tropic of Cancer

Lake Nasser

Lake Nasser

Wadi Halfa

Dongola

El Fasher

Nyala

El Geneina

D a r f u r

CHAD

SUDAN

Nubian Desert

N u b i a n   D e s e r t

Nile

Atbara

Port Sudan

Red Sea

(Halaʾib Triangle)

Kassala

Gedaref

Wad Medani

KHARTOUM

Omdurman

El Obeid

Dilling

Kadugli

Blue Nile

White Nile

SAUDI ARABIA

Persian Gulf

Gulf of Aden

YEMEN

Boosaaso

DJIBOUTI

DJIBOUTI

Aseb

Berbera

Mitsʾiwa

ASMARA

ERITREA

Keren

Mekʾele

Desē

Gonder

Bahir Dar

Tesenay

Himora

Gedaref

Tropic of Cancer

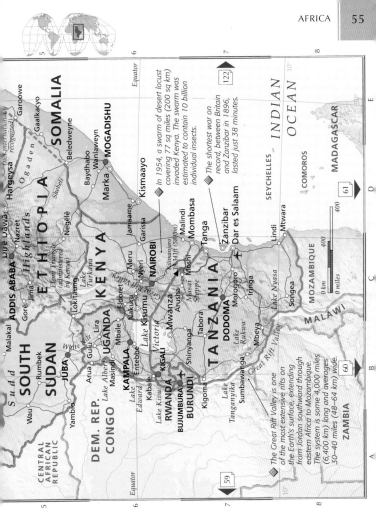

In 1954, a swarm of desert locust covering 77 sq miles (200 sq km) invaded Kenya. The swarm was estimated to contain 10 billion individual insects.

The shortest war on record, between Britain and Zanzibar in 1896, lasted just 38 minutes.

The Great Rift Valley is one of the most extensive rifts on the Earth's surface, extending from Jordan southward through eastern Africa to Mozambique. The system is some 4,000 miles (6,400 km) long and averages 30–40 miles (48–64 km) wide.

West Africa

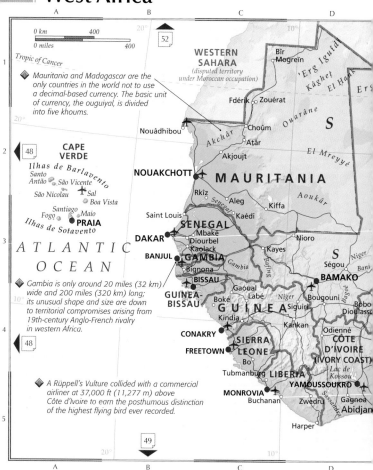

WESTERN SAHARA
(disputed territory under Moroccan occupation)

0 km 400
0 miles 400

Tropic of Cancer

◆ Mauritania and Madagascar are the only countries in the world not to use a decimal-based currency. The basic unit of currency, the ouguiya, is divided into five khoums.

Bîr Mogreïn

'Erg Iguîdi
Kâghet
El Hank
Ers

Fdérik • Zouérat

Ouârâne

S

Nouâdhibou

Choûm

Atâr

El Mreyyé

Akjoujt

48

CAPE VERDE

Ilhas de Barlavento
Santo
Antão • São Vicente
São Nicolau • Sal
Boa Vista
Fogo • Santiago • Maio
PRAIA
Ilhas de Sotavento

NOUAKCHOTT

MAURITANIA

Rkîz
Aleg
Kiffa

Aoukâr

Saint Louis

SENEGAL
Mbaké
Diourbel
Kaolack
DAKAR

Nioro

S

Kayes

Ségou

Niger

Bani

BAMAKO

ATLANTIC
OCEAN

BANJUL
GAMBIA
Bignona
BISSAU

Gambia

Boundou

◆ Gambia is only around 20 miles (32 km) wide and 200 miles (320 km) long; its unusual shape and size are down to territorial compromises arising from 19th-century Anglo-French rivalry in western Africa.

GUINEA-BISSAU

Boké
Labé

Gaoual

Niger

GUINEA

Siguiri

Bougouni

Bobo
Dioulasso

48

Kindia

Kankan

Odienné
**CÔTE
D'IVOIRE
(IVORY COAST)**

CONAKRY

**SIERRA
LEONE**
FREETOWN

Bo

Lac de
Kossou

◆ A Rüppell's Vulture collided with a commercial airliner at 37,000 ft (11,277 m) above Côte d'Ivoire to earn the posthumous distinction of the highest flying bird ever recorded.

Tubmanburg

LIBERIA

MONROVIA
Buchanan

YAMOUSSOUKRO

Zwedru

Gagnoa

Abidjan

Harper

ALGERIA   LIBYA

◆ The Niger River begins in Guinea just 150 miles
(240 km) from the Atlantic coast but then heads
inland on a 3000-mile (4100-km) journey
before finally reaching the Gulf of Guinea some
1200 miles (2000 km) to the east.

Tropic of Cancer

-hech

a   *h*   *a*   *r*   Ténéré
                                              du
Taoudenni   Tafassâsset

Tessalit   *Ténéré*

Araouane   Adrar des
            Ifoghas   Assamakka
Azaouâd   Massif
'Erg I-n-Sâkâne   de l'Aïr   Grand Erg de Bilma

MALI   Agadez

ac
guibine
Tombouctou   Gao
Lac   Ansongo
Niangay   Hombori   Tahoua   N I G E R   CHAD

opti   Zinder   Nguigmi
            h   *e*   *l*   Maradi   Gouré
BURKINA   NIAMEY
FASO   Sokoto   Katsina
OUAGADOUGOU   Sokoto   Kano   Maiduguri
                     Gusau
udougou   Fada-   Zaria
            Ngourma   Kaduna   Kumo
            Kandi   Kainji
BENIN   Natitingou Reservoir   Gongola

Wa   NIGERIA
            Parakou   Jos
Tamale   Sokodé Ilorin   Plateau   CAMEROON
GHANA   Oyo   Niger   ABUJA
                 Abomey Ogbomosho   Benue
kumasi Lake   Ede
          Volta   Ibadan Benin
Nsawan   Lagos   City   Enugu
mankese LOMÉ PORTO-   Onitsha
ACCRA   NOVO   Sapele   Aba   Calabar   C.A.R.
         Bight of Benin   Port Harcourt
         Gulf of Guinea   Mouths of the Niger

Lake Volta is one of the largest man-made
lakes in the world, covering 3283 sq miles   EQUATORIAL
(8502 sq km), or 3.6% of Ghana's area.   GUINEA

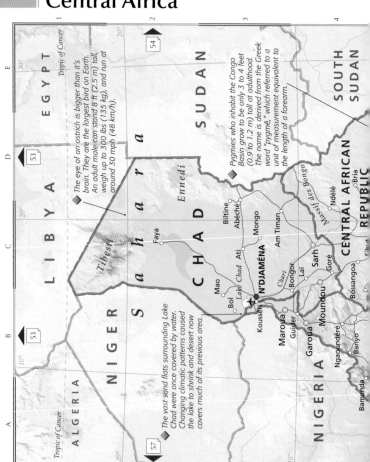

The eye of an ostrich is bigger than it's brain. They are the largest bird on Earth. An adult male can stand 8 ft (2.5 m) tall, weigh up to 300 lbs (135 kg), and run at around 30 mph (48 km/h).

Pygmies who inhabit the Congo Basin grow to be only 3 to 4 feet (0.9 to 1.2 m) tall at adulthood. The name is derived from the Greek word "pygmē," which referred to a unit of measurement equivalent to the length of a forearm.

The vast sand flats surrounding Lake Chad were once covered by water. Changing climatic patterns caused the lake to shrink and desert now covers much of its previous area.

Tropic of Cancer

E
D
C
B
A

LIBYA
EGYPT
SUDAN
SOUTH SUDAN
CENTRAL AFRICAN REPUBLIC
ALGERIA
NIGER
NIGERIA

Sahara
Tibesti
Ennedi
CHAD

Faya
Mao
Bol
Lake Chad
Kousséri
N'DJAMÉNA
Chari
Ati
Bongor
Laï
Bitine
Abéché
Mongo
Am Timan
Sarh
Goré
Moundou
Bossangoa
Bria
Ndélé

Maroua
Guider
Garoua
Ngaoundéré
Banyo
Bamenda

Massif des Bongo

Tropic of Cancer

53
53
54
57

1 2 3 4

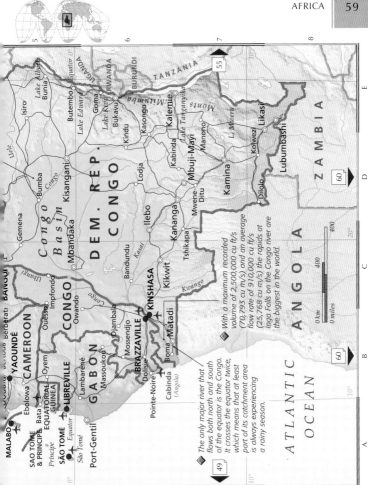

TANZANIA

55

UGANDA

Lake Albert

Butembo

Goma
Lake Kivu
RWANDA
BURUNDI

Bunia

Isiro

Lake Edward

Bukavu

Monts Mitumba

E

Kindu

Kasongo

Lake Tanganyika

Kalemie

Uele

Bumba

Congo

Kisangani

D E M . R E P .

Lodja

Kabinda

Manono

L. Mweru

Likasi

Kolwez

60

D

Gemena

Congo **Basin**

Mbandaka

C O N G O

Ilebo

Kananga

Mwene-
Ditu

Mbuji-Mayi

Kamina

Lubumbashi

Dilolo

Z A M B I A

Bangui

Berberati

Yaoundé

CAMEROON

Ebolowa

Oyem

Impfondo

Ouesso

Owando

CONGO

Djambala

Bandundu

Tshikapa

Kasai

Kikwit

Kwango

Kiango

A N G O L A

C

Ubangi

GABON

Lambaréné

Massoukou

Mossendjo

BRAZZAVILLE

KINSHASA

Matadi

60

0 km   400

0 miles   400

With a maximum recorded
volume of 2,500,000 cu ft/s
(70,793 cu m/s) and an average
flow rate of 910,000 cu ft/s
(25,768 cu m/s) the rapids at
Inga Falls on the Congo river are
the biggest in the world.

55

MALABO

SÃO TOMÉ
& PRÍNCIPE

EQUATORIAL
GUINEA

Bata

LIBREVILLE

Port-Gentil

Príncipe

São Tomé

SÃO TOMÉ

Equator

Pointe-Noire

Dolisie

Cabinda
(Angola)

Boma

B

49

The only major river that
flows both north and south
of the equator is the Congo.
It crosses the equator twice,
which means that at least
part of its catchment area
is always experiencing
a rainy season.

A T L A N T I C

O C E A N

A

Southern Africa

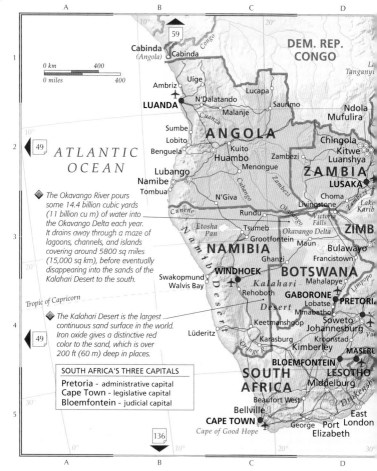

59

Cabinda
(Angola) ● Cabinda

DEM. REP. CONGO

Ambriz ● Uíge
● N'Dalatando ● Lucapa
LUANDA ●
● Malanje ● Saurimo ● Ndola
Sumbe ● Mufulira
Lobito ● Kuito **ANGOLA**
Benguela ● Huambo ● Zambezi Chingola ●
● Kitwe
Lubango ● Menongue **ZAMBIA** Luanshya ●
Namibe ● **LUSAKA** ●
Tombua ● N'Giva ● Choma ●
Livingstone ● Lake Karib

49

ATLANTIC OCEAN

◆ The Okavango River pours
some 14.4 billion cubic yards
(11 billion cu m) of water into
the Okavango Delta each year.
It drains away through a maze of
lagoons, channels, and islands
covering around 5800 sq miles
(15,000 sq km), before eventually
disappearing into the sands of the
Kalahari Desert to the south.

Rundu ● Victoria Falls
Etosha ● Tsumeb Okavango Delta **ZIMB**
Pan ● Grootfontein ● Maun
● Bulawayo
NAMIBIA ● Ghanzi ● Francistown
● **WINDHOEK** **BOTSWANA**
Swakopmund ● *Kalahari*
Walvis Bay ● ● Rehoboth **GABORONE** ● Mahalapye **PRETORI**
Desert ● Lobatse ●
Tropic of Capricorn ● Mmabatho ● Soweto ●
● Keetmanshoop ● Johannesburg ●

◆ The Kalahari Desert is the largest
continuous sand surface in the world.
Iron oxide gives a distinctive red
color to the sand, which is over
200 ft (60 m) deep in places.

Lüderitz ● ● Kroonstad **MASERU**
● Karasburg ● Kimberley **LESOTHO**
● **BLOEMFONTEIN**
SOUTH ● Middelburg
AFRICA
Beaufort West ● East
Bellville ● ● George ● Port London
CAPE TOWN ● Elizabeth
Cape of Good Hope

SOUTH AFRICA'S THREE CAPITALS
Pretoria - administrative capital
Cape Town - legislative capital
Bloemfontein - judicial capital

---

0 km   400
0 miles   400

136

◆ Coco de Mer, or the double coconut palm, produces some of the largest seeds in the plant kingdom. Weighing up to 60 lbs (27 kg), they take around 10 years to ripen.

TANZANIA

Mbala

Kasama

MALAWI

Mzuzu

Mpika   *Lake*   Mocímboa

               Nyasa   da Praia

LONGWE   Salima

               Zomba

Blantyre   Moçambique

Tete   *Rovuma*

HARARE   Nsanje   Nampula

hitungwiza   Mocuba

               Quelimane

WE

Chimoio   Beira

Inhambane

Xai-Xai

MAPUTO

BABANE

WAZILAND

Pietermaritzburg

urban

COMOROS

Grande Comore

MORONI ●

Mwali   *Anjouan*

Mamoudzou

Mayotte

(French overseas department)

Nacala

Moçambique

Aldabra Group

Farquhar Group

Amirante Islands

Inner Islands

VICTORIA ✈

Mahé

SEYCHELLES

Outer Islands

Antsirañana

Ambanja

Antsohihy

Antalaha

Mahajanga

MADAGASCAR

ANTANANARIVO

Fenoarivo Atsinanana

Toamasina

Morondava

Ambositra

Fianarantsoa

Mananjary

Ihosy

Saint-Denis

Réunion

(French overseas department)

Toliara

Farafangana

Vangaindrano

Amboasary

INDIAN OCEAN

MAURITIUS

PORT LOUIS

Mascarene Islands

Tropic of Capricorn

MOZAMBIQUE

Mozambique Channel

◆ In 1905, the world's largest rough diamond was discovered at the Cullinan Diamond Mine. Weighing 3106 carats, or about 1.3 pounds (0.6 kg), the diamond was cut into nine smaller stones, including the 530.2 carat "Cullinan I" or "Great Star of Africa," which forms part of the British Crown Jewels and is estimated to be worth over $400 million.

◆ Thought to have been extinct for 70 million years, a living coelacanth was netted in the Indian Ocean in 1938. They are powerful predators, averaging 5 feet (1.5 m) in length and weighing about 100 lbs (45 kg).

Europe

ATLANTIC

OCEAN

Arctic Circle

20°

Limit of winter pack ice

ICELAND

Lofo

Norwegian Sea

Faroe Islands
(Denmark)

Outer Hebrides

British Isles

Ireland

IRELAND

Britain

North Sea

DENMARK

Celtic Sea

UNITED KINGDOM

NETHERLANDS

English Channel

BELGIUM

Nort

Elbe

LUX.

GERMANY

Seine

CZEC REPUB

Loire

FRANCE

Rhine

LIECH

SWITZ.

AUSTRI

Massif Central

Mont Blanc 15,771ft (4807m)

Po

SLOVEN

Garonne

PORTUGAL

Duero

Pyrenees

Ebro

MONACO

SAN MARINO

CROAT BOS & HE

ANDORRA

Tagus

Iberian Peninsula

SPAIN

Corsica

ITALY

Bay of Biscay

VATICAN CITY

Balearic Islands

Sardinia

Strait of Gibraltar

Gibraltar
(UK)

Madeira (to Portugal)

Tyrrhenian Sea

Mediterr

a

Sicily

MALTA

Canary Islands (to Spain)

Atlas Mountains

AFRICA

20°

0°

0 km   800

0 miles   800

137

48

48

50

Barents Sea

North Cape

Ostrov Kolguyev

137

SWEDEN

FINLAND

Kola
Peninsula

White
Sea

Åland

Gulf of Bothnia

Lake Onega

Lake
Ladoga

Northern Dvina

Ural Mountains

RUSSIAN

FEDERATION

94

ESTONIA

LATVIA

LITHUANIA

RUSS.
FED.

European Plain

Central
Russian
Upland

Volga Uplands

Volga

Ural

Aral Sea

BELARUS

POLAND

Pripet
Marshes

Bug

Dnieper Lowlands

Dnieper

Don

94

Vistula

UKRAINE

Carpathian Mts.

Dniester

SLOVAKIA

MOLDOVA

Sea of
Azov

(the Ukrainian territory of
Crimea was annexed by
Russia in 2014)

Caspian Sea

HUNGARY

ROMANIA

Crimea

Caucasus

Elbrus
18,510ft
(5642m)

94

SERBIA

Danube

Black Sea

KOS.

BULGARIA

Balkan
Mts.

MACED.

TURKEY

ALBANIA

Aegean
Sea

Anatolia

ASIA

GREECE

Peloponnese

Cyprus

Sea

Crete

94

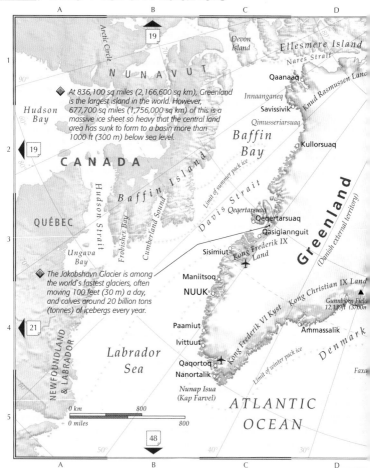

At 836,100 sq miles (2,166,600 sq km), Greenland is the largest island in the world. However, 677,700 sq miles (1,756,000 sq km) of this is a massive ice sheet so heavy that the central land area has sunk to form to a basin more than 1000 ft (300 m) below sea level.

The Jakobshavn Glacier is among the world's fastest glaciers, often moving 100 feet (30 m) a day, and calves around 20 billion tons (tonnes) of icebergs every year.

Arctic Circle

Devon Island

Ellesmere Island

Nares Strait

NUNAVUT

Qaanaaq

Innaanganeq

Knud Rasmussen Land

Hudson Bay

Savissivik

Qimusseriarsuaq

Baffin Bay

Kullorsuaq

CANADA

Limit of summer pack ice

QUÉBEC

Baffin Island

Hudson Strait

Frobisher Bay

Cumberland Sound

Davis Strait

Qeqertarsuaq

Qeqertarsuaq

Qasigiannguit

Sisimiut

Kong Frederik IX Land

Greenland
(Danish external territory)

Ungava Bay

Maniitsoq

NUUK

Kong Christian IX Land

Gunnbjørn Field
12,139ft (3700 m)

Paamiut

Kong Frederik VI Kyst

Ammassalik

Ivittuut

NEWFOUNDLAND & LABRADOR

Labrador Sea

Qaqortoq

Nanortalik

Denmark

Limit of winter pack ice

Faxa

Nunap Isua
(Kap Farvel)

ATLANTIC OCEAN

0 km   800
0 miles   800

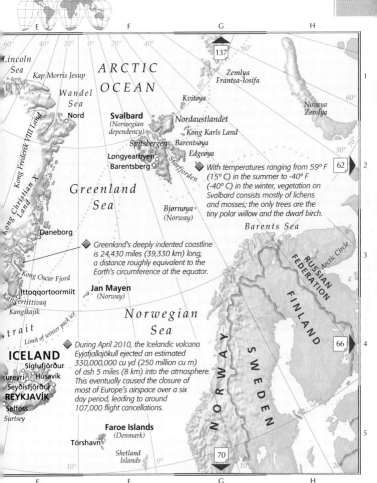

ARCTIC OCEAN

Lincoln Sea

Kap Morris Jesup

Wandel Sea

Nord

Svalbard
(Norwegian dependency)

Spitsbergen

Longyearbyen
Barentsberg

Kong Frederik VIII Land

Kong Christian X Land

Greenland Sea

Daneborg

Kong Oscar Fjord

Ittoqqortoormiit

Kangertittivaq

Kangikajik

strait

Limit of winter pack ice

ICELAND
Siglufjörður
Akureyri Húsavik
Seyðisfjörður
REYKJAVÍK
Selfoss
Surtsey

Zemlya
Frantsa-Iosifa

Kvitøya

Nordaustlandet
Kong Karls Land

Barentsøya
Edgeøya

Storfjorden

Novaya
Zemlya

Bjørnøya
(Norway)

Barents Sea

RUSSIAN FEDERATION

Arctic Circle

FINLAND

SWEDEN

NORWAY

Jan Mayen
(Norway)

Norwegian Sea

Faroe Islands
(Denmark)

Tórshavn

Shetland
Islands

◆ With temperatures ranging from 59° F (15° C) in the summer to -40° F (-40° C) in the winter, vegetation on Svalbard consists mostly of lichens and mosses; the only trees are the tiny polar willow and the dwarf birch.

◆ Greenland's deeply indented coastline is 24,430 miles (39,330 km) long, a distance roughly equivalent to the Earth's circumference at the equator.

◆ During April 2010, the Icelandic volcano Eyjafjallajökull ejected an estimated 330,000,000 cu yd (250 million cu m) of ash 5 miles (8 km) into the atmosphere. This eventually caused the closure of most of Europe's airspace over a six day period, leading to around 107,000 flight cancellations.

137

62

66

70

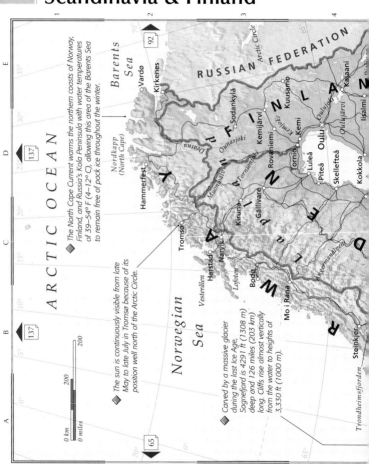

ARCTIC OCEAN

The North Cape Current warms the northern coasts of Norway, Finland, and Russia's Kola Peninsula with water temperatures of 39–54° F (4–12° C), allowing this area of the Barents Sea to remain free of pack ice throughout the winter.

The sun is continuously visible from late May to late July in Tromsø because of its position well north of the Arctic Circle.

Carved by a massive glacier during the last Ice Age, Sognefjord is 4291 ft (1308 m) deep and 126 miles (203 km) long. Cliffs rise almost vertically from the water to heights of 3,330 ft (1000 m).

Barents Sea

RUSSIAN FEDERATION

Arctic Circle

Vardø
Kirkenes
Sodankylä
Kuusamo
Kajaani
Iisalmi

Nordkapp (North Cape)
Hammerfest

Ounasjoki
Kemijärvi
Rovaniemi
Kemi
Oulu
Oulujärvi
Oulujoki

Tornio
Tornio
Kiruna
Gällivare
Luleå
Piteå
Skellefteå
Kokkola

Tromsø
Narvik
Harstad
Vesterålen
Lofoten
Bodø
Mo i Rana

Norwegian Sea

Steinkjer
Trondheimsfjorden

0 km   200
0 miles 200

Finns consume an average of 26.5 lbs (12 kg) of coffee each per year, which is over twice the amount of most other Europeans, and makes them among the biggest coffee drinkers in the world.

The 10 mile (16 km) bridge and tunnel link across the Øresund Sound is one of the largest infrastructure projects in European history. It connects the Danish capital Copenhagen to the Swedish port of Malmö.

Places on map:

E — Joensuu, Varkaus, Saimaa, Imatra, Kouvola, Kotka, Lappeenranta, Jyväskylä, Seinäjoki, Hämeenlinna, Riihimäki, Espoo, HELSINKI, Vantaa, Tampere, Turku, Pori, Rauma, Mariehamn, Åland, Gulf of Finland, ESTONIA, LATVIA, BELARUS, LITHUANIA, KALININGRAD (part of Russian Federation), Vaasa

88, 80

D — POLAND

Örnsköldsvik, Sundsvall, Gulf of Bothnia, Gävle, Falun, Borlänge, Uppsala, Västerås, STOCKHOLM, Örebro, Nyköping, Norrköping, Linköping, Skövde, Vättern, Jönköping, Växjö, Visby, Gotland, Kalmar, Öland, Karlskrona, Kristianstad, Malmö, Baltic Sea

C — Bornholm, Ronne

S W E D E N

B — Lillehammer, Hamar, Gjøvik, Mjøsa, Elverum, OSLO, Moss, Fredrikstad, Halden, Uddevalla, Trollhättan, Vänern, Borås, Göteborg, Halmstad, Helsingborg, Frederikshavn, Aalborg, Randers, Århus, Silkeborg, Herning, Vejle, Odense, Esbjerg, COPENHAGEN, Sjælland, Nykøbing, Aabenrå, GERMANY

76, 70

A — Ålesund, Hermansverk, Honefoss, Drammen, Porsgrunn, Arendal, Kristiansand, Skagerrak, Jylland, DENMARK, Sognefjorden, Bergen, Haugesund, Stavanger, N O R W A Y, Hjørring, Storebælt

The Low Countries

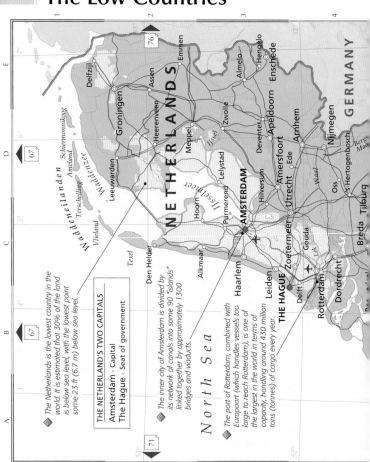

THE NETHERLAND'S TWO CAPITALS

Amsterdam - Capital
The Hague - Seat of government

◆ The Netherlands is the lowest country in the world. It is estimated that 30% of the land is below sea level, with the lowest point some 23 ft (6.7 m) below sea level.

◆ The inner city of Amsterdam is divided by its network of canals into some 90 "islands" linked together by approximately 1300 bridges and viaducts.

◆ The port of Rotterdam, combined with Europoort (which handles vessels too large to reach Rotterdam), is one of the largest in the world in terms of capacity, handling around 430 million tons (tonnes) of cargo every year.

GERMANY

NETHERLANDS

North Sea

Waddeneilanden

Schiermonnikoog
Ameland
Terschelling
Vlieland
Texel

Delfzijl
Emmen
Assen
Groningen
Heerenveen
Leeuwarden
Almelo
Hengelo
Enschede
Apeldoorn
Arnhem
Nijmegen
Zwolle
Meppel
Deventer
Ede
's-Hertogenbosch
IJssel
Lelystad
Amersfoort
Hilversum
Utrecht
Oss
Waal
Tilburg
Breda
IJsselmeer
Hoorn
Purmerend
AMSTERDAM
Zoetermeer
Gouda
Dordrecht
Haarlem
Leiden
THE HAGUE
Delft
Rotterdam
Lek
Alkmaar
Den Helder

GERMANY

BELGIUM

FRANCE

Ardennes

LUXEMBOURG

Over 80 percent of the world's rough-cut diamonds pass through Antwerp's (Antwerpen) Diamond Quarter every year, making it the largest diamond trading center in the world, with an annual turnover of over US$50 billion.

Echternach is the home of the only religious dancing procession remaining in the Western world. Every year since the 15th century, thousands of pilgrims have marched down the streets of the town performing a ritual dance involving specific movements, music, and prayers.

On August 23, 1914, three weeks after Britain entered World War I, the 70,000 strong British Expeditionary Force encountered the advancing German army for the first time at the battle of Mons.

Venlo
Heerlen
Maastricht
Eindhoven
Genk
Verviers
Liège
Hasselt
Seraing
Leuven
Tienen
Namur
Diekirch
LUXEMBOURG
Bastogne
Arlon
Esch-sur-Alzette
Turnhout
Antwerpen
Mechelen
Aalst
BRUSSELS
Charleroi
Dinant
Sint Niklaas
Gent
La Louvière
Mons
Zeebrugge
Brugge
Oostende
Roeselare
Kortrijk
Tournai
Mouscron
Ieper
Terneuzen

Flanders
Scheldt
Westerschelde
Vlissingen

Meuse
Ourthe
Sambre
Sûre
Our
Moselle

0 km 50
0 miles 50

The British Isles

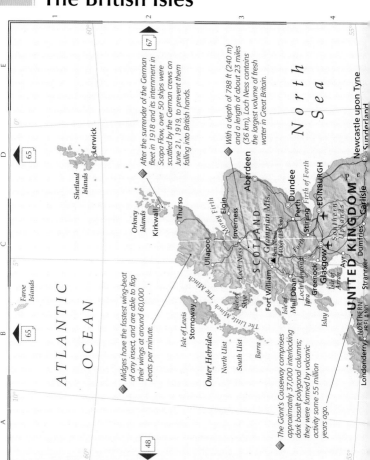

After the surrender of the German fleet in 1918 and its internment in Scapa Flow, over 50 ships were scuttled by the German crews on June 21, 1919, to prevent them falling into British hands.

With a depth of 788 ft (240 m) and a length of about 23 miles (36 km), Loch Ness contains the largest volume of fresh water in Great Britain.

Midges have the fastest wing-beat of any insect, and are able to flap their wings at around 60,000 beats per minute.

The Giant's Causeway comprises approximately 37,000 interlocking dark basalt polygonal columns; they were formed by volcanic activity some 55 million years ago.

North Sea

ATLANTIC OCEAN

Lerwick

Shetland Islands

Orkney Islands

Kirkwall

Thurso

Faroe Islands

Isle of Lewis

Stornoway

Outer Hebrides

North Uist

South Uist

Barra

The Minch

The Little Minch

Isle of Skye

Ullapool

Loch Ness

Inverness

Moray Firth

Elgin

SCOTLAND

Aberdeen

Dundee

Perth

Grampian Mts.

Ben Nevis

4406 (1343 m)

Fort William

Loch Lomond

Isle of Mull

Oban

Jura

Islay

Greenock

Glasgow

Isle of Arran

Ayr

Stirling

Firth of Forth

EDINBURGH

Southern Uplands

Dumfries

Stranraer

Carlisle

UNITED KINGDOM

NORTHERN IRELAND

Londonderry

Newcastle upon Tyne

Sunderland

Pe

With evidence of teaching beginning as early as 1096, Oxford University is the second oldest university in the world.

The River Severn has the second highest tidal range in the world, as much as 50 ft (15 m), often giving rise to a tidal bore. In September 1996, one such wave carried a surfer for 5.7 miles (9 km).

Every year over 1.5 billion pints (850 million litres) of Guinness® Irish stout are consumed in over 120 countries around the world.

IRELAND

DUBLIN

BELFAST

Sligo
Galway
Tralee
Killarney
Cork
Limerick
Ennis
Athlone
Waterford
Wexford
Fishguard
Milford Haven
Swansea
CARDIFF
Newport
Barnstaple
Penzance
Land's End
Isles of Scilly
Plymouth
Exeter
Bournemouth
Taunton
Salisbury
Bath
Bristol
Gloucester
Worcester
Birmingham
Stratford-upon-Avon
Coventry
Swindon
Oxford
Reading
Southampton
Isle of Wight
Portsmouth
Brighton
Canterbury
Dover
LONDON
Colchester
Southend-on-Sea
Ipswich
Cambridge
Peterborough
Norwich
Leicester
Nottingham
Derby
Stoke-on-Trent
Shrewsbury
Chester
Liverpool
Manchester
Bolton
Preston
Blackpool
Bradford
Leeds
Sheffield
Lincoln
York
Grimsby
Kingston upon Hull
Middlesbrough
Lancaster
Bangor
Isle of Man
(British Crown Dependency)
Douglas
Holyhead
Anglesey
Aberystwyth
Newry
Dundalk
Wexford

WALES
ENGLAND

Irish Sea

Brecon Beacons
Cambrian Mts.
Cardigan Bay
The Wash
The Fens
Dartmoor
Exmoor
Severn
Thames
Wicklow Mts.
Lough Ree
Lough Derg
Lough Corrib
Barrow
Blackwater
Shannon
Bantry Bay

Channel Islands
St. Peter Port Guernsey
(British Crown Dependency)
St. Helier Jersey
(British Crown Dependency)

FRANCE

English Channel
Channel Tunnel

ATLANTIC OCEAN

0 km   100
0 miles   100

68
72
74
48

France, Andorra & Monaco

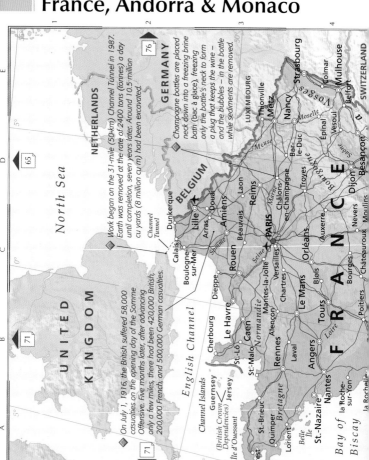

Champagne bottles are placed neck down into a freezing brine bath (bac à glace), freezing only the bottle's neck to form a plug that keeps the wine – and the bubbles – in the bottle while sediments are removed.

Work began on the 31-mile (50-km) Channel Tunnel in 1987. Earth was removed at the rate of 2400 tons (tonnes) a day until completion, seven years later. Around 10.5 million cu yards (8 million cu m) had been excavated.

On July 1, 1916, the British suffered 58,000 casualties on the opening day of the Somme Offensive. Five months later, after advancing only a few miles, there had been 420,000 British, 200,000 French, and 500,000 German casualties.

NORTH SEA

UNITED KINGDOM

NETHERLANDS

GERMANY

BELGIUM

LUXEMBOURG

SWITZERLAND

English Channel

Channel Islands
Guernsey
(British Crown
Dependencies) Jersey

Île d'Ouessant

Bay of Biscay

Dunkerque
Calais
Boulogne-sur-Mer
Lille
Douai
Arras
Amiens
Somme
Dieppe
Le Havre
Rouen
Beauvais
PARIS
Versailles
Laon
Reims
Marne
Châlons-en-Champagne
Troyes
Auxerre
Nevers
Moulins

Cherbourg
St-Lô
Caen
Normandie
Alençon
Mantes-la-Jolie
Chartres
Le Mans
Blois
Orléans
Bourges
Châteauroux

St-Malo
Rennes
Laval
Angers
Tours
Poitiers

St-Brieuc
Quimper
Lorient
Belle Île
St-Nazaire
Nantes
La Roche-sur-Yon
La Rochelle
Brest
Bretagne
Loire

Strasbourg
Colmar
Mulhouse
Belfort
Vosges
Thionville
Metz
Nancy
Moselle
Meuse
Bar-le-Duc
Épinal
Vesoul
Besançon
Dijon
Côte d'Or
Saône

Channel Tunnel

North Sea

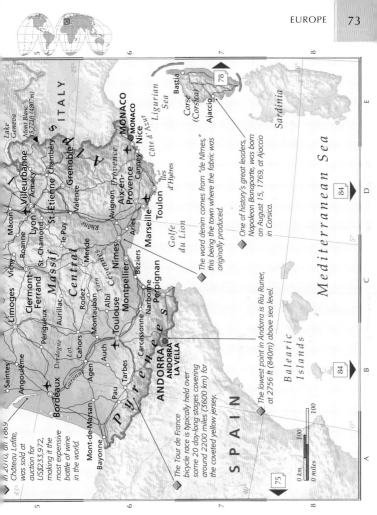

In 2010, an 1869 Château Lafite, was sold at auction for US$233,972, making it the most expensive bottle of wine in the world.

The Tour de France bicycle race is typically held over some 20 day-long stages covering around 2200 miles (3600 km) for the coveted yellow jersey.

The lowest point in Andorra is Riu Runer, at 2756 ft (840m) above sea level.

The word denim comes from "de Nîmes," this being the town where the fabric was originally produced.

One of history's great leaders, Napoleon Bonaparte, was born on August 15, 1769, at Ajaccio in Corsica.

ITALY

Mont Blanc 15,771ft (4807m)
Lake Geneva
Villeurbanne
Annecy
Chambéry
Grenoble
Lyon
Mâcon
St. Étienne
St. Chamond
Vichy
Roanne
Le Puy
Valence
Avignon
Aix-en-Provence
Provence
Cannes
Nice
MONACO
MONACO
Côte d'Azur
Îles d'Hyères
Toulon
Marseille
Arles
Rhône
Nîmes
Cévennes
Béziers
Montpellier
Narbonne
Perpignan
Golfe du Lion

Ligurian Sea

Bastia
78
Corse (Corsica)
Ajaccio

Sardinia

Mediterranean Sea

Limoges
Clermont-Ferrand
Aurillac
Mende
Rodez
Tarn
Lot
Cahors
Albi
Montauban
Toulouse
Carcassonne
Massif Central

Saintes
Angoulême
Périgueux
Dordogne
Bordeaux
Garonne
Agen
Auch
Tarbes
Pau
Mont-de-Marsan
Bayonne

P y r e n e e s

ANDORRA
ANDORRA LA VELLA

SPAIN

Balearic Islands

84
84
75

0 km   100
0 miles 100

Spain & Portugal

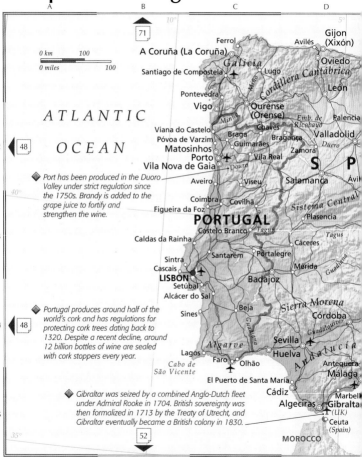

71

0 km   100

0 miles   100

ATLANTIC

OCEAN

Gijon
(Xixón)
Ferrol
Avilés
A Coruña (La Coruña)
Oviedo
Galicia
Lugo
Cordillera Cantábrica
Santiago de Compostela
León
Pontevedra
Vigo
Ourense
(Orense)
Emb. de
Ricobayo
Palencia
Minho
Chaves
Valladolid
Viana do Castelo
Braga
Bragança
Zamora
Duero
Póvoa de Varzim
Guimarães
Matosinhos
Vila Real
S P
Porto
Duoro
Viseu
Vila Nova de Gaia
Salamanca
Aveiro
Ávi
Coimbra
Covilhã
Sistema Central
Figueira da Foz
Plasencia
PORTUGAL
Castelo Branco
Tagus
Caldas da Rainha
Cáceres
Tagus
Santarém
Portalegre
Sintra
Mérida
Cascais
Guadiana
LISBON
Badajoz
Setúbal
Alcácer do Sal
Beja
Sierra Morena
Sines
Córdoba
Guadiana
Guadalquivir
Algarve
Sevilla
Andalucía
Lagos
Huelva
Faro
Olhão
Antequera
*Cabo de
São Vicente*
El Puerto de Santa María
Málaga
Cádiz
Marbel
Algeciras
Gibralta
Gibraltar
(UK)
Ceuta
(Spain)
MOROCCO

◆ Port has been produced in the Duoro
Valley under strict regulation since
the 1750s. Brandy is added to the
grape juice to fortify and
strengthen the wine.

48

◆ Portugal produces around half of the
world's cork and has regulations for
protecting cork trees dating back to
1320. Despite a recent decline, around
12 billion bottles of wine are sealed
with cork stoppers every year.

48

◆ Gibraltar was seized by a combined Anglo-Dutch fleet
under Admiral Rooke in 1704. British sovereignty was
then formalized in 1713 by the Treaty of Utrecht, and
Gibraltar eventually became a British colony in 1830.

52

Bay of Biscay

FRANCE

Santander
Bilbao
Donostia/San Sebastián

Vitoria-Gasteiz
Miranda de Ebro
Pamplona (Iruña)
Pyrenees
ANDORRA
Figueres
Girona (Gerona)
Costa Brava

Burgos
Logroño
Huesca
Cataluña
Soria
Lleida
Terrassa
Mataró

S A I N
Sistema
Zaragoza
Sabadell
Barcelona
L'Hospitalet de Llobregat

egovia
Ibérico
Reus
Tarragona

◆ Work continues on the Sagrada Família, Gaudí's unfinished cathedral. Begun in 1882, construction passed the mid-point in 2010 and is now due to be completed in around 2025.

MADRID
Getafe
Teruel
Tortosa

oledo
Cuenca
Castellón de la Plana
Palma
Menorca

Mallorca

Albacete
Valencia
Gandia
Ibiza
Islas Baleares
(Balearic Islands)

iudad Real
País Valenciano
Elda
Benidorm
Formentera

Linares
Cieza
Alicante (Alacant)
Elche (Elx)
Costa Blanca

◆ Seat of many great civilizations throughout history, the name Mediterranean translates as "sea between the lands."

Jaén
Murcia
Lorca

Granada
Cartagena

Sierra Nevada
Almería

Motril
Costa del Sol

M e d i t e r r a n e a n   S e a

◆ Spain produces just under half of the world's olive oil, which amounted to around 1.2 million tons (tonnes) in 2009. Of this, roughly 30 percent is the highest quality "extra-virgin" olive oil.

Golfe du Lion

A L G E R I A

Germany & The Alpine States

The Kiel Canal is 61 miles (98 km) long and one of the busiest canals in the world, with around 45,000 ships a year passing between the Baltic and the North Sea.

Early in the morning of Sunday, August 13, 1961, work began on the Berlin Wall, which would eventually run for 66 miles (107 km) between east and west Berlin, cutting through 192 streets.

During what became known as "The Berlin Airlift" a total of 2,326,406 tons (tonnes) of supplies were flown into Berlin over an 18-month period to break a Soviet blockade of the city.

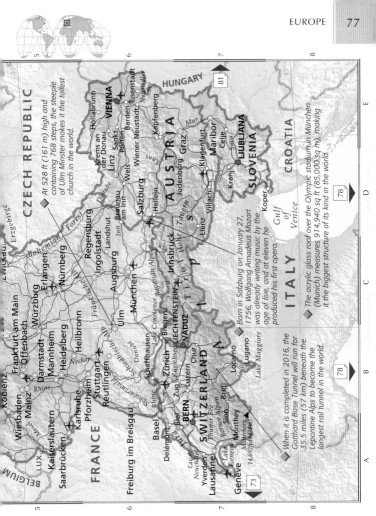

At 528 ft (161 m) high, and containing 768 steps, the steeple of Ulm Minster makes it the tallest church in the world.

Born in Salzburg on January 27, 1756, Wolfgang Amadeus Mozart was already writing music by the age of five, and at eleven he produced his first opera.

The acrylic glass roof over the Olympic stadium in München (Munich) measures 914,940 sq ft (85,000 sq m), making it the biggest structure of its kind in the world.

When it is completed in 2016, the Gotthard Base Tunnel will run for 35.5 miles (57 km) beneath the Lepontine Alps to become the longest rail tunnel in the world.

CZECH REPUBLIC

HUNGARY

Erzgebirge

Bohemian Forest

VIENNA
Krems an der Donau
Sankt Pölten
Linz
Wels
Rollabrunn
Barden
Wiener Neustadt
Eisenstadt
Kapfenberg

A U S T R I A

Mur
Judenburg
Graz
Klagenfurt
Maribor
Celje
LJUBLJANA
SLOVENIA
Villach
Kranj
Koper

CROATIA

Salzburg
Braunau am Inn
Hallein
Lienz
Hohe Tauern

A L P S

Gulf of Venice

ITALY

Frankfurt am Main
Koblenz
Offenbach
Würzburg
Mainz
Wiesbaden
Darmstadt
Mannheim
Heidelberg
Heilbronn
Kaiserslautern
Karlsruhe
Pforzheim
Saarbrücken
Stuttgart
Reutlingen

Regensburg
Erlangen
Nürnberg
Ingolstadt
Landshut
Augsburg
München
Inn

Ulm
Danube

Lake Constance
Austrian Alps
Tirol
Innsbruck
Bregenz
LIECHTENSTEIN
VADUZ
Chur
Bodensee

Schaffhausen
Zürich
Zugersee
Zug
Luzern
St. Gallen

Freiburg im Breisgau
Rhine

Basel
Biel
Delémont
Bern
Berner Alpen
Thuner See
Sion

SWITZERLAND

Lago Maggiore
Lake Lugano
Locarno
Lugano

FRANCE

Neckar
Schwarzwald

Mosel

BELGIUM
LUX

Matterhorn
14,692 ft (4478 m)

Lake Neuchâtel
Yverdon
Lausanne
Lake Geneva
Genève
Monthey
Brig

73

78

78

81

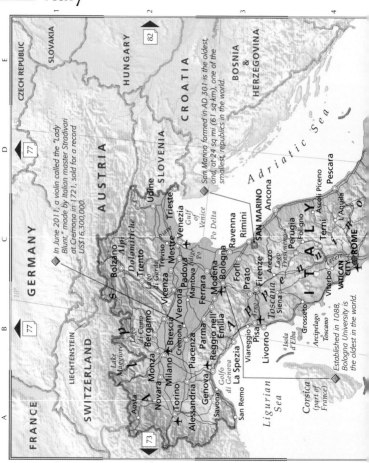

In June 2011, a violin called the "Lady Blunt," made by Italian master Stradivari at Cremona in 1721, sold for a record US$16,300,000.

San Marino formed in AD 301 is the oldest, and, at 24 sq mi (61 sq km), one of the smallest republics in the world.

Established in 1088, Bologna University is the oldest in the world.

SLOVAKIA

CZECH REPUBLIC

HUNGARY

AUSTRIA

SLOVENIA

CROATIA

BOSNIA & HERZEGOVINA

GERMANY

LIECHTENSTEIN

SWITZERLAND

FRANCE

Aosta
Novara
Torino
Alessandria
Monza
Milano
Bergamo
Brescia
Cremona
Piacenza
Savona
San Remo
Genova
Golfo di Genova
La Spezia
Reggio nell'Emilia
Parma
Modena
Bologna
Ferrara
Padova
Vicenza
Verona
Mantova
Treviso
Mestre
Venezia
Bolzano
Trento
Udine
Trieste
Gulf of Venice
Po Delta
Ravenna
Rimini
SAN MARINO
Ancona
Ascoli Piceno
Pescara
L'Aquila
Terni
Foligno
Perugia
Arezzo
Viterbo
VATICAN CITY
ROME
Siena
Grosseto
Livorno
Pisa
Viareggio

Dolomitiche
Alpi
Lago di Garda
Lago di Como
Lago Maggiore
Lake Maggiore

Po
Lago Trasimeno

Tevere

Arno

Toscana

Isola d'Elba
Arcipelago Toscano

Corsica (part of France)

Ligurian Sea

Adriatic Sea

ITALY

APPENNINO

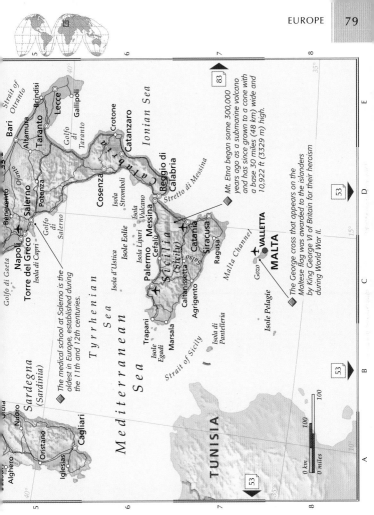

Strait of Otranto

Brindisi
Gallipoli
Bari
Altamura
Lecce
Taranto
Golfo di Taranto
Crotone
Ofanto
Benevento
Catanzaro
Salerno
Potenza
Cosenza
Ionian Sea
Golfo di Salerno
Isola Stromboli
Reggio di Calabria
Napoli
Torre del Greco
Isola di Capri
Stretto di Messina
Isola Vulcano
Isole Eolie
Messina
Isola Lipari
Cetalù
Catania
Isola d'Ustica
Palermo
Siracusa
Sicilia (Sicily)
Salso
Ragusa
Caltanissetta
Enna
Agrigento
Trapani
Isole Egadi
Marsala
Isola di Pantelleria

Tyrrhenian Sea

Mediterranean Sea

Sardegna (Sardinia)
Nuoro
Cagliari
Oristano
Iglesias
Alghero

Strait of Sicily

Gozo
VALLETTA
MALTA

Isole Pelagie

Malta Channel

TUNISIA

Mt. Etna began some 300,000 years ago as a submarine volcano and has since grown to a cone with a base 30 miles (48 km) wide and 10,922 ft (3329 m) high.

The medical school at Salerno is the oldest in Europe, established during the 11th and 12th centuries.

The George cross that appears on the Maltese flag was awarded to the islanders by King George VI of Britain for their heroism during World War II.

83

53

53

53

53

0 km   100
0 miles   100

Central Europe

Built between 1747 and 1795, the Zaluski Library in Warsaw was one of the world's first public libraries.

Founded in Gdansk shipyard in 1980, the Solidarity trade union, and its leader Lech Walesa, played a key role in the downfall of communism across much of eastern Europe.

In November 1989, the so-called "Velvet Revolution" saw Czechoslovakia split into the Czech Republic and Slovakia.

LATVIA

LITHUANIA

BELARUS

KALININGRAD (part of Russian Federation)

Baltic Sea

SWEDEN

DENMARK

Bornholm (part of Denmark)

GERMANY

POLAND

WARSAW

Białystok

Lublin

Ostrowiec Świętokrzyski

Radom

Kielce

Łódź

Warta

Wrocław

Oder

Legnica

Zielona Góra

Poznań

Kalisz

Włocławek

Toruń

Bydgoszcz

Płock

Ostrołęka

Olsztyn

Elbląg

Grudziądz

Piła

Noteć

Czluchów

Koszalin

Słupsk

Gdańsk

Gdynia

Gulf of Danzig

Wisła

Courland Lagoon

Mazury

Narew

Bug

Wisła

Gorzów Wielkopolski

Szczecin

Zalew Szczeciński

Pomeranian Bay

Warta

Oder

Dĕčín

Liberec

Tepli

67

88

89

76

25°

20°

15°

55°

50°

0 km   100
0 miles   100

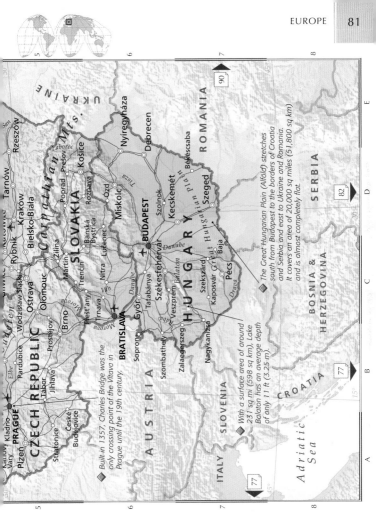

EUROPE

UKRAINE

Sán

Rzeszów

Tarnów

Carpathian Mts.

Laborec

Prešov Košice

Poprad

Bielsko-Biała

Kraków

Rožňava

Ózd

Ondava

Nyíregyháza

Debrecen

Tisza

ROMANIA

Békéscsaba

Miskolc

Kecskemét

Szeged

SLOVAKIA

Žilina

Banská
Bystrica

Martin

Trenčín

Nitra

Lučenec

Szolnok

Rybnik

Wodzisław Śląski

Ostrava

Olomouc

Prostějov

Brno

Přerov

Morava

Trnava

Tatabánya

Györ

Székesfehérvár

BUDAPEST

Danube

Great Hungarian Plain

The Great Hungarian Plain (Alföld) stretches
south from Budapest to the borders of Croatia
and Serbia and east to Ukraine and Romania.
It covers an area of 20,000 sq miles (51,800 sq km)
and is almost completely flat.

SERBIA

CZECH REPUBLIC

Pardubice

Jihlava

Tábor

BRATISLAVA

Sopron

Vah

Danube

Balaton

Veszprém

Pécs

Szekszárd

Baja

Dráva

HUNGARY

BOSNIA &
HERZEGOVINA

Pilzeň PRAGUE

Vltava

Kladno

Vary

Stříbro

Strakonice

České
Budějovice

Built in 1357, Charles Bridge was the
only crossing point of the Vltava in
Prague until the 19th century.

Szombathely

Zalaegerszeg

Nagykanizsa

Kapošvár

AUSTRIA

SLOVENIA

CROATIA

ITALY

With a surface area of around
231 sq mi (598 sq km), Lake
Balaton has an average depth
of only 11 ft (3.25 m).

Adriatic
Sea

Raba

Morava

Elbe

Rába

Mura

90

82

77

77

Labe

Southeast Europe

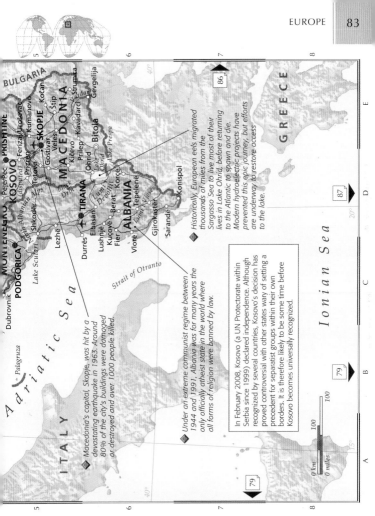

Macedonia's capital, Skopje, was hit by a devastating earthquake in 1963. Around 80% of the city's buildings were damaged or destroyed and over 1000 people killed.

Under an extreme communist regime between 1944 and 1991, Albania was for many years the only officially atheist state in the world where all forms of religion were banned by law.

Historically, European eels migrated thousands of miles from the Sargasso Sea to live most of their lives in Lake Ohrid, before returning to the Atlantic to spawn and die. Modern hydroelectric projects have prevented this epic journey, but efforts are underway to restore lake access to the lake.

In February 2008, Kosovo (a UN Protectorate within Serbia since 1999) declared independence. Although recognized by several countries, Kosovo's decision has proved controversial with other states wary of setting a precedent for separatist groups within their own borders. It is therefore likely to be some time before Kosovo becomes universally recognized.

The Mediterranean

UNITED KINGDOM

Thames

NETHERLANDS

BELGIUM

Rhine

GERMANY

LUX.

English Channel

Seine

E U

Danube

Loire

FRANCE

LIECH.

SWITZ.

L. Geneva

A l p s

A T L A N T I C

Bay of Biscay

Massif Central

Dordogne

Rhône

Po

SAN MARINO

O C E A N

Garonne

Pyrenees

Genoa

Apennine

MONACO

Marseille

Livorno

ANDORRA

Golfe du Lion

Corsica

VATICAN CITY

Iberian

SPAIN

Ebro

Barcelona

PORTUGAL

Tagus

Balearic Is.

Sardinia

M e d i t e

Peninsula

Valencia

Tyrrhenian Sea

Guadalquivir

Algiers

Tunis

Gibraltar (UK)

Gibraltar

Oran

TUNISIA

Strait of Gibraltar

Sfax

MOROCCO

A t l a s   M o u n t a i n s

Chott el Jerid

Trip

Madeira (Portugal)

ALGERIA

Grand Erg Occidental

Grand Erg Oriental

Canary Is. (Spain)

A F R I

Saha

a

0 km 400

0 miles 400

POLAND

CZECH REP.

E U R O P E

SLOVAKIA

AUSTRIA

HUNGARY

SLOVENIA

Hungarian
Plain

CROATIA

BOS. &
HERZ.

ROMANIA

SERBIA

KOSOVO
(disputed)

MON.

Dinaric Alps

ITALY

Naples

ALBANIA

MACEDONIA

Pindus Mts.

Adriatic Sea

Ionian
Sea

Sicily

MALTA

r r a n e a n

GREECE

Piraeus

Peloponnese

Carpathian
Mountains

MOLDOVA

Danube
Delta

Danube

BULGARIA

Balkan Mts.

Rhodope Mts.

Aegean
Sea

Lesbos

Kos

Rhodes

Crete

UKRAINE

(the Ukrainian territory of
Crimea was annexed by
Russia in 2014)

Dnieper

Crimea

Sea
of Azov

Black Sea

Bosporus

T U R K E Y

Anatolia

Taurus Mts.

Cyprus

S e a

Don

RUSSIAN
FEDERATION

Caucasus

GEORGIA

Lake
Van

Euphrates

SYRIA

LEBANON

Anti-Lebanon

IRAQ

Tigris

Gulf of Sirte

L I B Y A

C   A

r a

EGYPT

Libyan

Desert

Nile
Delta

Nile

Port Said

Haifa

ISRAEL

Suez Canal

Syrian Desert

JORDAN

Red Sea

A S I A

SAUDI
ARABIA

Arabian
Peninsula

Bulgaria & Greece

Bulgaria is one of the few countries in the world where locals shake their heads from side to side to mean "yes" and nod up and down for "no."

Sofia's skyline is dominated by the gold domes of the Alexander Nevski Memorial Church, which took craftsmen and artists some thirty years to build between 1882 and 1912.

Built between 447 and 438 BCE, the Parthenon survived almost unscathed for over 2000 years until, in 1687, the building exploded beneath a gunpowder magazine, causing considerable damage.

ROMANIA

SERBIA

KOSOVO (disputed)

ALBANIA

MACEDONIA

BULGARIA

TURKEY

G R E E C E

Black Sea

Marmara Denizi

Thracian Sea

Vóreíes Sporádes

Danube

Balkan Mountains

Rhodope Mountains

Píndos

Vidin
Vratsa
SOFIA
Pernik
Blagoevgrad
Kyustendil
Petrich
Serres
Kilkís
Thessaloníki
Kateríni
Véroia
Kozáni
Flórina
Ioánnina
Kérkyra
Préveza
Tríkala
Kardítsa
Lárisa
Vólos
Kalamariá
Thermaïkós Kólpos
Akrotírio Palioúri
Akrotírio Drépano
Akrotírio Pínes
Thásos
Samothráki
Límnos
Lésvos
Mitilíni
Alexandroúpoli
Komotiní
Xánthi
Kaválá
Dráma
Khaskovo
Kŭrdzhali
Svilengrad
Orestiáda
Momchilgrad
Plovdiv
Pazardzhik
Velingrad
Kazanlŭk
Stara Zagora
Sliven
Yambol
Burgas
Varna
Dobrich
Razgrad
Shumen
Ruse
Pleven
Loveč
Gabrovo
Veliko Tŭrnovo
Maritsa
Struma
Vardar
Píneios
Aliákmon
Lake Prespa

90
90
93
83

1 · 2 · 3 · 4

20°E · 25°E

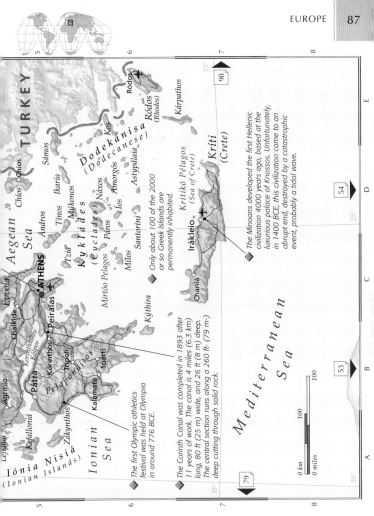

TURKEY

Chíos

Sámos

Ikaría

Kos

Dodekánisa
(Dodecanese)

Ródos
(Rhodes)

Kárpathos

Astypálaia

Kríti
(Crete)

Kritikó Pélagos
(Sea of Crete)

Irákleio

Chaniá

The Minoans developed the first Hellenic
civilization 4000 years ago, based at the
luxurious palace of Knossos. Unfortunately,
in 1400 BCE, this civilization came to an
abrupt end, destroyed by a catastrophic
event, probably a tidal wave.

Aegean
Sea

Évvoia

ATHENS

Peiraías

Chalkída

Kykládes
(Cyclades)

Ándros

Tínos

Mýkonos

Náxos

Íos

Amorgós

Páros

Santoríni

Tziá

Mílos

Mirtóo Pélagos

Kýthira

Korinthiakós
Kólpos

Kórinthos

Trípoli

Pátra

Peloponnisos

Kalámata

Spárti

Agrínio

Kefalloniá

Zákynthos

Iónia Nisiá
(Ionian Islands)

Ionian
Sea

Mediterranean Sea

Only about 100 of the 2000
or so Greek Islands are
permanently inhabited.

The Corinth Canal was completed in 1893 after
11 years of work. The canal is 4 miles (6.3 km)
long, 80 ft (25 m) wide, and 26 ft (8 m) deep.
The central section runs along a 260 ft (79 m)
deep cutting through solid rock.

The first Olympic athletics
festival was held at Olympia
in around 776 BCE.

0 km   100
0 miles   100

35°

25°

20°

The Baltic States & Belarus

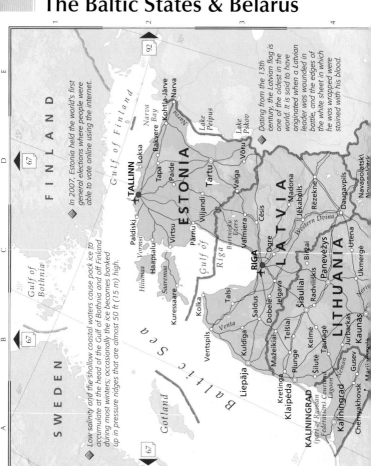

In 2007, Estonia held the world's first general elections where people were able to vote online using the internet.

Low salinity and the shallow coastal waters cause pack ice to accumulate at the head of the Gulf of Bothnia and off Finland during most winters; occasionally the ice becomes banked up in pressure ridges that are almost 50 ft (15 m) high.

Dating from the 13th century, the Latvian flag is one of the oldest in the world. It is said to have originated when a Latvian leader was wounded in battle, and the edges of the white sheet in which he was wrapped were stained with his blood.

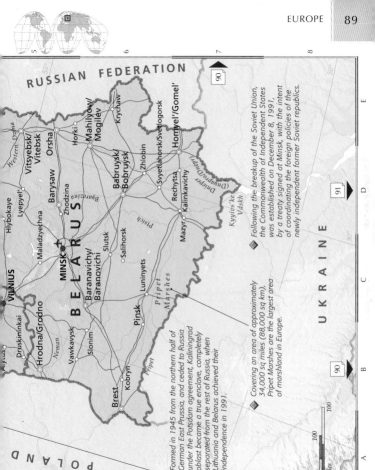

RUSSIAN FEDERATION

BELARUS

Western Dvina

Hlybokaye
Lyepyel'

Vitsyebsk/Vitebsk
Orsha
Barysaw
Zhodzina

Horki

Mahilyow/Mogilev
Krychaw

Byarezina

MINSK

Maladzyechna

VILNIUS

Druskininkai

Hrodna/Grodno

Neman

Slonim

Vawkavysk

Baranavichy/Baranovichi

Slutsk

Salihorsk

Ptsich

Babruysk/Bobruysk

Zhlobin

Svyetlahorsk/Svetlogorsk

Homyel'/Gomel'

Redhytsa

Mazyr
Kalinkavichy

Dnieper (Dnipro/Dnyapro)

Kyyivs'ke Vdskh.

Pripet Marshes

Luninyets

Pinsk

Kobryn

Brest

Pripet

POLAND

UKRAINE

Formed in 1945 from the northern half of German East Prussia, and ceded to Russia under the Potsdam agreement, Kaliningrad oblast became a true enclave, completely separated from the rest of Russia, when Lithuania and Belarus achieved their independence in 1991.

Covering an area of approximately 34,000 sq miles (88,000 sq km), Pripet Marshes are the largest area of marshland in Europe.

Following the breakup of the Soviet Union, the Commonwealth of Independent States was established on December 8, 1991, by a treaty signed at Minsk, with the intent of coordinating the foreign policies of the newly independent former Soviet republics.

0 km   100

0 miles   100

Ukraine, Moldova & Romania

POLAND

BELARUS

Pripet

Pripet Marshes

Kovel'

◆ On April 25, 1986, engineers accidentally initiated an uncontrolled chain reaction in the number 4 reactor of the Chornobyl' nuclear power plant. The resulting explosion released 8 tons (tonnes) of radioactive material in the world's worst-ever nuclear accident.

Luts'k

Korosten

Rivne

Zhytomyr

L'viv

SLOVAKIA

Ternopil'

U K R

Ivano-Frankivs'k

Khmel'-nyts'kyy

Vinnytsya

◆ Vlad Dracula or Vlad the Impaler was the real-life prince upon whom Bram Stoker based his famous Count Dracula. Dracula was born in Transylvania in 1431 in the town of Sighisoara.

Uzhhorod

Kam"yanets'-Podil's'kyy

Dniester

Transnistri

Satu Mare

Chernivtsi

Baia Mare

Suceava

Botoşani

Ribniţa

Bălţi

MOLDOVA

HUNGARY

Oradea

Dej

Transylvania

Piatra-Neamţ

Dubăsari

CHIŞINĂ

Cluj-Napoca

Iaşi

Tiraspol

Arad

Târgu Mureş

Bacău

Tighina (Bendery)

Alba Iulia

Sighişoara

Siret

Timişoara

Deva

R O M A N I A

Basarabeasca

SERBIA

Reşiţa

Sibiu

Carpaţii Meridionali

Focşani

Braşov

Galaţi

Reni

Râmnicu Vâlcea

Buzău

Brăila

Tulcea

◆ In 1889, Timisoara became the first city in Europe to have electric street lighting.

Drobeta-Turnu Severin

Piteşti

Târgovişte

Ploieşti

BUCHAREST

Craiova

Olt

Danube

Constanţa

Corabia

Giurgiu

Eforie Sud

Mangalia

BULGARIA

RUSSIAN
FEDERATION

93

◆ A monument in central Kiev stands as testament to the 7–12 million Ukrainian peasants who died during the Great Famine, or Holodomor, of 1932–33.

E
30°
35°
40°

Shostka

Chernihiv

Chornobyl'

Kyyivs'ke Vdskh.

Sumy

◆ KIEV

Kaniys'ke Vdskh.

Bila Tserkva

Lubny

Kharkiv

A I N E

Cherkasy

Kremenchuts'ke Vdskh.

Poltava

Donets

Syeverodonets'k

Oleksandriya

Kremenchuk

Slov"yans'k

Luhans'k

Kirovohrad

Dnipropetrovs'k

Pavlohrad

Horlivka

Kostyantynivka

Yenakiyeve

Makiyivka

Krasnyy Luch

Kryvyy Rih

Nikopol'

Zaporizhzhya

Donets'k

Piydennyy Buh

Mariupol'

Mykolayiv

Kakhovs'ka Vdskh.

Melitopol'

Berdyans'k

Kherson

Dnieper

Kakhovka

Sea of Azov

(the Ukrainian territory of Crimea was annexed by Russia in 2014)

◆ Odesa

Karkinits'ka Zatoka

Kryms'kyy Pivostriv

Kerch

◆ In 1872, an iron foundry was established at Donets'k by British industrialist John Hughes (from whom the town's pre-Revolutionary name Yuzovka was derived) to produce rails for the growing Russian transportation network.

93

RUSSIAN FEDERATION

Yevpatoriya

Simferopol'

Sevastopol'

Yalta

Black Sea

0 km   100
0 miles   100

◆ Odesa was one of the major flashpoints in the Russian Revolution of 1905, and was the scene of the mutiny on the warship Potemkin, when sailors protesting against the serving of rotten meat eventually killed several of the ship's officers.

98

E
F
G
35°
H

1

2
50°

3

4
45°

5

European Russia

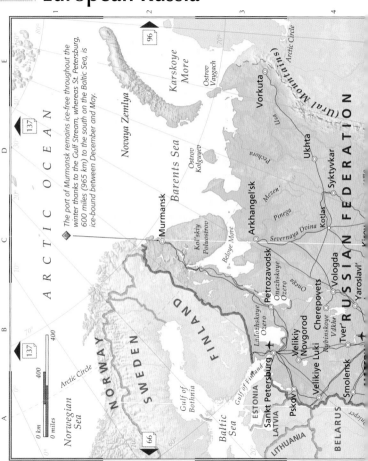

The port of Murmansk remains ice-free throughout the winter thanks to the Gulf Stream, whereas St. Petersburg, 600 miles (965 km) to the south on the Baltic Sea, is ice-bound between December and May.

96
137
137
66
66

ARCTIC OCEAN

Karskoye More

Novaya Zemlya

Ostrov Vaygach

Vorkuta

Ural Mountains

Usa

Ukhta

Pechora

Syktyvkar

RUSSIAN FEDERATION

Barents Sea

Ostrov Kolguyev

Mezen'

Kotlas

Murmansk

Kol'skiy Poluostrov

Arkhangel'sk

Pinega

Severnaya Dvina

Beloye More

Onega

Vologda

Petrozavodsk

Onezhskoye Ozero

Cherepovets

Yaroslavl'

NORWAY

FINLAND

Ladozhskoye Ozero

Velikiy Novgorod

Rybinskoye Vdkhr.

Tver'

Norwegian Sea

SWEDEN

Velikiye Luki

Smolensk

Gulf of Bothnia

Gulf of Finland

Sankt Peterburg

Pskov

ESTONIA

BELARUS

Baltic Sea

LATVIA

LITHUANIA

Arctic Circle

0 km 400
0 miles 400

The Ural Mountains form the traditional boundary between Europe and Asia, extending some 1550 miles (2500 km). They were formed over 280 million years ago as the East European and Siberian plates moved together.

From August 1942 to February 1943, German armies laid siege to Volgograd, formerly known as Stalingrad. The Germans themselves were eventually surrounded and lost almost 250,000 men.

Caviar is the processed eggs, or roe, of sturgeon that live in the Caspian Sea and Volga River. Overfishing and poaching in recent years have seen the price of the finest caviar, a pearly white variety called Gawad, rise to around US$25,000 for 2.2 lbs (1 kg).

Running from the Black Sea to the Caspian Sea, the Caucasus Mountains include Mt Elbrus, which at 18,510 ft (5642 m) is the highest point in Europe, and still uplifting at the rate of 0.4 inches (1 cm) every year.

KAZAKHSTAN

UZBEKISTAN

TURKMENISTAN

Ural'skiye Gory

Perm'

Izhevsk
Naberezhnyye
Chelny
Ufa
Orenburg
Orsk

Nizhniy Novgorod
Cheboksary
Kazan'
Vyatka
Ul'yanovsk
Saransk
Tol'yatti
Samara
Penza
Balakovo
Tambov
Saratov
Ryazan'
Voronezh
Mikhaylovka
Orël
Volgograd
Belgorod
Rostov-na-Donu
Elista
Astrakhan'

Bryansk
Don
Krasnodar
Stavropol'
Cherkessk
El'brus
18,510 ft (5642m)
Nal'chik
Vladikavkaz
Groznyy
Makhachkala

Sochi
Caucasus
GEORGIA
AZERBAIJAN
ARMENIA
Az

Black
Sea

Sea of
Azov

Donets

Kuma

Volga

Volga

UKRAINE
(the Ukrainian territory of Crimea was annexed by Russia in 2014)

Caspian
Sea

Aral Sea

TURKEY

IRAN

IRAQ

96

99

99

104

99

99

North & West Asia

OCEAN

137

New Siberian Islands

Laptev Sea

East Siberian
Sea

berian Lowland

Anabar *Olenёk* *Lena* *Yana* *Indigirka* *Kolyma*

Wrangel Island

Chukchi
Sea

Long Strait

Bering Strait

berian Plateau

Arctic Circle

16

FEDERATION

Velikaya

b *e* *r* *i* *a*

Lena   Amga

Bering
Sea

Vitim

Lake
Baikal

Amur Zeya

Sea of
Okhotsk

Kamchatka

Aleutian Islands

Argun

Sakhalin

I   A

Kuril Islands

Gobi

(administered by
Russian Federation,
claimed by Japan.)

Sea of
Japan
(East Sea)

PACIFIC

16

Yellow River

East
China
Sea

OCEAN

Yangtze

Tropic of Cancer

0 km   800

0 miles   800

Mekong

South
China
Sea

80°100°120° 140°   160°

OCEAN

Ostrov
Vrangelya

18

◈ Also known as the "Road of Bones," construction
of the 1262 mile (2031 km) road between Yakutsk
and Magadan took over twenty years and cost
the lives of a huge number of prisoners from
Stalin's notorious Gulag camps.

Pevek

Anadyr'

*Vostochno-
Sibirskoye More*

Ambarchik

*Bering
Sea*

180°

*Severnaya
Zemlya*

*Novosibirskiye
Ostrova*

*More
Laptevykh*

Ossora

*Poluostrov
Taymyr*

*Ozero
Taymyr*

Tiksi

Ust'-Kamchatsks

*Poluostov
Kamchatka*

Petropavlovsk
-Kamchatskiy

Oleněk

Magadan

50°

160°

*rednesibirskoye
Ploskogor'ye*

Okhotsk

Verkhoyanskiy Khrebet

Lena

Yakutsk

I A N

Suntar

*Sea of
Okhotsk*

i b i r
(S i b e r i a)

A T I O N

Sakhalin

Lena

Komsomol'sk-
na-Amure

Kuril Islands

Kansk

Bratsk

Skovorodno

Yuzhno-
Sakhalinsk

*Ozero
Baykal*

Blagoveshchensk

134

Irkutsk

Chita

Amur

Khabarovsk

40°

Ulan-Ude

C H I N A

JAPAN

Vladivostok

◈ The Trans-Siberian Railroad, completed in 1916, runs
5578 miles (9297 km) between Moscow and Vladivostok.
Crossing eight time zones, the journey takes six days.

0 km   500

0 miles   500

MONGOLIA

100°   110°   120°   130°

110

E   F   G   H

Black Sea

An average of 50,000 commercial ships pass through the Bosporus a year, along with thousands of ferries and smaller passenger boats. The strait is three times busier than the Suez Canal and four times as busy as the Panama Canal.

ROMANIA

BULGARIA

GREECE

Edirne
Kırklareli
Tekirdağ
Çanakkale Boğazı (Dardanelles)
Bosporus
İstanbul
Marmara Denizi
Bursa
İzmit
Adapazarı
Çanakkale
Balıkesir
Ayvalık
Lésvos
Mánisa
Chíos
İzmir
Afyon
Sámos
Uşak
Aydın
Muğla
Denizli
Isparta
Bodrum
Dalaman
Antalya
Ródos
Megísti
Antalya Körfezi
Kríti
Kárpathos
Toros Dağları

Zonguldak
Küre Dağları
Sinop
Kastamonu
Samsun
Karabük
Çankırı
Kızıl Irmak
Canik Dağları
Örc
Eskişehir
ANKARA
Çorum
Tokat
Kütahya
Kırıkkale
Sivas
Kayseri
Tuz Gölü
Nevşehir
Niğde
Kahramanmaraş
Konya
Ereğli
Osmaniye
Adana
Mersin
Tarsus
Gaziantep
İskenderun
Antakya

A n a t o l i a

T U R K

TURKISH REPUBLIC OF NORTHERN CYPRUS
(recognized only by Turkey)
Girne (Kyrenia)
Gazimağusa (Famagusta)
NICOSIA
Paphos
Larnaca
Limassol
CYPRUS
LEBANON

Mediterranean Sea

RUSSIAN FEDERATION

[93]

◆ The Spitak earthquake struck Armenia in 1988, killing at least 25,000 people and devastating the country's infrastructure.

Caspian Sea

Caucasus

Gagra
Sokhumi
Ochamchire
Enguri
Kutaisi
Poti
GEORGIA
Batumi
Hopa
TBILISI Rustavi
Trabzon   Rize
Vanadzor
Kura
Quba
Vanadzor   Gäncä   Mingäçevir
Sumqayıt [104]
Doğu Karadeniz Dağları   Gyumri   **ARMENIA**   **AZERBAIJAN**   **BAKU**
Kars   Sevana Lich
Erzurum   **YEREVAN**   *Nagorno-Karabakh*
Erzincan   *Aras*   Xankändi
E Y   *Büyükağrı Dağı (Mount Ararat) 16,853ft (5137m)*   Naxçıvan
AZERBAIJAN   *Aras*   • Länkäran
Elazığ   *Van Gölü*
Tigris   Muş   ◆ Azerbaijan has substantial oil reserves located in and around the Caspian Sea. They were some of the earliest oilfields in the world to be exploited.
Malatya   *Güney Doğu Toroslar*   Van
Diyarbakır   Siirt   **I R A N**
Adıyaman   Batman
Mardin
Şanlıurfa   *Kurdistan*
◆ The salty water of Lake Van inhibits all animal life except the Pearl Mullet, a small fish that has adapted to the harsh conditions.
[102]

◆ Atatürk Dam, one of the largest dams in the world, was completed in 1990. The reservoir behind the dam covers an area of 315 sq miles (816 sq km) and often requires interruptions in the flow of the Euphrates River to maintain water levels.

S Y R I A   I R A Q

0 km   200
0 miles   200

E   F   G   H
40°   45°   50°

[102]

The Near East

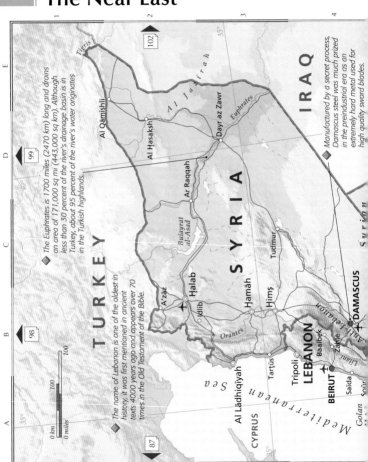

The Euphrates is 1700 miles (2470 km) long and drains an area of 171,000 sq mi (443,000 sq km). Although less than 30 percent of the river's drainage basin is in Turkey, about 95 percent of the river's water originates in the Turkish highlands.

Manufactured by a secret process, Damascus steel was much prized in the preindustrial era as an extremely hard metal used for high quality sword blades.

The name of Lebanon is one of the oldest in history; it was first mentioned in ancient texts 4000 years ago and appears over 70 times in the Old Testament of the Bible.

TURKEY

IRAQ

SYRIA

LEBANON

CYPRUS

DAMASCUS

BEIRUT

Tigris

Al Qāmishlī

Al Hasakah

Dayr az Zawr

Euphrates

Al Jazirah

Ar Raqqah

Buḥayrat al-Asad

Tudmur

A'zāz

Ḥalab

Idlib

Ḥamāh

Ḥimṣ

Orontes

Tartūs

Tripoli

Baalbek

Zahlé

Saïda

Şūr

Al Lādhiqīyah

Mediterranean Sea

Litani

Anti-Lebanon

Golan

Syrian

0 km   100
0 miles 100

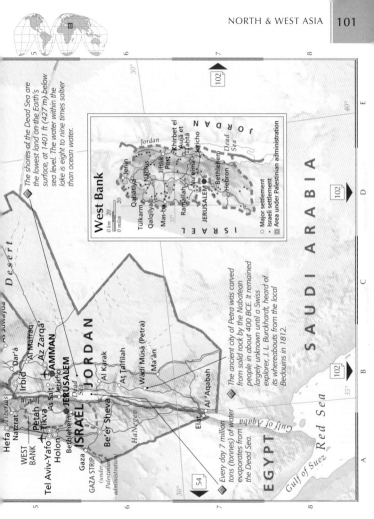

The shores of the Dead Sea are the lowest land on the Earth's surface, at 1401 ft (427 m) below sea level. The water within the lake is eight to nine times saltier than ocean water.

West Bank

0 km 20
0 miles 20

Jordan

Jenin

Qabatiya

Tulkarm   Nablus   Khirbet el Lahta

Qalqiliya   Jiftlik Post   Jericho

Mas-ha   Nu'eima   Dead Sea

Ramallah   Bethlehem

JERUSALEM   Hebron

ISRAEL   JORDAN

○ Major settlement
• Israeli settlement
◎ Area under Palestinian administration

The ancient city of Petra was carved from solid rock by the Nabatean people in about 400 BCE. It remained largely unknown until a Swiss explorer, J. L. Burckhardt, heard of its whereabouts from the local Bedouins in 1812.

Every day 7 million tons (tonnes) of water evaporates from the Dead Sea.

SAUDI ARABIA

JORDAN

As Suwayda'
Dar'a
Al Matraq
Irbid
Az Zarqa'
AMMAN
As Salt
Jericho
Dead Sea
Al Karak
At Tafilah
Wadi Musa (Petra)
Ma'an

Desert

Hefa
Nazerat
Petah Tikva
Tel Aviv-Yafo
Holon
WEST BANK
JERUSALEM
Bethlehem
Gaza
GAZA STRIP
(under Palestinian administration)
Be'er Sheva
HaNegev
ISRAEL

Elat
Al 'Aqabah

Gulf of Aqaba

Gulf of Suez

Red Sea

EGYPT

54

102

The Middle East

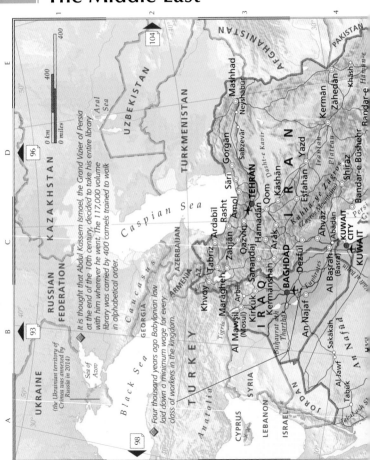

400

400

0 km   400

0 miles

It is thought that Abdul Kassem Ismael, the Grand Vizier of Persia at the end of the 10th century, decided to take his entire library with him wherever he went. The 117,000 volume library was carried by 400 camels trained to walk in alphabetical order.

Four thousand years ago Babylonian law laid down a minimum wage for every class of workers in the kingdom.

(the Ukrainian territory of Crimea was annexed by Russia in 2014)

PAKISTAN

AFGHANISTAN

UZBEKISTAN

TURKMENISTAN

KAZAKHSTAN

RUSSIAN FEDERATION

UKRAINE

GEORGIA

ARMENIA

AZERBAIJAN

Caucasus

Black Sea

Sea of Azov

Caspian Sea

Aral Sea

Mashhad

Neyshābūr

Gorgān

Sārī

Āmol

Rasht

Ardabil

Tabriz

Khvoy

Marāgheh

Arbīl

Kirkūk

Al Mawşil (Mosul)

Zanjān

Qazvīn

TEHRĀN

Qom

Kāshān

Eşfahān

Yazd

Kermān

Zāhedān

Khāsh

Bandar-e Ḥamīm

Bandar-e Būshehr

Shīrāz

Ahvāz

Ābādān

KUWAIT CITY

KUWAIT

Al Başrah (Basra)

Al Najaf

BAGHDAD

Dezfūl

Kermānshāh

Sanandaj

Hamadān

Arāk

Sabzevār

Dasht-e Kavir

Dasht-e Lut

Iranian Plateau

Kūh-e Zāgros (Zagros Mountains)

I R A N

I R A Q

T U R K E Y

SYRIA

LEBANON

ISRAEL

JORDAN

CYPRUS

Anatolia

Tigris

Euphrates

Buḩayrat ath Tharthār

As Sakākah

Al Jawf

An Nafūd

Tabūk

Qāsim

Persian Gulf

104

96

93

98

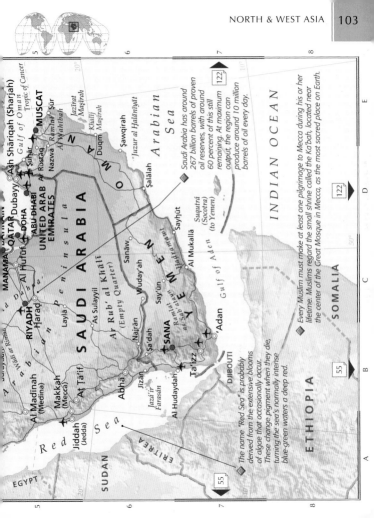

Saudi Arabia has around 267 billion barrels of proven oil reserves, with around 60 percent of this still remaining. At maximum output, the region can produce around 10 million barrels of oil every day.

Every Muslim must make at least one pilgrimage to Mecca during his or her lifetime. Muslims regard the small shrine called the Ka'bah, located near the center of the Great Mosque in Mecca, as the most sacred place on Earth.

The name "Red Sea" is probably derived from the extensive blooms of algae that occasionally occur. These change pigment when they die, turning the sea's normally intense blue-green waters a deep red.

Tropic of Cancer

MUSCAT
Gulf of Oman
Aṣḥ Shāriqah (Sharjah)
Dubayy Dubai
DOHA
QATAR
MANAMA
BAHRAIN
ABU-DHABI
UNITED ARAB EMIRATES
Ṣuḥār
Ar Rustāq
Nazwā
'Ibrī
Sūr
Jazīrat Maṣīrah
Khalīj Maṣīrah
Rīmah
Al Wahībah
Sawqirah
Duqm
'Juzur al Ḥalāniyāt
Ṣalālah
Al Ḥufūf
Al Buraymī
O M A N

Arabian Sea

INDIAN OCEAN

RIYADH
Harad
Laylā
As Sulayyil
Najrān
SAUDI ARABIA
Ar Rub' al Khālī
(Empty Quarter)
Ramlat as Sab'atayn
Wudayʻah
Sayʼūn
Shabwah
Sana
Sayḥūt
Al Mukallā
Ḥaḍramawt
SANA
Saʻdah
Y E M E N
Ta'izz
'Adan
Gulf of Aden

Suqutrā
(Socotra)
(to Yemen)

Makkah (Mecca)
Al Madīnah (Medina)
At Ṭāʼif
Abhā
Jīzān
Jazā'ir Farāsān
Al Hudaydah

SANA

Red Sea

EGYPT
SUDAN
ERITREA
DJIBOUTI
ETHIOPIA
SOMALIA

Arabian Peninsula
Dahnā
Wādī al Rimah
Wadīa

122
55

◆ Since 1960, the Aral Sea has shrunk by 90 percent, becoming extremely saline and consequently losing all but one of its once-abundant fish species.

◆ The desert of Kara Kum (Garagum) occupies over 70 percent of Turkmenistan, severely limiting human settlement across much of the country.

◆ The Kara Kum (Garagum) Canal, the world's longest irrigation canal, stretches some 850 miles (1375 km) and is known as the "River of Life," since it irrigates large areas of arid land.

Aral Sea

Ustyurt Plateau

Turan Lowland

Nukus

UZBEKISTAN

Köneürgenç
Daşoguz
Urganch
To'rtko'l
Uchquduq
Zarafshon
Aydarko'l Ko'li

Türkmenbaşy
Hazar
Balkanabat
Bereket

TURKMENISTAN

Garagum

Buxoro
Seýdi
Navoiy
Samarqan
Qarshi

Caspian Sea

Serdar

Baharly
Gökdepe
Abadan
AŞGABAT
Kaka
Tejen

Türkmenabat
Mary
Saýat
Bayramaly
Atamyrat

Garagum Kanaly

Murgap

Aqchah
Shibirghān
Mazar-e Sharī
Maimanah

Bāla Murghāb
Darya-ye Murghāb

Serhetabat

IRAN

Herāt

Harīrūd

AFGHANISTAN

Farāh

Gereshk
Qalāt

0 km   200

0 miles   200

Zaranj
Dasht-e Mārgow
Kandahār

Darya-ye Helmand

97

E F G H

80°

KAZAKHSTAN

Kara-Balta **BISHKEK** Tyup
Tokmak Ozero Karakol
Talas Issyk-Kul

ASHKENT Chirchiq **KYRGYZSTAN** Tien Shan

Namangan Dzhalal-Abad Naryn
Olmaliq Angren Andijon Kokshaal-Tau
Qo'qon Khŭjand Osh
oteppa Farg'ona
Sulyukta Khaydarkan
Zeravshan

♦ The "Epic of Manas" is a verbally transmitted
poem of close to 500,000 lines that tells the
story of Kyrgyz hero Manas and his
descendants and followers.

USHANBE **TAJIKISTAN** Pamirs

Norak Surkhob
urghon' Danghara Bartang Murghob
eppa Kŭlob
rmez Farkhor Khorugh **C H I N A**
hŭlm Faizābād Pamir
Kunduz
Baghlān

Pul-e Hindu Kush
umri

♦ Until recent years, people living in remote areas of
Afghanistan were immunized against smallpox by
having dried powdered scabs from victims of the
disease blown up their noses. This treatment was
invented by the Chinese in the 11th century, and is
thought to be the oldest form of vaccination.

arikār Asadābād

ABUL Jalālābād

Ghaznī
Gardēz

♦ Despite an area of 251,771 sq miles (652,090 sq km),
Afghanistan has a limited road network and no
railroads whatsoever, making access to much
of the country extremely difficult.

P A K I S T A N **I N D I A**

70°

E F G H

1

2

108

40°

3

4

108

5

30°

116

97

South & East Asia

Black Sea

Caspian Sea

Aral Sea

Syr Darya

Lake Balkhash

Irtysh

Yenisey

Lake Baikal

Hovsgol Nuur

Uvs Nuur

Altai Mountains

MONGOLIA

A S I A

Tien Shan

Gobi

Iranian Plateau

Hindu Kush

Takla Makan Desert

Altun Shan

Kunlun Mountains

Plateau of Tibet

C H I N A

Yellow River

Persian Gulf

Gulf of Oman

PAKISTAN

Indus

Sutlej

Yamuna

Ganges

Himalayas

Brahmaputra

NEPAL

▲ Mount Everest 29,029ft (8848m)

BHUTAN

Salween

Mekong

Yangi

Thar Desert

Rann of Kachchh

BANGLADESH

INDIA

Gulf of Khambhat

Deccan

Western Ghats

Eastern Ghats

MYANMAR (BURMA)

Irrawaddy

Red River

Xi Jia

VIETNA

LAOS

Arabian Sea

Bay of Bengal

THAILAND

Mekong

Hai

CAMBOI

Tônlé Sa

Laccadive Islands (to India)

Andaman Islands (to India)

Andaman Sea

Gulf of Thailand

Gulf of Mannar

SRI LANKA

Nicobar Islands (to India)

MALDIVES

Equator

MAL

SINGAPORE

Sumatra

I N D I A N

O C E A N

Ja

40°

60°

80°

100°

20°

60°

80°

100°

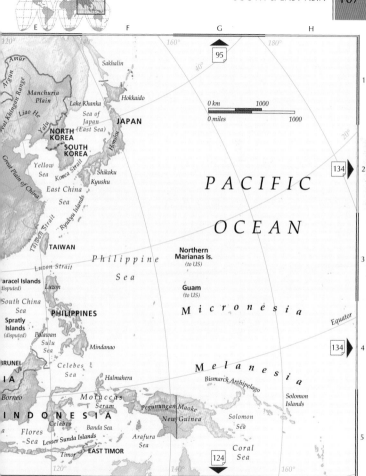

E   F   G   H

120°   140°   160°   180°

95

Amur

Sakhalin

Argun

Manchuria
Plain

Lesser Khingan Range

Liao He

Hokkaido

Lake Khanka

Sea of
Japan
(East Sea)

JAPAN

NORTH
KOREA

SOUTH
KOREA

Yalu

Honshu

Korea Strait

Yellow
Sea

Great Plain of China

East China
Sea

Shikoku
Kyushu

Ryukyu Islands

TAIWAN

Taiwan Strait

Luzon Strait

Caracel Islands
(disputed)

South China
Sea

Luzon

Philippine

Sea

PHILIPPINES

Spratly
Islands
(disputed)

Palawan

Sulu
Sea

Mindanao

BRUNEI

IA

Celebes
Sea

Halmahera

A

Borneo

Moluccas
Seram

Celebes

INDONESIA

Flores
Sea

Banda Sea

Lesser Sunda Islands

Timor

EAST TIMOR

Pegunungan Maoke
New Guinea

Arafura
Sea

PACIFIC

OCEAN

Northern
Marianas Is.
(to US)

Guam
(to US)

M i c r o n e s i a

M e l a n e s i a

Bismarck Archipelago

Solomon
Islands

Solomon
Sea

Coral
Sea

40°

20°

Equator

134

134

124

1

2

3

4

5

0 km   1000

0 miles   1000

120°   140°   160°

E   F   G   H

Western China & Mongolia

◆ The Altai Mountains provide one of the last refuges for the endangered snow leopard. There are thought to be only a few thousand animals left in the wild.

◆ The Turpan Depression is the lowest and hottest place in China. Temperatures can exceed 117°F (47°C) around the lake of Aydingkol Hu, which lies 505 ft (154 m) below sea level.

◆ Although forming around 20 percent of China's landmass, Tibet is sparsely populated, supporting only 1 percent of China's 1.3 billion population.

RUSSIAN FED

KAZAKHSTAN

MONG

KYRGYZSTAN

TAJIKISTAN

AFGH.

PAKISTAN

INDIA

NEPAL

BHUTAN

INDIA

Altai Mountains
Ulaangom
Uvs Nuur
Ölgiy
Hyargas Nuur
Har-Us Nuur
Hövsgöl Nuur
Mörön
Altay
Hovd
Tsetserle
Altay
Bayanhongor

Karamay
Ulungur Hu
Junggar Pendi
Kuytun
Yining
Shihezi
ÜRÜMQI
Qitai
Hami
Go

Tien Shan
Korla
Bosten Hu
Turpan
Dalain Hot

Kashi
Tarim He
Tarim Basin
Lop Nur
Xingxingxia

Yengisar
XINJIANG UYGUR
GANSU
Qilian Shan
Shache
ZIZHIQU
Ruoqiang
Yecheng
(claimed by India)
Taklimakan Shamo
Altun Shan
Qaidam Pendi
Qinghai H

Moyu
Qira
Kunlun Shan
Golmud
Dulan

Karakoram Range
Aksai Chin
(administered by China, claimed by India)
Qinghai
Tongtian He
Bayan Har Shan
Yushu

Demchok/Dêmqog
(administered by China, claimed by India)
Rutog
Qingzang Gaoyuan
(Plateau of Tibet)
C H I
Mekong

Gar
(Shiquanhe)
XIZANG
ZIZHIQU
(Tibet)
Tanggula Shan
Amdo
Qamdo

Zanda
Tangra Yumco
Nyima
Siling Co
Nam Co
Damxung
Nagqu
Salween

Himalayas
Brahmaputra
thaze
Nyainqêntanglha Shan
LHASA
Amdo
Arunachal Pradesh
(claimed by China)

Mount Everest
29,029ft (8848m)
Gyangzê

◆ 96

◆ 96

◆ 116

◆ 117

0 km   400
0 miles   400

96
116
117

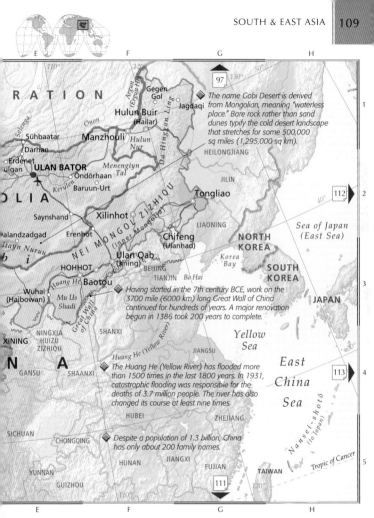

RATION

97

The name Gobi Desert is derived from Mongolian, meaning "waterless place." Bare rock rather than sand dunes typify the cold desert landscape that stretches for some 500,000 sq miles (1,295,000 sq km).

Argun (Ergun He)

Gegen Gol

Jagdaqi

Onon

Hulun Buir (Hailar)

Selenga

Da Hinggan Ling

Sühbaatar

Manzhouli

Hulun Nur

HEILONGJIANG

Darhan

Erdenet

ulgan

ULAN BATOR

Menengiyn Tal

JILIN

Ondörhaan

Baruun-Urt

Kerulen

Tongliao

112

OLIA

Saynshand

Xilinhot

NEI MONGOL ZIZHIQU

LIAONING

Sea of Japan (East Sea)

alandzadgad

Erenhot

(Inner Mongolia)

Chifeng (Ulanhad)

NORTH KOREA

Ilayn Nuruu

Ulan Qab (Jining)

Korea Bay

SOUTH KOREA

b   i

HOHHOT

BEIJING

JAPAN

Baotou

TIANJIN   Bo Hai

Wuhai (Haibowan)

Huang He (Yellow River)

Mu Us Shadi

Having started in the 7th century BCE, work on the 3700 mile (6000 km) long Great Wall of China continued for hundreds of years. A major renovation begun in 1386 took 200 years to complete.

Yellow Sea

XINING

NINGXIA HUIZU ZIZHIQU

SHANXI

Great Wall of China

East China Sea

113

GANSU

SHAANXI

Huang He (Yellow River)

JIANGSU

The Huang He (Yellow River) has flooded more than 1500 times in the last 1800 years. In 1931, catastrophic flooding was responsible for the deaths of 3.7 million people. The river has also changed its course at least nine times.

NA

SICHUAN

CHONGQING

Despite a population of 1.3 billion, China has only about 200 family names.

HUBEI

ZHEJIANG

Nansei-shotō (to Japan)

Tropic of Cancer

YUNNAN

GUIZHOU

HUNAN

JIANGXI

FUJIAN

TAIWAN

111

Eastern China & Korea

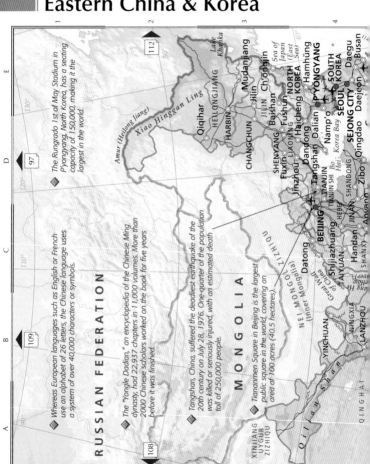

Whereas European languages such as English or French use an alphabet of 26 letters, the Chinese language uses a system of over 40,000 characters or symbols.

The *Rungrado 1st of May Stadium* in Pyongyang, North Korea, has a seating capacity of 150,000, making it the largest in the world.

The "Yongle Dadian," an encyclopedia of the Chinese Ming dynasty, had 22,937 chapters in 11,000 volumes. More than 2000 Chinese scholars worked on the book for five years before it was finished.

Tangshan, China, suffered the deadliest earthquake of the 20th century on July 28, 1976. One-quarter of the population was killed or seriously injured, with an estimated death toll of 250,000 people.

Tiananmen Square in Beijing is the largest public square in the world, covering an area of 100 acres (40.5 hectares).

RUSSIAN FEDERATION

MONGOLIA

Amur (Heilong Jiang)

Xiao Hinggan Ling

Lake Khanka

Sea of Japan (East Sea)

Qiqihar
HEILONGJIANG
Mudanjiang
HARBIN
JILIN Jilin Ch'ŏngjin
CHANGCHUN Baishan
JILIN
Fushun NORTH
SHENYANG Haicheng KOREA Hamhŭng
LIAONING PYONGYANG
Fuxin Dandong Namp'o
Jinzhou Tangshan Dalian Korea Bay
TIANJIN SHI
BEIJING TIANJIN Bo Hai
HEBEI
Datong Shijiazhuang HANDAN SHANDONG
TAIYUAN
SHANXI JINAN Zibo Qingdao
NINGXIA Anyang
YINCHUAN

SEJONG CITY
SEOUL
SOUTH
KOREA
Daegu
Daejeon Busan

NEI MONGOL (Inner Mongolia)

NINGXIA

Great Wall

Huang He (Yellow River)

QINGHAI

Qilian Shan

XINJIANG UYGUR ZIZHIQU

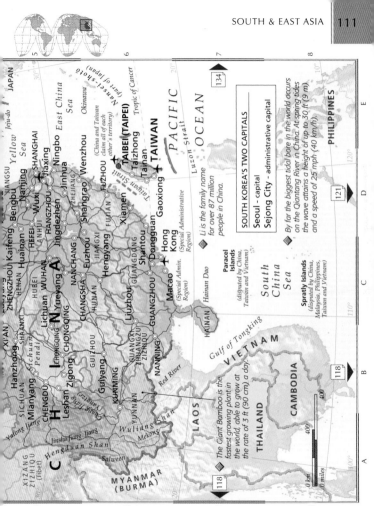

SOUTH KOREA'S TWO CAPITALS

Seoul - capital
Sejong City - administrative capital

By far the biggest tidal bore in the world occurs on the Qiantang River in China. At spring tides the wave attains a height of up to 30 ft (9 m) and a speed of 25 mph (40 km/h).

Li is the family name for over 87 million people in China.

The Giant Bamboo is the fastest growing plant in the world, able to grow at the rate of 3 ft (90 cm) a day.

JAPAN
Iejima (to Japan)
Yellow Sea
East China Sea
SHANGHAI
Ningbo
Wenzhou
Nansei-shoto
Okinawa
Tropic of Cancer
China and Taiwan claim all of each other's territory)
TAIBEI (TAIPEI)
TAIWAN
Taizhong
Tainan
Gaoxiong
PACIFIC
OCEAN
Luzon Strait
Taiwan Strait
PHILIPPINES

GUANGXU
ZHENGZHOU Kaifeng
HENAN
Bengbu
Huainan
ANHUI
HEFEI
Nanjing
Wuxi
JIANGSU
Jiaxing
ZHEJIANG
HANGZHOU
Jingdezhen
Shangrao
JIANGXI
FUJIAN
Fuzhou
Xiamen
Hengyang
Dongguan
Hong Kong (Special Administrative Region)
GUANGDONG
Shantou
GUANGZHOU
Macao (Special Admin. Region)

XIAN
SHAANXI
Hanzhong
Lichuan
Yueyang
HUNAN
CHANGSHA
NANCHANG
WUHAN
HUBEI
CHONGQING
GUIZHOU
GUIYANG
GUANGXI ZIZHIQU
NANNING
Liuzhou

XIZANG ZIZHIQU (Tibet)
SICHUAN
Mianyang
CHENGDU
Leshan
Zigong
Hanzhong
Sichuan Pendi
Yalong Jiang
Jinsha Jiang
Jinsha Jiang
Hengduan Shan
Salween
Wuliang Shan
Mekong
YUNNAN
KUNMING
C H I N A

Paracel Islands (disputed by China, Taiwan and Vietnam)
Hainan Dao
HAINAN
South China Sea
Spratly Islands (disputed by China, Malaysia, Philippines, Taiwan and Vietnam)

Gulf of Tongking
Red River
VIETNAM
LAOS
THAILAND
CAMBODIA
MYANMAR (BURMA)

134
121
118
118

0 km   400
0 miles   400

Japan

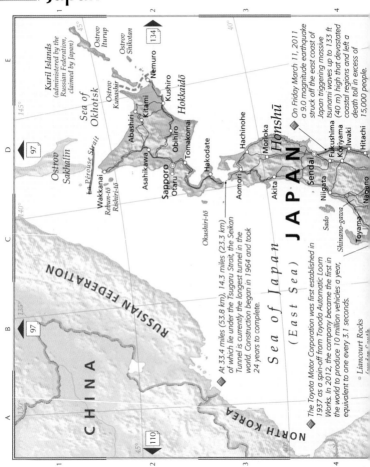

Kuril Islands
(administered by the
Russian Federation,
claimed by Japan)

Ostrov Iturup

Ostrov Shikotan

134

Ostrov Kunashir

Sea of
Okhotsk

Nemuro

Kushiro

Kitami

Hokkaidō

Abashiri

Ostrov
Sakhalin

97

La Pérouse Strait

Wakkanai

Tomakomai

Obihiro

Hachinohe

Honshū

Rebun-tō
Rishiri-tō

Asahikawa

Sapporo

Hakodate

Morioka

Fukushima

Otaru

Aomori

Kōriyama

Hitachi

On Friday March 11, 2011
a 9.0 magnitude earthquake
struck off the east coast of
Japan triggering massive
tsunami waves up to 133 ft
(40 m) high that devastated
coastal regions and left a
death toll in excess of
15,000 people.

Okushiri-tō

Akita

Sendai

Niigata

J A P A N

Sado

Nagano

Shinano-gawa

Toyama

Sea of Japan

(East Sea)

At 33.4 miles (53.8 km), 14.3 miles (23.3 km)
of which lie under the Tsugaru Strait, the Seikan
Tunnel is currently the longest tunnel in the
world. Construction began in 1964 and took
24 years to complete.

97

RUSSIAN FEDERATION

The Toyota Motor Corporation was first established in
1937 as a spin-off from Toyota Automatic Loom
Works. In 2012, the company became the first in
the world to produce 10 million vehicles a year,
equivalent to one every 3.1 seconds.

Liancourt Rocks

CHINA

NORTH KOREA

110

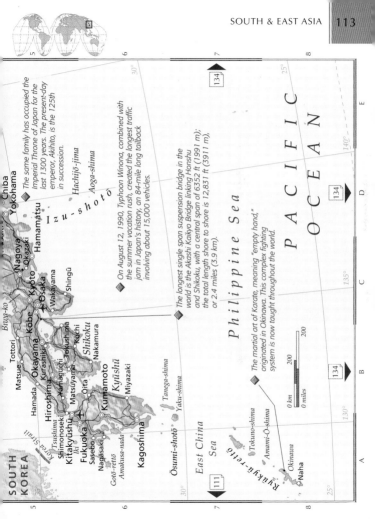

The same family has occupied the Imperial Throne of Japan for the last 1300 years. The present-day emperor, Akihito, is the 125th in succession.

On August 12, 1990, Typhoon Winona, combined with the summer vacation rush, created the longest traffic jam in Japan's history, an 84-mile long tailback involving about 15,000 vehicles.

The longest single span suspension bridge in the world is the Akashi Kaikyo Bridge linking Honshu and Shikoku, with a central span of 6352 ft (1991 m); the total length shore to shore is 12,831 ft (3911 m), or 2.4 miles (3.9 km).

The martial art of Karate, meaning "empty hand," originated in Okinawa. This complex fighting system is now taught throughout the world.

SOUTH KOREA

Kyŏra Strait

Tsushima

Matsue   Tottori

Hamada   Okayama   Okazaki

Hiroshima   Kurashiki

Shimonoseki   Yamaguchi   Tokushima

Kitakyūshū   Matsuyama   Kōchi

Fukuoka   Ōita   Shikoku   Nakamura

Iki   Kumamoto

Sasebo   Kyūshū   Tanega-shima

Nagasaki   Miyazaki   Yaku-shima

Gotō-rettō   Amakusa-nada

Kagoshima

Ōsumi-shotō

East China Sea

Tokuno-shima

Amami-O-shima

Okinawa

Naha

Ryūkyū-rettō

Matsue   Nagoya   Yokohama   Chiba

Kyōto   Hamamatsu

Ōsaka   Kōbe

Wakayama   Shingū

Biwa-ko

Izu-shotō

Hachijō-jima

Aoga-shima

Philippine Sea

PACIFIC OCEAN

0 km   200

0 miles   200

30°

25°

25°

140°

135°

130°

111   134

134

134

134

111

Southern India & Sri Lanka

A B C D

116

70° Mumbai ○ Kalyān ○ Nānded ○
(Bombay) ○ Nizāmābād ○
Pune ○
Solāpur ○
Hyderābād ○
Belgaum ○ Hubli ○ Kurnool ○
Pānāji ○
Dāvangere ○ Anantāpur
Karnātaka
Mangalore ○ Bangalore ○ Vell
Mysore ○ Tam
Salem Nac

The Mumbai (Bombay) movie industry, known as Bollywood, makes around 900 films each year, compared to Hollywood's 100, making it the most prolific film-producing country in the world.

A r a b i a n

103

S e a

10°

Amīndīvi Is.

Lakshadweep
(part of India)

Kozhikode / Calicut
Kavaratti I. Coimbatore ○ Tiruchi
Kalpeni I. Ernākulam ○ rāpp
Kochi / Cochin ○ Madu
Thiruvananthapuram /
Tivandrum ○ G
Minicoy I. Nāgercoil Man

The word ghats, literally "stairs that descend to a river," refers to the stair-like appearance of the slopes of the Western Ghats mountain range, as they descend to the coastal plain.

Ihavananthapuram
Atoll

MALDIVES

55

✈
● **MALE'**

There are over 1300 islands in the Maldives but only about 200 are inhabited. All the islands are low-lying, none rising more than 6 ft (1.8m) above sea level.

I N D I A N

| 0 km | | 300 | |
| 0 miles | | | 300 |

Kolhumadulu Atoll

Equator

O C E A N

Huvadhu Atoll

123

70°

A B C D

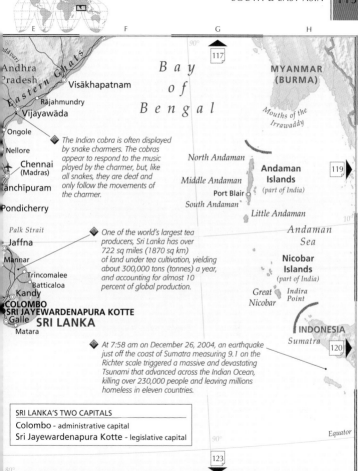

E F G H

117

Bay

of

Bengal

Andhra
Pradesh • Visākhapatnam

Eastern Ghats

• Rājahmundry
• Vijayawāda

MYANMAR
(BURMA)

*Mouths of the
Irrawaddy*

• Ongole

• Nellore

Chennai
(Madras)

The Indian cobra is often displayed
by snake charmers. The cobras
appear to respond to the music
played by the charmer, but, like
all snakes, they are deaf and
only follow the movements of
the charmer.

North Andaman

Andaman
Islands
(part of India)

119

Middle Andaman

ānchīpuram

Port Blair
South Andaman

• Pondicherry

Little Andaman

Palk Strait

*Andaman
Sea*

• Jaffna

One of the world's largest tea
producers, Sri Lanka has over
722 sq miles (1870 sq km)
of land under tea cultivation, yielding
about 300,000 tons (tonnes) a year,
and accounting for almost 10
percent of global production.

Nicobar
Islands
(part of India)

• Mannar

• Trincomalee
• Batticaloa

Great
Nicobar

Indira
Point

Kandy

COLOMBO
SRI JAYEWARDENAPURA KOTTE
Galle SRI LANKA
• Matara

INDONESIA
Sumatra

120

At 7:58 am on December 26, 2004, an earthquake
just off the coast of Sumatra measuring 9.1 on the
Richter scale triggered a massive and devastating
Tsunami that advanced across the Indian Ocean,
killing over 230,000 people and leaving millions
homeless in eleven countries.

Equator

SRI LANKA'S TWO CAPITALS

Colombo - administrative capital
Sri Jayewardenapura Kotte - legislative capital

123

E F G H

North India & Pakistan

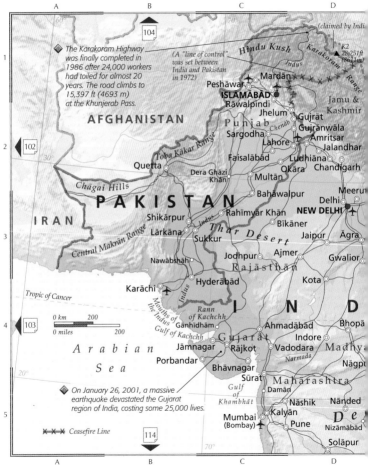

◆ The Karakoram Highway was finally completed in 1986 after 24,000 workers had toiled for almost 20 years. The road climbs to 15,397 ft (4693 m) at the Khunjerab Pass.

(A "line of control" was set between India and Pakistan in 1972)

(claimed by Indi

104

AFGHANISTAN

Hindu Kush

Indus

K2 28,251ft
8611(m)

Karakoram Range

Mardān

Peshāwar

ISLĀMĀBĀD

Rāwalpindi

Jamu & Kashmir

Jhelum

Gujrāt

102

30°

Punjab

Sargodha

Gujrānwāla

Chenāb

Lahore

Amritsar

Jalandhar

Toba Kākar Range

Quetta

Faisalābād

Ludhiāna

Okāra

Chandigarh

Dera Ghāzi Khān

Multān

P A K I S T A N

Chāgai Hills

Bahāwalpur

Meeru

Delhi

I R A N

Shikārpur

Rahīmyār Khān

Bīkāner

NEW DELHI

Lārkāna

Indus

Thar Desert

Jaipur

Āgra

Sukkur

Jodhpur

Ajmer

Gwalior

Nawābshāh

Rājasthān

Kota

Karāchi

Hyderābād

I

N

D

Tropic of Cancer

0 km   200

0 miles   200

103

Mouths of the Indus

Rann of Kachchh

Gānghidhām

Ahmadābād

Bhopā

Gujarāt

Gulf of Kachchh

Indore

Madhy

A r a b i a n

Jāmnagar

Rājkot

Vadodara

Narmada

Nāgpr

Porbandar

Bhāvnagar

Sūrat

M a h a r a s h t r a

20°

S e a

Gulf of Khambhāt

Daman

Nāshik

Nānded

◆ On January 26, 2001, a massive earthquake devastated the Gujarat region of India, costing some 25,000 lives.

D

e

Kalyān

Mumbai (Bombay)

Pune

Nizāmābād

✕✕✕ Ceasefire Line

114

Solāpur

70°

E F G H

80° 90°

XINJIANGUYGUR
ZIZHIQU

108

◆ The northern ranges of the Himalayas contain the highest
mountains in the world, with average heights of more than
23,000 ft (7000 m) and many peaks higher
than 26,000 ft (8000m).

1

ksai Chin
administered by China,
laimed by India)

C H I N A QINGHAI

emchok/Dêmqog
administered by China,
laimed by India)

◆ Cherrapunji, 4872 ft (1484 m) above sea level, has an average
annual rainfall of 450 inches (1143 cm), although most of this
falls during the monsoon – the winter is a virtual drought.
The highest-ever seasonal rainfall was 904 inches (2298 cm).

XIZANG ZIZHIQU
(Tibet)

◆ The Kingdom of Bhutan is
the only country in the world
to measure the happiness
of its citizens.

108

2

Arunachal Pradesh
(claimed by China)

H i m a l a y a s

Bareilly

N E P A L

Mount Everest
29,029ft (8848m)

Uttar
Pradesh

KATHMANDU   THIMPHU
Gangtok   BHUTAN

Lucknow

Birātnagar

Guwāhāti

Dispur   Kohima

3

Saidpur

Kānpur   Varānasi   Patna

Jamālpur

Imphāl

Brahmaputra

Ganges

amuna

Allahābād

Bihār

BANGLADESH

Sylhet

Rājshāhi

Tropic of Cancer

I A

Gaya

Dhanbād   West   DHAKA

Jabalpur   Rānchi   Bengal   Comilla

Chittagong

118

4

Raipur

Kolkata
(Calcutta)

Khulna

Pradesh

Mahānadi

MYANMAR
(BURMA)

an

Orissa

Cuttack

Mouths of the Ganges

B a y

20°

◆ The heaviest hailstones
on record, weighing about
2.25 lbs (1 kg), are
reported to have killed 92
people in the Gopalganj
area of Bangladesh on
April 14, 1986.

Eastern Ghats

averi

Warangal

Visākhapatnam

o f

B e n g a l

115

5

90° G

One of the world's largest "books" is a collection of sacred Buddhist texts engraved onto 730 polished marble slabs. Each "page" is 1.5 m (5.0 ft) high, 1.0 m (3.5 ft) wide, and 12.7 cm (5 in) thick

Every year around 300–500 million people worldwide are infected with malaria from the bite of female Anopheles Mosquitos, of which between 1 and 3 million die, making this the deadliest animal in the world

Around 60 percent of Myanmar's cultivated land is given over to growing rice, producing almost 20 million tons (tonnes) each year.

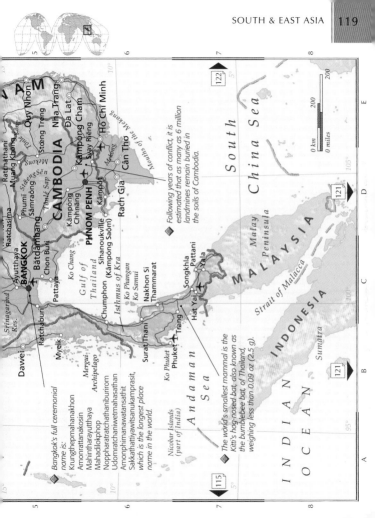

VIETNAM

Quy Nhon

Nha Trang

Da Lat

Ratthathani
Muang Khong

Ratchasima

Stoeng Treng

Kampong Cham

Ho Chi Minh

Mekong

Phumi
Samraong

Stoeng Sen

Sw'y Rieng

Tonlé Sap

CAMBODIA

PHNOM PENH

Kampong
Chhnang

Kampôt

Can Tho

Mouth of the Mekong

Ayutthaya

Bătdâmbâng

Kampong
Saôm (Kâmpông Saôm)

Rach Gia

Chon Buri

BANGKOK

Pattaya

Sihanoukville

South China Sea

Srinagarind
Res.

Ratchaburi

Ko Chang

Gulf of
Thailand

Malay Peninsula

MALAYSIA

Isthmus of Kra

Chumphon

Ko Phangan
Ko Samui

Pattani

Songkhla

Yala

Strait of Malacca

Dawei

Nakhon Si
Thammarat

Hat Yai

INDONESIA

Myeik

Surat Thani

Trang

Sumatra

Mergui
Archipelago

Ko Phuket
Phuket

Andaman
Sea

Following years of conflict, it is estimated that as many as 6 million landmines remain buried in the soils of Cambodia.

Bangkok's full ceremonial name is:
Krungthepmahanakhon
Amonrattanakosin
Mahintharayutthaya
Mahadilokphop
Noppharatratchathaniburirom
Udomratchaniwetmahasathan
Amonphimanawatansathit
Sakkathattiyawitsanukamprasit, *which is the longest place name in the world.*

The world's smallest mammal is the Kitti's hog-nosed bat, also known as the bumblebee bat of Thailand, weighing less than 0.09 oz (2.5 g).

Nicobar Islands
(part of India)

INDIAN OCEAN

0 km   200
0 miles   200

MYANMAR (BURMA)

THAILAND

LAOS

VIETNAM

CAMBODIA

Gulf of Tonkin

Paracel Islands
(disputed by China, Taiwan, and Vietnam)

South China Sea

Spratly Islands
(disputed by China, Malaysia, Philippines, Taiwan, and Vietnam)

0 km 400
0 miles 400

MALAYSIA'S TWO CAPITALS
Kuala Lumpur - Capital
Putrajaya - Administrative capital

Andaman Sea

Gulf of Thailand

Nicobar Islands *(to India)*

◆ *The Rafflesia plant has the largest single flower in the world. The bloom, 3 ft (90 cm) in diameter, attracts insects by imitating the foul smell of rotting flesh.*

Bandaaceh

Isthmus of Kra

George Town
Kota Bharu
Kuala Terengganu
Kota Kinabalu
BANDAR SERI BEGAWAN
BRUNEI

Taiping
Ipoh
Kuantan

Medan
Klang
KUALA LUMPUR

Pematangsiantar
Sibu
Sarawak

Pulau Simeulue
Danau Toba
PUTRAJAYA
MALAYSIA

Kuching
Pegunungan

Sibolga
Pulau Nias
Johor Bahru
SINGAPORE

Equator

Sumatera (Sumatra)
Pekanbaru
Pontianak
Kapuas
Borneo

Padang
Pulau Siberut
Jambi
Kalimantan
Samarinda
Balikpapan

Kepulauan Mentawai
Batang Hari
Bangka
Palembang
Banjarmasin

Pulau Belitung

Bengkulu
Selat Karimata
INDONESIA

INDIAN OCEAN

Bandar Lampung
Java Sea

Tegal
Pekalongan
Semarang
Surabaya

◆ *In August 1883, a devastating volcanic eruption destroyed most of the island of Krakatau and triggered a tsunami that claimed around 35,000 lives.*

Selat Sunda
JAKARTA
Kudus
Mataram
Jember

Bogor
Sukabumi
Bandung
Cilacap
Magelang
Yogyakarta
Surakarta
Jawa (Java)
Kediri
Madiun
Malang
Denpasar
Bali

E   F   G   H

112

Luzon Strait
Babuiani Channel   130°   140°
Tuguergarao
Ilagan   *Luzon*
aguio   **Dagupan**   *Philippine*
ngeles   **Cabanatuan**
ANILA   **Lucena**   *Sea*
atangas   **Naga**
Mindoro   **Legazpi City**
Sibuyan   **Calbayog**
Roxas City   *Sea*   **Tacloban**
Iloilo   **Cadiz**
Bacolod   **Cebu**
City   **Butuan**
Bohol Sea   **Cagayan de Oro**
Iligan   *Mindanao*
Davao

◆ The Philippines take their name from Philip II
of Spain, who was king when the islands were
colonized during the 16th century.

PHILIPPINES

Northern
Mariana
Islands
(to US)

Guam *(to US)*

P A C I F I C

126

Yap

O C E A N

Babeldaob
PALAU   **MICRONESIA**

◆ Indonesia is the world's largest archipelago,
with over 17,500 islands stretching
3100 miles (5000 km) between the Indian
and Pacific oceans.

General
Santos
Kepulauan
Talaud

Kepulauan
Sangir

Celebes Sea

Manado
Gorontalo   *Pulau Morotai*

Moluccu Sea   *Pulau*
Halmahera

Pulau
Biak

Equator

Sungai Mamberamo

Jayapura   126

alu   *Gulf of*
Tomini   *Halmahera*
Sea   **Sorong**

alu   *Kepulauan*
Banggai   *Jazirah*
Doberai

Sulawesi
(Celebes)

Kepulauan
Sula   *Ceram Sea*   **Wahai**
Pulau
Seram

Pegunungan Maake

N   **Kendari**   **Ambon**
Pulau
Buru   **Papua**
(Irian Jaya)

PAPUA
NEW
GUINEA

Parepare   *Kepulauan*
Kai   *New Guinea*

Makassar   *Banda Sea*   *Kepulauan*
Aru   *Digul*

Flores   *T e n g g a r a*   *Kepulauan*
Sea   *Wetar*   *Tanimbar*
Strait
usa   *Kepulauan Alor*   *Pulau Yamdena*
Sumba   *Flores*   *Kepulauan Leti*
Savu Sea   **DILI**
Pulau   **EAST TIMOR**
Sumba   *Timor*   **Kupang**

A r a f u r a   S e a
Torres Strait   10°

Timor Sea

130   **AUSTRALIA**

120°   130°   140°

1

2

3

4

5

120°   130°   140°

The Indian Ocean

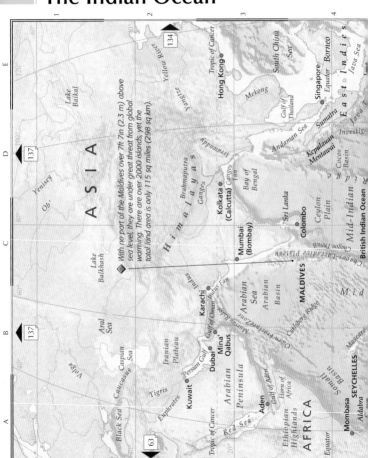

With no part of the Maldives over 7ft 7in (2.3 m) above sea level, they are under great threat from global warming. There are over 2000 islands, yet the total land area is only 115 sq miles (298 sq km).

ASIA

Lake Baikal

Lake Balkhash

Yenisey

Ob'

Yangtze

Yellow River

Tropic of Cancer

Hong Kong

Mekong

South China Sea

Singapore   Equator   Borneo

Sumatra   East Indies

Java Sea

Andaman Sea   Investigator

Kepulauan Mentawai

Cocos Basin

Himalayas

Brahmaputra   Ganges   Ganges Fan

Irrawaddy

Kolkata (Calcutta)

Bay of Bengal

Sri Lanka   Colombo   Ceylon Plain

Mid-Indian   Ridge

Mumbai (Bombay)

Chagos-Laccadive Plateau   Chagos Trench

Indus   Indus Fan

Karachi

Murray Ridge   Arabian Sea   Arabian Basin

MALDIVES

Owen Fracture Zone   Carlsberg Ridge

Mid

Aral Sea

Caspian Sea

Iranian Plateau

Gulf of Oman

Dubai   Mina' Qabus

Persian Gulf

Kuwait

Tigris   Euphrates

Volga

Caucasus

Black Sea

Arabian Peninsula

Gulf of Aden   Aden

Red Sea

Horn of Africa

Ethiopian Highlands

AFRICA

Mombasa   Aldabra

SEYCHELLES

Somali Basin

Mascarene

Tropic of Cancer

Equator

134

137

137

63

AUSTRALASIA

Tropic of Capricorn

Exmouth
Plateau

Australian
Basin

(to Australia)

Perth
Basin

Fremantle

Naturaliste
Plateau

134

Antarctic Circle

Cocos Islands
(to Australia)

Wharton
Basin

East Indian Ridge

Diamantina Fracture Zone

Southeast Indian Ridge

South Indian Basin

Limit of winter pack ice

Limit of summer pack ice

A N T A R C T I C A

136

Osborn
Plateau

Ninetyeast Ridge

Amsterdam Island

Île St-Paul

◆ Every cubic mile (4.3 cu km) of
seawater holds over 150 million
tons (tonnes) of minerals.

Kerguelen Plateau

Banzart
Seamounts

136

I N D I A N   O C E A N

Central Indian Ridge

MAURITIUS

Réunion
(to France)

Crozet
Basin

French Southern &
Antarctic Lands
(to France)

Crozet
Islands

Heard & Mcdonald Islands
(to Australia)

Southwest Indian Ridge

Madagascar
Basin

Farafangana

Mayotte
(to France)

Madagascar
Plateau

Natal
Basin

Mascarene
Plain

◆ The largest animal ever seen alive was a 110-ft (34 m)
170-ton (tonne) female blue whale.

E n d e r b y   P l a i n

49

Atlantic-Indian
Basin

Antarctic Circle

Mozambique
Channel

Indian Ridge

0 km   1500

0 miles   1500

1500

Australasia & Oceania

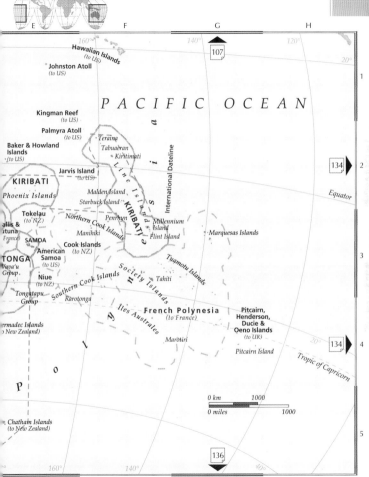

E   F   G   H

160°   140°   120°

107

Hawaiian Islands
(to US)   20°

Johnston Atoll
(to US)   1

P A C I F I C   O C E A N

Kingman Reef
(to US)   a

Palmyra Atoll
(to US)   Teraina
Tabuaeran
Kiritimati

**Baker & Howland
Islands**
(to US)   i

134   2

Jarvis Island
(to US)   Equator

KIRIBATI   Line Islands   International Dateline

Phoenix Islands   Malden Island
Starbuck Island   s

Tokelau
(to NZ)   **KIRIBATI**

**Wallis &
Futuna**
(to France)   Northern Cook Islands   Penrhyn   Millennium
Island   **Marquesas Islands**

SAMOA   Manihiki   Flint Island

**American
Samoa**
(to US)   **Cook Islands**
(to NZ)   3

TONGA

*Vava'u
Group*   **Niue**
(to NZ)   Society Islands   Tuamotu Islands

*Tongatapu
Group*   Rarotonga   Tahiti   o

Southern Cook Islands   n

Îles Australes   **French Polynesia**
(to France)   **Pitcairn,
Henderson,
Ducie &
Oeno Islands**
(to UK)   y   20°

Kermadec Islands
(to New Zealand)   Marotiri   134   4

Pitcairn Island   Tropic of Capricorn

P   o   l

0 km   1000

0 miles   1000

Chatham Islands
(to New Zealand)   5

160°   140°   40°

136

E   F   G   H

The Southwest Pacific

Guam
(US unincorporated territory) HAGÅTÑA

Yap

Marianas Trench

134

MARSHALL ISLANDS

Ratak Chain

Caroline Islands

Chuuk Is.

Pohnpei
PALIKIR

Ralik Chain Majuro

MICRONESIA

Kosrae

NGERULMUD

PALAU

121

Equator

◆ The Pitohui bird has a poison on its feathers and skin similar to the poison arrow tree frog, making it the only known example of a poisonous bird.

BAIRIKI
Tarawa

NAURU

Banaba

PAPUA NEW GUINEA

Bismarck Archipelago New Ireland

INDONESIA

Mt. Wilhelm
14,793ft (4509m) ▲ Madang

New Guinea

Bougainville I.

Lae New Britain

PORT MORESBY

Solomon Sea

New Georgia Islands

Santa Cruz Islands

HONIARA

Melanesia

Arafura Sea

Torres Strait

10°

Gulf of Carpentaria

128

◆ Found only in the rainforest of New Guinea, Queen Alexandra's Birdwing, with a wingspan of 11 inches (280 mm), is the largest butterfly in the world.

Coral Sea

SOLOMON ISLANDS

Banks Is.

VANUATU

PORT VILA

Coral Sea Islands
(Australian external territory)

New Caledonia
(French special collectivity)

20°

AUSTRALIA

NOUMÉA

Îles Loyauté

Great Barrier Reef

Tropic of Capricorn

131

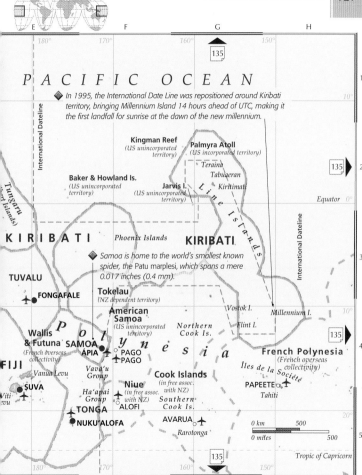

PACIFIC OCEAN

◆ In 1995, the International Date Line was repositioned around Kiribati territory, bringing Millennium Island 14 hours ahead of UTC, making it the first landfall for sunrise at the dawn of the new millennium.

International Dateline

180°   170°   160°   150°

10°N

1

Kingman Reef
(US unincorporated territory)

Palmyra Atoll
(US incorporated territory)

Teraina

Tabuaeran

Baker & Howland Is.
(US unincorporated territory)

Jarvis I.
(US unincorporated territory)

Kiritimati

Line Islands

135

2

Equator 0°

Tungaru (Gilbert Islands)

KIRIBATI

Phoenix Islands

KIRIBATI

International Dateline

3

◆ Samoa is home to the world's smallest known spider, the Patu marplesi, which spans a mere 0.017 inches (0.4 mm).

TUVALU

FONGAFALE ✈●

Tokelau
(NZ dependent territory)

American Samoa
(US unincorporated territory)

Northern Cook Is.

Vostok I.

Flint I.

Millennium I.

10°S

135

4

Wallis & Futuna
(French overseas collectivity)

P o l y n e s i a

SAMOA
ÁPIA ✈●

✈ **PAGO PAGO**

Vava'u Group

Cook Islands
(in free assoc. with NZ)

Îles de la Société

French Polynesia
(French overseas collectivity)

PAPEETE ✈

Tahiti

FIJI

Vanua Levu

✈ **SUVA**

Viti Levu

Niue
(in free assoc. with NZ)

ALOFI ○

Southern Cook Is.

Ha'apai Group

TONGA

● **NUKU'ALOFA**

AVARUA ✈

Rarotonga

20°S

0 km   500

0 miles   500

5

135

Tropic of Capricorn

180°   170°   160°   150°

E   F   G   H

Western Australia

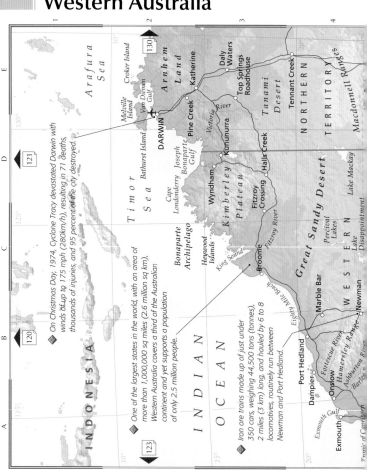

Arafura Sea

Croker Island

Arnhem Land

Daly Waters

Katherine

Melville Island

Van Diemen Gulf

Top Springs Roadhouse

Tanami Desert

Tennant Creek

Bathurst Island

DARWIN

Pine Creek

Victoria River

Kununurra

Halls Creek

NORTHERN

TERRITORY

Macdonnell Ranges

Lake Mackay

On Christmas Day, 1974, Cyclone Tracy devastated Darwin with winds of up to 175 mph (280 km/h), resulting in 71 deaths, thousands of injuries, and 95 percent of the city destroyed.

121

Timor Sea

Cape Londonderry

Joseph Bonaparte Gulf

Wyndham

Kimberley Plateau

Fitzroy Crossing

Fitzroy River

Great Sandy Desert

Percival Lakes

Lake Disappointment

One of the largest states in the world, with an area of more than 1,000,000 sq miles (2.6 million sq km), Western Australia covers a third of the Australian continent and yet supports a population of only 2.5 million people.

120

Bonaparte Archipelago

Heywood Islands

King Sound

Broome

WESTERN

Marble Bar

Newman

123

INDIAN

OCEAN

Eighty Mile Beach

Port Hedland

Fortescue River

Hamersley Range

Ashburton River

Onslow

Dampier

Exmouth Gulf

Exmouth

Iron ore trains made up of just under 350 cars, weighing 44,500 tons (tonnes), 2 miles (3 km) long, and hauled by 6 to 8 locomotives, routinely run between Newman and Port Hedland.

INDONESIA

Tropic of Capricorn

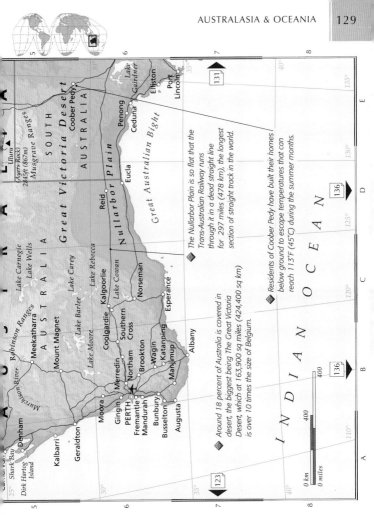

AUSTRALIA

SOUTH

AUSTRALIA

Great Victoria Desert

Nullarbor Plain

Great Australian Bight

Uluru
(Ayers Rock)
2845ft (867m)
Musgrave Ranges

Coober Pedy

Lake
Gairdner

Penong
Ceduna
Ellison
Port
Lincoln

131

Eucla

Reid

136

Lake Rebecca

Lake Cowan

Kalgoorlie

Norseman

Esperance

Coolgardie

Southern
Cross

Albany

Katanping

Wagin

Brookton

Mahjimup

Merredin

Northam

PERTH

Fremantle

Mandurah

Bunbury

Busselton

Augusta

Moora

Gingin

Robinson Ranges

Lake Wells

Lake Carnegie

Meekatharra

Mount Magnet

Lake Barlee

Lake Carey

Lake Moore

Murchison River

Kalbarri

Geraldton

Denham

Shark Bay

Dirk Hartog
Island

INDIAN OCEAN

136

123

The Nullarbor Plain is so flat that the Trans-Australian Railway runs through it in a dead straight line for 297 miles (478 km), the longest section of straight track in the world.

Residents of Coober Pedy have built their homes below ground to escape temperatures that can reach 113°F (45°C) during the summer months.

Around 18 percent of Australia is covered in desert, the biggest being The Great Victoria Desert, which at 163,900 sq miles (424,400 sq km) is over 10 times the size of Belgium.

0 km   400
0 miles   400

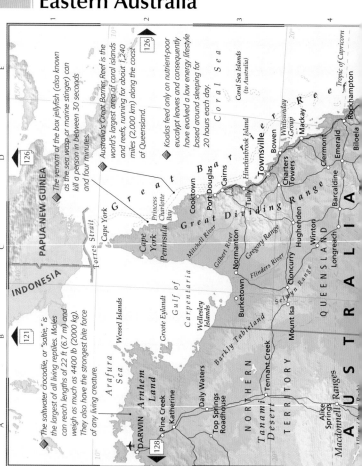

The venom of the box jellyfish (also known as the sea wasp or marine stinger) can kill a person in between 30 seconds and four minutes.

Australia's Great Barrier Reef is the world's largest area of coral islands and reefs, running for about 1,240 miles (2,000 km) along the coast of Queensland.

Koalas feed only on nutrient-poor eucalypt leaves and consequently have evolved a low energy lifestyle based around sleeping for 20 hours each day.

The saltwater crocodile, or "saltie," is the largest of all living reptiles. Males can reach lengths of 22 ft (6.7 m) and weigh as much as 4400 lb (2000 kg). They also have the strongest bite force of any living creature.

PAPUA-NEW GUINEA

INDONESIA

Coral Sea

Coral Sea Islands (to Australia)

Tropic of Capricorn

Rockhampton

Biloela

Emerald

Clermont

Barcaldine

Mackay

Whitsunday Group

Bowen

Charters Towers

Townsville

Hinchinbrook Island

Hughenden

Winton

Longreach

QUEENSLAND

Cairns

Tully

Port Douglas

Cooktown

Great Barrier Reef

Great Dividing Range

Gregory Range

Selwyn Range

Cloncurry

Mount Isa

Flinders River

Gilbert River

Mitchell River

Normanton

Burketown

Princess Charlotte Bay

Cape York

Cape York Peninsula

Torres Strait

Gulf of Carpentaria

Wellesley Islands

Groote Eylandt

Barkly Tableland

Arafura Sea

Wessel Islands

Arnhem Land

DARWIN

Pine Creek

Katherine

Daly Waters

Top Springs Roadhouse

Tennant Creek

NORTHERN TERRITORY

Tanami Desert

Alice Springs

Macdonnell Ranges

AUSTRALIA

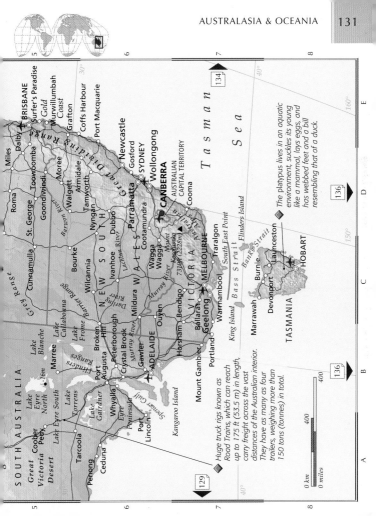

T a s m a n   S e a

The platypus lives in an aquatic environment, suckles its young like a mammal, lays eggs, and has webbed feet and a bill resembling that of a duck.

Huge truck rigs known as Road Trains, which can reach up to 175 ft (53.5 m) in length, carry freight across the vast distances of the Australian interior. They have as many as four trailers, weighing more than 150 tons (tonnes) in total.

BRISBANE
Surfer's Paradise
Gold Coast
Murwillumbah
Grafton
Coffs Harbour
Port Macquarie
Dalby
Toowoomba
Roma
St. George
Goondiwindi
Cunnamulla
Bourke
Walgett
Moree
Narrabri
Tamworth
Armidale
Inverell
Newcastle
Gosford
SYDNEY
Parramatta
Wollongong
CANBERRA
AUSTRALIAN
CAPITAL TERRITORY
Cooma
Cootamundra
Wagga Wagga
Dubbo
Nyngan
Ivanhoe
Wilcannia
Broken Hill
Mildura
Ouyen
Bendigo
Ballarat
MELBOURNE
Geelong
Warrnambool
Traralgon
Portland
Mount Gambier
Horsham
Peterborough
Crystal Brook
ADELAIDE
Gawler
Whyalla
Port Augusta
Port Lincoln
Ceduna
Penong
Tarcoola
Marree
Cooper Pedy

NEW SOUTH WALES
VICTORIA
SOUTH AUSTRALIA

Great Dividing Range
Grey Range
Barrier Range
Flinders Ranges

Barwon River
Darling River
Lachlan River
Murrumbidgee River
Murray River

Lake Callabonna
Lake Blanche
Lake Frome
Lake Gairdner
Lake Torrens
Lake Eyre North
Lake Eyre South
Great Victoria Desert
Eyre Peninsula
Spencer Gulf
Kangaroo Island

Mount Kosciuszko
7310 ft (2228m)

South East Point
Flinders Island
King Island
Bass Strait
Banks Strait
Marrawah
Burnie
Devonport
Launceston
HOBART
TASMANIA

129
134
136
136

New Zealand

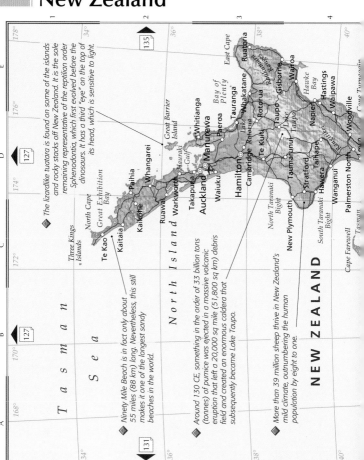

The lizardlike tuatara is found on some of the islands and rocky stacks off New Zealand. It is the sole remaining representative of the reptilian order Sphenodontia, which first evolved before the dinosaurs. It has a third "eye" on the top of its head, which is sensitive to light.

Ninety Mile Beach is in fact only about 55 miles (88 km) long. Nevertheless, this still makes it one of the longest sandy beaches in the world.

Around 130 CE, something in the order of 33 billion tons (tonnes) of pumice was ejected in a massive volcanic eruption that left a 20,000 sq mile (51,800 sq km) debris field and created an enormous caldera that subsequently became Lake Taupo.

More than 39 million sheep thrive in New Zealand's mild climate, outnumbering the human population by eight to one.

Tasman Sea

NEW ZEALAND

North Island

Three Kings Islands

North Cape

Great Exhibition Bay

Te Kao

Kaitaia

Kaikohe

Paihia

Whangarei

Ruawai

Warkworth

Takapuna

Auckland

Waiuku

Hamilton

Cambridge

Te Kuiti

Taumarunui

New Plymouth

Stratford

Hawera

Wanganui

Palmerston North

Taihape

Hauraki Gulf

Matamata

Whitianga

Great Barrier Island

Pheroa

Te Awamutu

Rotorua

Lake Rotorua

Taupo

Lake Taupo

Tauranga

Bay of Plenty

Whakatane

Opotiki

East Cape

Ruatoria

Gisborne

Hawke Bay

Napier

Hastings

Waipawa

Woodville

North Taranaki Bight

South Taranaki Bight

Tasman

Cape Farewell

Cape Turnagain

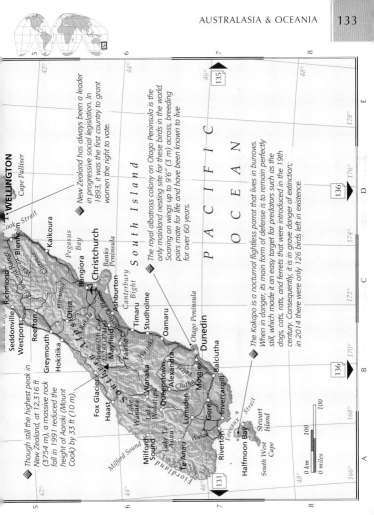

New Zealand has always been a leader in progressive social legislation. In 1893, it was the first country to grant women the right to vote.

The royal albatross colony on Otago Peninsula is the only mainland nesting site for these birds in the world. Soaring on wings up to 9'6" (3 m) across, breeding pairs mate for life and have been known to live for over 60 years.

The kakapo is a nocturnal flightless parrot that lives in burrows. When in danger, its main form of defense is to remain perfectly still, which made it an easy target for predators such as the dogs, cats, rats, and ferrets that were introduced in the 19th century. Consequently, it is in grave danger of extinction; in 2014 there were only 126 birds left in existence.

Though still the highest peak in New Zealand, at 12,316 ft (3754 m), a massive rock fall in 1991 reduced the height of Aoraki (Mount Cook) by 33 ft (10 m).

PACIFIC OCEAN

South Island

WELLINGTON

Cape Palliser

Cook Strait

Blenheim

Kaikoura

Pegasus Bay

Richmond

Richmond Range

Seddonville

Westport

Reefton

Greymouth

Hokitika

Otira

Christchurch

Banks Peninsula

Rangiora

Ashburton

Canterbury Bight

Timaru

Studholme

Oamaru

Fox Glacier

Haast

Mayfield

Fairlie

Aoraki/Mt Cook 12,283 ft/3744 m

Wanaka

Lake Wanaka

Lake Wakatipu

Queenstown

Alexandra

Lumsden

Lake Te Anau

Te Anau

Milford Sound

Fiordland

Dunedin

Otago Peninsula

Mosgiel

Balclutha

Gore

Clutha

Waitaki

Rakaia

Waimakariri

Waiau

Invercargill

Riverton

Halfmoon Bay

Stewart Island

South West Cape

Foveaux Strait

0 km 100

0 miles 100

The Pacific Ocean

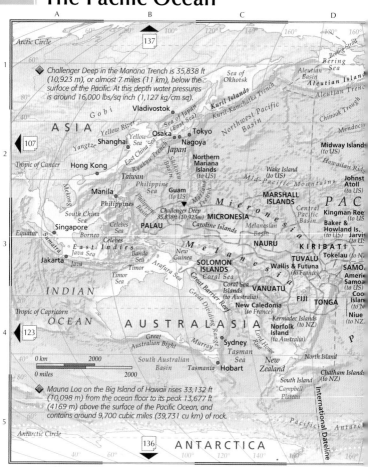

Arctic Circle

◆ *Challenger Deep in the Mariana Trench is 35,838 ft (10,923 m), or almost 7 miles (11 km), below the surface of the Pacific. At this depth water pressures is around 16,000 lbs/sq inch (1,127 kg/cm sq).*

Bering Sea

Aleutian Basin

Aleutian Islands

Aleutian Trench

Chinook Trough

Gobi

Vladivostok

Sea of Okhotsk

Kuril Islands

Kuril-Kamchatka Trench

Northwest Pacific Basin

Mendocino

ASIA

Yellow River

Sea of Japan (East Sea)

Tokyo

Yangtze

Yellow Sea

Osaka

Shanghai

East China Sea

Nagoya

Japan

Midway Islands (to US)

107

Tropic of Cancer

Hong Kong

Taiwan

Ryukyu Trench

Northern Mariana Islands (to US)

Wake Island (to US)

Mid-Pacific Mountains

Hawaiian Ridge

Johnston Atoll (to US)

Manila

Philippines

Philippine Sea

Philippine Basin

Guam (to US)

Challenger Deep 35,838ft (10,923m)

MICRONESIA

Micronesia

MARSHALL ISLANDS

Central Pacific Basin

PAC

Kingman Ree (to US)

South China Sea

Singapore

Sumatra

Celebes Sea

Borneo

PALAU

Caroline Islands

Melanesian Basin

Baker & Howland Is. (to US)

Jarvis (to US)

Equator

East Indies

Java Sea

Jakarta

Java

Celebes

Banda Sea

New Guinea

Melanesia

NAURU

KIRIBATI

Tokelau (to N.

SAMO.

Ameri Samoa (to

Timor Sea

Arafura Sea

SOLOMON ISLANDS

Wallis & Futuna (to France)

TUVALU

Coo Islan (to

Coral Sea

Great Barrier Reef

Coral Sea Islands (to Australia)

VANUATU

FIJI

TONGA

Niue (to NZ)

INDIAN

Great Dividing Range

New Caledonia (to France)

Kermadec Islands (to NZ)

Tropic of Capricorn

Norfolk Island (to Australia)

OCEAN

AUSTRALASIA

Lord Howe Rise

123

Great Australian Bight

Murray

Sydney

Tasman Sea

North Island

0 km   2000

South Australian Basin

Tasmania

Hobart

New Zealand

0 miles   2000

Chatham Islands (to NZ)

South Island

Campbell Plateau

International Dateline

◆ *Mauna Loa on the Big Island of Hawaii rises 33,132 ft (10,098 m) from the ocean floor to its peak 13,677 ft (4169 m) above the surface of the Pacific Ocean, and contains around 9,700 cubic miles (39,731 cu km) of rock.*

Pacific O

Antarc

Antarctic Circle

ANTARCTICA

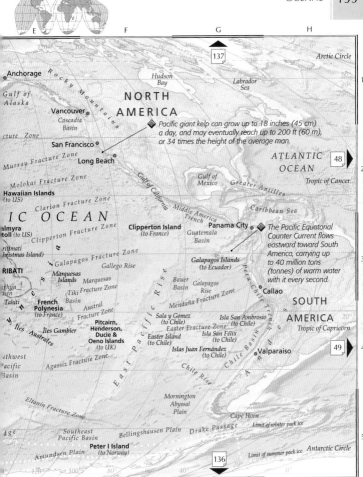

E F G H

137

Arctic Circle

1

Anchorage

Rocky Mountains

Gulf of
Alaska

Hudson
Bay

Labrador
Sea

Vancouver

Cascadia
Basin

**NORTH
AMERICA**

Pacific giant kelp can grow up to 18 inches (45 cm)
a day, and may eventually reach up to 200 ft (60 m),
or 34 times the height of the average man.

cture Zone

San Francisco

Long Beach

Murray Fracture Zone

**ATLANTIC
OCEAN**

48

2

Molokai Fracture Zone

Gulf of California

Gulf of
Mexico

Greater Antilles

Tropic of Cancer

Hawaiian Islands
(to US)

Clarion Fracture Zone

Caribbean Sea

IC OCEAN

almyra
toll (to US)

Clipperton Fracture Zone

Middle America Trench

ritimati
hristmas Island)

Clipperton Island
(to France)

Panama City

The Pacific Equatorial
Counter Current flows
eastward toward South
America, carrying up
to 40 million tons
(tonnes) of warm water
with it every second.

RIBATI

Galapagos Fracture Zone

Gallego Rise

Guatemala
Basin

thin
sin

Marquesas
Islands

Marquesas

Galapagos Islands
(to Ecuador)

Peru-Chile Trench

3

Tahiti

Tiki Fracture Zone

Bauer
Basin

Galapagos
Rise

Callao

SOUTH

**French
Polynesia**
(to France)

Austral
Fracture Zone

Mendaña Fracture Zone

AMERICA

Iles
Australes

Iles Gambier

**Pitcairn,
Henderson,
Ducie &
Oeno Islands**
(to UK)

Sala y Gomez
(to Chile)

Easter Fracture Zone

Isla San Ambrosio
(to Chile)

Tropic of Capricorn

East Pacific Rise

Easter Island
(to Chile)

Isla San Félix
(to Chile)

Chile Basin

49

Islas Juan Fernández
(to Chile)

Valparaiso

4

thwest
acific
Basin

Agassiz Fracture Zone

Chile Rise

Eltanin Fracture Zone

Mornington
Abyssal
Plain

Cape Horn

Limit of winter pack ice

ge

Southeast
Pacific Basin

Bellingshausen Plain

Drake Passage

Antarctic Circle

Peter I Island
(to Norway)

Limit of summer pack ice

Amundsen Plain

136

E F G H

Antarctica

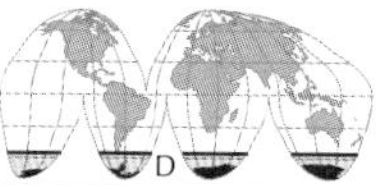

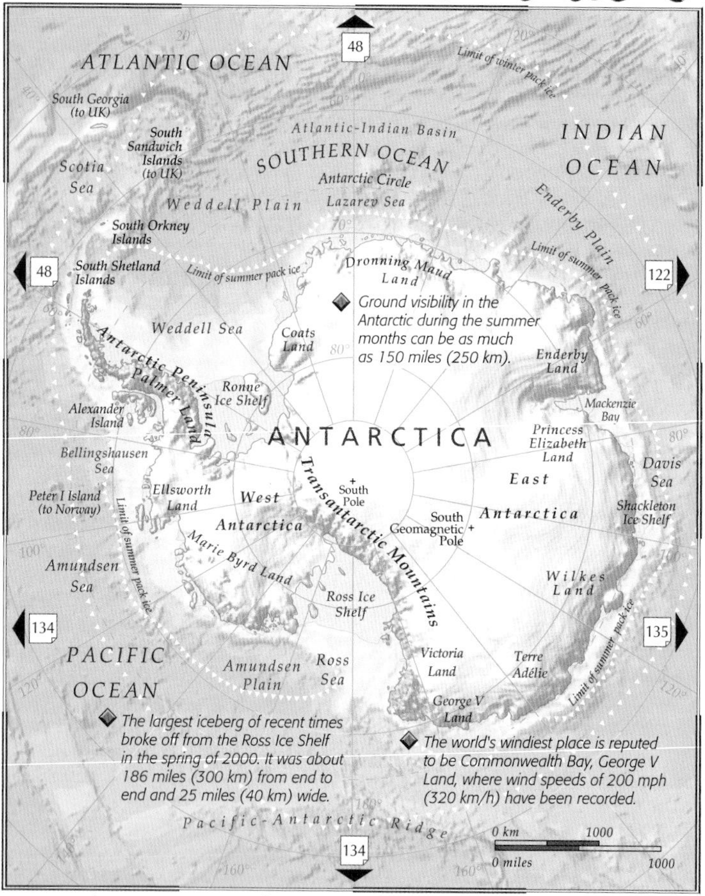
A
B
C
D
1
2
3
4
5
48
122
135
134
ATLANTIC OCEAN
South Georgia (to UK)
South Sandwich Islands (to UK)
Scotia Sea
Atlantic-Indian Basin
SOUTHERN OCEAN
Antarctic Circle
Lazarev Sea
Weddell Plain
South Orkney Islands
South Shetland Islands
Limit of summer pack ice
Limit of winter pack ice
INDIAN OCEAN
Enderby Plain
Dronning Maud Land
Ground visibility in the Antarctic during the summer months can be as much as 150 miles (250 km).
Weddell Sea
Coats Land
Antarctic Peninsula
Palmer Land
Ronne Ice Shelf
Enderby Land
Mackenzie Bay
Alexander Island
ANTARCTICA
Princess Elizabeth Land
Bellingshausen Sea
Transantarctic Mountains
South Pole
Davis Sea
East Antarctica
Peter I Island (to Norway)
Ellsworth Land
West Antarctica
South Geomagnetic Pole
Shackleton Ice Shelf
Marie Byrd Land
Amundsen Sea
Wilkes Land
Ross Ice Shelf
PACIFIC OCEAN
Amundsen Plain
Ross Sea
Victoria Land
Terre Adélie
George V Land
The largest iceberg of recent times broke off from the Ross Ice Shelf in the spring of 2000. It was about 186 miles (300 km) from end to end and 25 miles (40 km) wide.
The world's windiest place is reputed to be Commonwealth Bay, George V Land, where wind speeds of 200 mph (320 km/h) have been recorded.
Pacific-Antarctic Ridge
0 km 1000
0 miles 1000

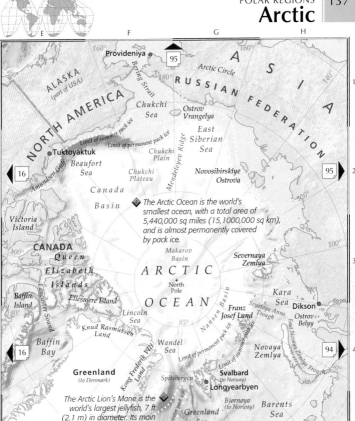

The Arctic Ocean is the world's smallest ocean, with a total area of 5,440,000 sq miles (15,1000,000 sq km), and is almost permanently covered by pack ice.

The Arctic Lion's Mane is the world's largest jellyfish, 7 ft (2.1 m) in diameter. Its main body trails tentacles up to 180 ft (55 m) in length.

Scale:
0 km   500
0 miles   500

The world factfiles

North & Central America

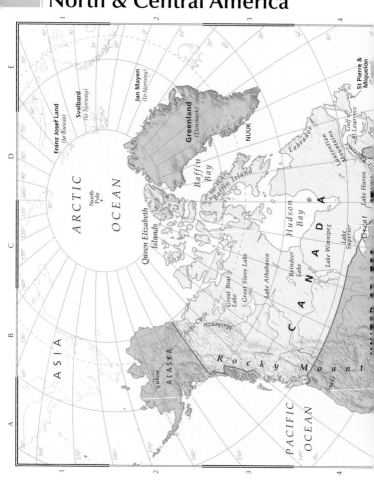

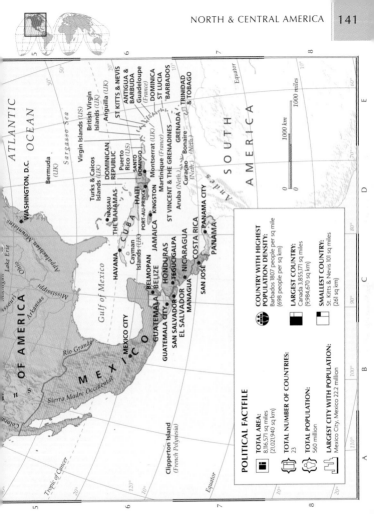

ATLANTIC OCEAN

Sargasso Sea

Bermuda (UK)

Virgin Islands (US)
British Virgin Islands (UK)
Anguilla (UK)
ST KITTS & NEVIS
ANTIGUA & BARBUDA
Guadeloupe (France)
DOMINICA
ST LUCIA
BARBADOS
GRENADA
TRINIDAD & TOBAGO

Turks & Caicos Islands (UK)
Puerto Rico (US)
SANTO DOMINGO
Montserrat (UK)
Martinique (France)
ST VINCENT & THE GRENADINES
Curaçao (Neth.)
Bonaire (Neth.)
Aruba (Neth.)

DOMINICAN REPUBLIC
HAITI
PORT-AU-PRINCE
NASSAU
THE BAHAMAS
HAVANA
CUBA
Cayman Islands (UK)
KINGSTON
JAMAICA
BELMOPAN
BELIZE
HONDURAS
TEGUCIGALPA
NICARAGUA
MANAGUA
SAN JOSÉ
COSTA RICA
PANAMA CITY
PANAMA

SOUTH AMERICA

Andes

Equator

1000 miles
1000 km

WASHINGTON, D.C.

Appalachian Mountains

Lake Erie
Ohio
Missouri
Mississippi
Arkansas
Rio Grande
Colorado

Gulf of Mexico

UNITED STATES OF AMERICA

MEXICO
MEXICO CITY
GUATEMALA CITY
GUATEMALA
SAN SALVADOR
EL SALVADOR

Sierra Madre Occidental

Clipperton Island (French Polynesia)

Tropic of Cancer

POLITICAL FACTFILE

TOTAL AREA:
8,116,571 sq miles
(21,021,940 sq km)

TOTAL NUMBER OF COUNTRIES:
23

TOTAL POPULATION:
560 million

LARGEST CITY WITH POPULATION:
Mexico City, Mexico 22.2 million

COUNTRY WITH HIGHEST POPULATION DENSITY:
Barbados 1807 people per sq mile (698 people per sq km)

LARGEST COUNTRY:
Canada 3,855,171 sq miles (9,984,670 sq km)

SMALLEST COUNTRY:
St Kitts & Nevis 101 sq miles (261 sq km)

South America

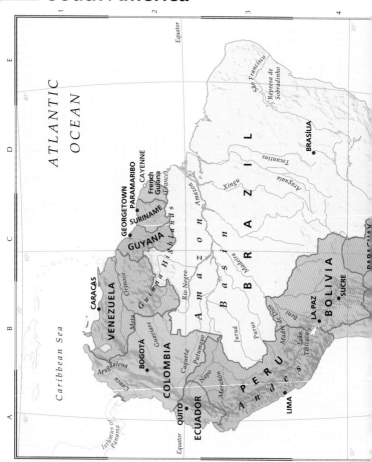

ATLANTIC OCEAN

Equator

Caribbean Sea

Isthmus of Panama

VENEZUELA

CARACAS

COLOMBIA

BOGOTÁ

ECUADOR

QUITO

Equator

GEORGETOWN

GUYANA

SURINAME

PARAMARIBO

CAYENNE

French Guiana (France)

Guiana Highlands

Orinoco

Meta

Guaviare

Caquetá

Putumayo

Napo

Marañón

Magdalena

Cauca

A n d e s

PERU

LIMA

A m a z o n   B a s i n

B R A Z I L

BRASÍLIA

São Francisco

Represa de Sobradinho

Tocantins

Araguaia

Xingu

Amazon

Rio Negro

Madeira

Juruá

Purus

Madre de Dios

Beni

BOLIVIA

LA PAZ

SUCRE

Lake Titicaca

PARAGUAY

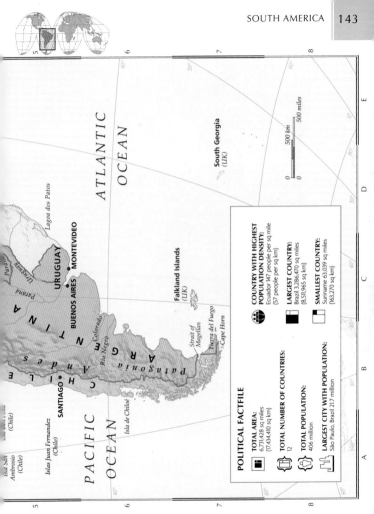

ATLANTIC

OCEAN

South Georgia
(UK)

0   500 km

0   500 miles

Lagoa dos Patos

URUGUAY

MONTEVIDEO

BUENOS AIRES

Paraná

Uruguay

Falkland Islands
(UK)

A R G E N T I N A

Colorado

Río Negro

Strait of
Magellan

Tierra del Fuego

Cape Horn

C H I L E

A n d e s

P a t a g o n i a

SANTIAGO

Isla de Chiloé

Islas Juan Fernandez
(Chile)

Isla San
Ambrosio
(Chile)

PACIFIC

OCEAN

POLITICAL FACTFILE

TOTAL AREA:
6,731,428 sq miles
(17,434,410 sq km)

TOTAL NUMBER OF COUNTRIES:
12

TOTAL POPULATION:
406 million

LARGEST CITY WITH POPULATION:
São Paulo, Brazil 217 million

**COUNTRY WITH HIGHEST
POPULATION DENSITY:**
Ecuador 147 people per sq mile
(57 people per sq km)

LARGEST COUNTRY:
Brazil 3,286,470 sq miles
(8,511,965 sq km)

SMALLEST COUNTRY:
Suriname 63,039 sq miles
(163,270 sq km)

Africa

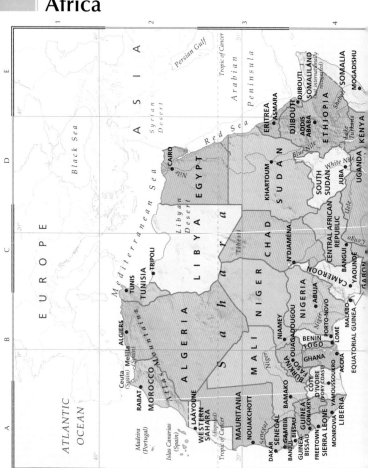

E

ASIA

Persian Gulf

Tropic of Cancer

SOMALILAND *(not internationally recognized)*

SOMALIA

MOGADISHU

Arabian Peninsula

ERITREA

ASMARA

DJIBOUTI

DJIBOUTI

Syrian Desert

ADDIS ABABA

ETHIOPIA

Black Sea

Red Sea

SUDAN

KHARTOUM

Blue Nile

White Nile

Lake Turkana

KENYA

D

CAIRO

EGYPT

Nile

SOUTH SUDAN

JUBA

UGANDA

Ubangi

EUROPE

Mediterranean Sea

Libyan Desert

Tibesti

CENTRAL AFRICAN REPUBLIC

BANGUI

Congo

C

TUNIS

TUNISIA

TRIPOLI

LIBYA

CHAD

N'DJAMÉNA

CAMEROON

YAOUNDÉ

GABON

ALGIERS

S a h a r a

NIGER

NIGERIA

ABUJA

Atlas Mountains

ALGERIA

NIAMEY

BENIN

PORTO-NOVO

LOMÉ

MALABO

B

Ceuta *(Spain)* Melilla *(Spain)*

MALI

Niger

BURKINA FASO

OUAGADOUGOU

TOGO

GHANA

ACCRA

EQUATORIAL GUINEA

ATLANTIC OCEAN

RABAT

MOROCCO

LAAYOUNE

WESTERN SAHARA *(disputed)*

MAURITANIA

NOUAKCHOTT

Senegal

SENEGAL

DAKAR

GAMBIA

BANJUL

GUINEA-BISSAU

BISSAU

GUINEA

CONAKRY

CÔTE D'IVOIRE *(IVORY COAST)*

YAMOUSSOUKRO

LIBERIA

MONROVIA

Tropic of Cancer

Madeira *(Portugal)*

Islas Canarias *(Spain)*

FREETOWN

SIERRA LEONE

A

BAMAKO

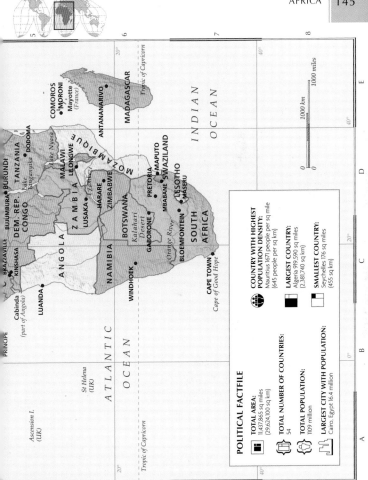

Europe

POLITICAL FACTFILE

TOTAL AREA:
3,739,678 sq miles
(9,685,756 sq km)

TOTAL NUMBER OF COUNTRIES:
46

TOTAL POPULATION:
721 million

LARGEST CITY WITH POPULATION:
Moscow, European Russia 16.7 million

**COUNTRY WITH HIGHEST
POPULATION DENSITY:**
Monaco 48,181 people per sq mile
(18,531 people per sq km)

LARGEST COUNTRY:
European Russia 1,527,341 sq miles
(3,955,818 sq km)

SMALLEST COUNTRY:
Vatican City, Italy 0.17 sq miles
(0.44 sq km)

REYKJAVÍK
ICELAND

Arctic Circle

Faroe Islands
(Denmark)

Norwegian
Sea

N O R W A

Shetland Islands

Outer
Hebrides

Orkney Islands

OSLO

British
Isles

North
Sea

DENMARK

IRELAND

COPENHAGEN

DUBLIN

UNITED
KINGDOM

Elbe

BERLIN

LONDON

AMSTERDAM

NETH.
THE
HAGUE

GERMANY

PRAG

BELGIUM
BRUSSELS

LUXEMBOURG

CZECH REPUB

PARIS

LUXEMBOURG

BRATISLA

Loire

Rhine

LIECH.

VIENNA
AUSTRIA

Bay of Biscay

FRANCE

BERN

SLOVEN

SWITZERLAND

LJUBLJANA

ZAGREB
CROATIA

Garonne

MONACO

SAN MARINO

SARAJ
BOS
& HE

ATLANTIC
OCEAN

PORTUGAL

Ebro

ANDORRA

Corsica

LISBON

MADRID

VATICAN CITY

Tagus

SPAIN

ROME

Madeira
(Portugal)

Guadalquivir

I T A L Y

Gibraltar
(UK)

Balearic Islands

Sardinia

M e d i t e r r a n e a

Canary Islands
(Spain)

Ceuta
(Spain)

Melilla
(Spain)

Sicily

VALLETTA

MALTA

A F R I C A

20° 40° 60° 80° 80°

DENMARK

FINLAND

Ural Mountains

Ob'

Irtysh

80°

Northern Dvina

Lake Onega

Lake Ladoga

R U S S I A N

HELSINKI

STOCKHOLM

TALLINN

ESTONIA

Baltic Sea

LATVIA

RIGA

F E D E R A T I O N

MOSCOW

50°

LITHUANIA

VILNIUS

KALININGRAD

(Russ.Fed.)

MINSK

BELARUS

Volga

Ural

Aral Sea

WARSAW

POLAND

KIEV

UKRAINE

Don

Dnieper

40°

SLOVAKIA

BUDAPEST

HUNGARY

MOLDOVA

CHIŞINĂU

(the Ukrainian territory of
Crimea was annexed by
Russia in 2014)

Caspian Sea

ROMANIA

SERBIA

BELGRADE

BUCHAREST

Danube

Caucasus

A

MONTENEGRO

PODGORICA

PRISHTINË

Black Sea

60°

SOFIA

BULGARIA

SKOPJE

MACED.

TURKEY

S

TIRANA

ALBANIA

I

GREECE

A

ATHENS

0   1000 km

Sea

Crete

Cyprus

0   1000 miles

30°

40°

E   F   G   H

Asia

ASIA

A B C D

ARCTIC OCEAN

Franz Josef
Land

Severnaya
Zemlya

Kara Sea

Laptev Sea

1

RUSSIAN FEDERATION

2

EUROPE

Ob
Irtysh
Yenisey

Lake Baikal

Black
Sea

ASTANA

KAZAKHSTAN

ULAN BATOR

ANKARA
TURKEY
GEORGIA
TBILISI
MONGOLIA
CYPRUS
NICOSIA
ARMENIA
YEREVAN
AZERBAIJAN
BAKU
UZBEKISTAN
BISHKEK
SYRIA
DAMASCUS
TURKMENISTAN
KYRGYZSTAN
LEBANON
BEIRUT
AŞGABAT
TASHKENT
JERUSALEM
AMMAN
TEHRĀN
DUSHANBE
TAJIKISTAN
CHINA

3

ISRAEL
JORDAN
BAGHDAD
KABUL
IRAQ
IRAN
AFGHANISTAN
ISLAMABAD
KUWAIT
KUWAIT
MANAMA
BAHRAIN
QATAR
RIYADH
DOHA
PAKISTAN
NEW
DELHI
NEPAL
THIMPHU
ABU DHABI
KATHMANDU
BHUTAN
SAUDI
ARABIA
U.A.E.
MUSCAT
Ganges
BANGLADESH
DHAKA
VIETNAM
HANOI
SANA
OMAN
MYANMAR
(BURMA)
LAOS

4

YEMEN
Arabian
Sea
INDIA
NAY PYI TAW
VIENTIANE
THAILAND
BANGKOK
Socotra
(Yemen)
Bay of
Bengal
CAMBODIA
PHN
PEN
Laccadive
Islands
(India)
Andaman &
Nicobar Islands
(India)

AFRICA

Tropic of Cancer

Red Sea

Indus

Yan

Equator

COLOMBO
SRI JAYEWARDENAPURA
KOTTE
MAL
MALE
SRI
LANKA
KUALA LUMPUR
PUTRAJAYA
SINGAPORE
IN

5

MALDIVES

INDIAN OCEAN

JAKAR

A B C D

40° 20° 60° 80° 80°
40° 60° 100°

POLITICAL FACTFILE

TOTAL AREA:
17,006,354 sq miles
(44,046,472 sq km)

TOTAL NUMBER OF COUNTRIES:
49

TOTAL POPULATION:
4309 million

LARGEST CITY WITH POPULATION:
Tokyo, Japan 39.4 million

COUNTRY WITH HIGHEST POPULATION DENSITY:
Singapore 22,881 people per sq mile
(8852 people per sq km)

LARGEST COUNTRY:
Asiatic Russia 5,065,394 sq miles
(13,119,382 sq km)

SMALLEST COUNTRY:
Maldives 116 sq miles
(300 sq km)

Sea of Okhotsk

Kuril Islands

NORTH KOREA

PYONGYANG

BEIJING

SEOUL

SEJONG CITY

SOUTH KOREA

JAPAN

TOKYO

Ryukyu Islands

Tropic of Cancer

TAIBEI (TAIPEI)

TAIWAN

PACIFIC OCEAN

MANILA

PHILIPPINES

Equator

BRUNEI

BANDAR SERI BEGAWAN

INDONESIA

AUSTRALASIA & OCEANIA

DILI

EAST TIMOR

0   1000 km
0   1000 miles

Australasia & Oceania

E F G H

160° 140° 120° 20° 1

Johnston Atoll
(US)

POLITICAL FACTFILE

TOTAL AREA:
3,244,632 sq miles (8,403,608 sq km)

TOTAL NUMBER OF COUNTRIES:
14

TOTAL POPULATION:
37.5 million

LARGEST CITY WITH POPULATION:
Sydney, Australia 4.8 million

COUNTRY WITH HIGHEST POPULATION DENSITY:
Nauru 1165 people per sq mile
(449 people per sq km)

LARGEST COUNTRY:
Australia 2,967,893 sq miles
(7,686,850 sq km)

SMALLEST COUNTRY:
Nauru 8.1 sq miles (21 sq km)

Baker & Howland
Islands
(US)

Jarvis Island
(US)

n *a*

P A C I F I C

2

KIRIBATI

Phoenix Islands

KIRIBATI

s *i*

O C E A N

Equator

Tokelau
(NZ)

Vallis &
Futuna
(Fr.)

SAMOA

American
Samoa
(US)

Cook Islands
(NZ)

e

Marquesas Islands

3

MATĀ'UTU

ĀPIA

PAGO PAGO

TONGA

Niue
(NZ)

PAPEETE

Society Islands

n

**NUKU'
ALOFA**

AVARUA

French Polynesia
(France)

y

Pitcairn,
Henderson,
Ducie &
Oeno Islands
(UK)

Iles Australes

20° 4

Kermadec Islands
(New Zealand)

l

o

Tropic of Capricorn

P

n

0 1000 km

0 1000 miles

5

Chatham Islands
(New Zealand)

International Dateline

30° 160° 140° 40° 120° 100°

E F G H

Key to factfile maps

FOREWORD

This factfile is intended as a guide to a world that is continually changing as political fashions and personalities come and go. Nevertheless, all the material in these factfiles has been researched from the most up-to-date and authoritative sources to give an incisive portrait of the geographical, social, and economic characteristics that make each country unique.

KEY TO MAP SYMBOLS

ELEVATION

- 4000m/13,124ft
- 3000m/9843ft
- 2000m/6562ft
- 1000m/3281ft
- 500m/1640ft
- 200m/656ft
- 0
- Below sea level

BORDERS

- ———— Full international
- ----- Disputed de facto
- ·········· Territorial claim
- ✳✳✳✳✳✳ Cease-fire line
- ———— State/Province

DRAINAGE FEATURES

- ———— River
- ·········· Seasonal river
- ⊔⊔⊔⊔⊔ Canal
- ⬭ Lake
- ⬭ Seasonal lake

SYMBOLS

- ● Capital city
- ○ Major town
- ✈ International airport
- ▲ Mountain

The asterisk in the Factfile denotes the country's official language(s)

Date of formation denotes the date of political origin or independence of a state, i.e. its emergence as a recognizable entity in the modern political world

The area figure denotes total land area

Afghanistan

About 75% of this landlocked Asian country is inaccessible. The Islamist *Taliban*, ousted in 2001, continue to fight a guerrilla war against Afghan and NATO-led forces.

 GEOGRAPHY
Predominantly mountainous. Highest range is the Hindu Kush. Mountains are bordered by fertile plains. Desert plateau in the south.

 CLIMATE
Harsh continental. Hot, dry summers. Cold winters with heavy snow, especially in the Hindu Kush.

 PEOPLE & SOCIETY
Mujahideen factions fought first against Soviet invaders (from 1979), and then against each other (after 1989). *Taliban* insurgents won control in 1996 and imposed a strict Islamist regime: women were denied all rights and ethnic tensions were exacerbated. In 2001, a US-led intervention justified as a "war on terrorism" helped install an elected anti-*Taliban* regime. NATO troops led the anti-insurgency campaign, but aimed ultimately to hand over and withdraw.

$ THE ECONOMY
Mainly agricultural, severely disrupted by war. Illicit opium trade is big cash earner. Natural gas pipeline planned from the Caspian Sea to Pakistan.

◆ **INSIGHT:** *The UN estimates that it could take 100 years to remove the 10 million landmines laid since 1979*

3000m/9843ft	
2000m/6562ft	
1000m/3281ft	
500m/1640ft	
200m/656ft	

0   100 km
0   100 miles

FACTFILE

OFFICIAL NAME: Islamic Republic of Afghanistan
DATE OF FORMATION: 1919
CAPITAL: Kabul
POPULATION: 30.6 million
TOTAL AREA: 250,000 sq. miles (647,500 sq. km)

DENSITY: 122 people per sq. mile
LANGUAGES: Pashtu*, Tajik, Dari*, other
RELIGIONS: Sunni Muslim 80%, Shi'a Muslim 19%, other 1%
ETHNIC MIX: Pashtun 38%, Tajik 25%, Hazara 19%, Uzbek and Turkmen 15%, other 3%
GOVERNMENT: Nonparty system
CURRENCY: Afghani = 100 puls

Albania

Lying at the southeastern end of the Adriatic Sea, Albania was the last east European country to liberalize its economy. The regional strife of the 1990s has left a difficult legacy.

GEOGRAPHY

Narrow coastal plain. Interior is mostly hills and mountains. Forest and scrub cover over 40% of the land.

CLIMATE

Mediterranean coastal climate, with warm summers and cool winters. Mountains receive heavy rains or snows in winter.

PEOPLE & SOCIETY

The pace of economic reform remains a major issue. Albania's application for EU membership reached candidate status in 2014. Mosques and churches have reopened in what was once the world's only officially atheist state. The Greek minority in the south suffers much discrimination.

INSIGHT: *The Albanians' name for their country, Shqipërisë, means "Land of the Eagles"*

THE ECONOMY

Oil and natural gas reserves have potential to offset rudimentary infrastructure and lack of foreign investment. Organized crime problem.

2000m/6562ft
1000m/3281ft
500m/1640ft
200m/656ft
Sea Level

MONTENEGRO
KOSOVO (disputed)
Lake Scutari
42°
Shkodër
Kukës
Adriatic Sea
MACEDONIA
Durrës
★ TIRANA
41°
Elbasan
Lake Ohrid
Lushnjë
Fier
Berat
Lake Prespa
Vlorë
Korçë
40°
Delvinë
Ionian Sea
GREECE
20°

0   50 km
0   50 miles

FACTFILE

OFFICIAL NAME: Republic of Albania
DATE OF FORMATION: 1912
CAPITAL: Tirana
POPULATION: 3.2 million
TOTAL AREA: 11,100 sq. miles (28,748 sq. km)
DENSITY: 302 people per sq. mile

LANGUAGES: Albanian*, Greek
RELIGIONS: Sunni Muslim 70%, Albanian Orthodox 20%, Roman Catholic 10%
ETHNIC MIX: Albanian 98%, Greek 1%, other 1%
GOVERNMENT: Parliamentary system
CURRENCY: Lek = 100 qindarka (qintars)

Algeria

On the Mediterranean coast, and independent from France since 1962, Algeria is now Africa's largest country. Its regime used the army to keep Islamists from power in 1992.

GEOGRAPHY

85% of the country lies within the Sahara Desert. Fertile coastal region with plains and hills rises to meet the Atlas Mountains.

CLIMATE

Coastal areas are warm and temperate, with most rainfall during the mild winters. The south is very hot, with negligible rainfall.

PEOPLE & SOCIETY

Algerians are predominantly Arab, under 35 years of age, and urban. Berbers consider the mountainous Kabylia region in the northeast to be their homeland. They have been granted greater ethnic rights in recent years. The Sahara sustains just 500,000 people, mainly oil workers or Tuareg nomads herding goats and camels. A national reconciliation process has followed the suppression of the Islamist challenge to the regime.

THE ECONOMY

Oil and natural gas exports. Political turmoil has led to exodus of skilled foreign labor. Limited agriculture.

INSIGHT: *Some of the world's highest dunes are located in the deserts of east central Algeria*

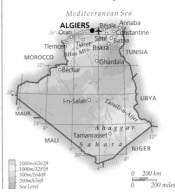

FACTFILE

OFFICIAL NAME: People's Democratic Republic of Algeria

DATE OF FORMATION: 1962

CAPITAL: Algiers

POPULATION: 39.2 million

TOTAL AREA: 919,590 sq. miles (2,381,740 sq. km)

DENSITY: 43 people per sq. mile

LANGUAGES: Arabic*, Tamazight, French

RELIGIONS: Sunni Muslim 99%, Christian and Jewish 1%

ETHNIC MIX: Arab 75%, Berber 24%, European and Jewish 1%

GOVERNMENT: Presidential system

CURRENCY: Algerian dinar = 100 centimes

Andorra

A tiny landlocked principality, Andorra lies high in the eastern Pyrenees between France and Spain. It held its first full elections in 1993. Tourism is the main source of income.

GEOGRAPHY

High mountains, with six deep, glaciated valleys that drain into the Valira River as it flows into Spain.

CLIMATE

Cool, wet springs followed by dry, warm summers. Mountain snows linger until March.

PEOPLE & SOCIETY
Immigration is strictly monitored and restricted by quota to French and Spanish nationals seeking employment in Andorra. Low taxes attract wealthy expatriates. A referendum in 1993 ended 715 years of semifeudal status, but Andorran society remains conservative.

◆ **INSIGHT:** *Andorra's coprincipality status dates from the 13th century. The "princes" are the president of France and the bishop of Urgel in Spain.*

THE ECONOMY
Tourism and duty-free sales dominate the economy. Banking secrecy laws and low consumer taxes promote investment and commerce. France and Spain effectively decide economic policy. The country is dependent on imported food and raw materials.

FACTFILE

OFFICIAL NAME: Principality of Andorra
DATE OF FORMATION: 1278
CAPITAL: Andorra la Vella
POPULATION: 85,293
TOTAL AREA: 181 sq. miles (468 sq. km)
DENSITY: 474 people per sq. mile

LANGUAGES: Spanish, Catalan*, French, Portuguese
RELIGIONS: Roman Catholic 94%, other 6%
ETHNIC MIX: Spanish 46%, Andorran 28%, other 18%, French 8%
GOVERNMENT: Parliamentary system
CURRENCY: Euro = 100 cents

Angola

Located in southwest Africa, Angola suffered a civil war following independence from Portugal in 1975, until a 2002 peace deal. Hundreds of thousands of people died.

GEOGRAPHY
Most of the land is hilly and grass-covered. Desert in the south. Mountains in the center and north.

CLIMATE
Varies from temperate to tropical. Rainfall decreases north to south. Coast is cooler and dry.

PEOPLE & SOCIETY
Civil war pitched the ruling Kimbundu-dominated MPLA against UNITA, representing the Ovimbundu. Multiparty elections in 1991–1992, after the MPLA had abandoned Marxism, failed to stall the war for long. Power-sharing from 2002 ended when the MPLA won the 2008 election. In 2006, separatists in the Cabinda exclave agreed a peace deal.

INSIGHT: *Angola has the greatest number of amputees (caused by landmines) in the world*

THE ECONOMY
Potentially one of Africa's richest countries, but long civil war hampered economic development. Oil and diamonds are exported.

FACTFILE

OFFICIAL NAME: Republic of Angola

DATE OF FORMATION: 1975

CAPITAL: Luanda

POPULATION: 21.5 million

TOTAL AREA: 481,351 sq. miles (1,246,700 sq. km)

DENSITY: 45 people per sq. mile

LANGUAGES: Portuguese*, Umbundu, Kimbundu, Kikongo

RELIGIONS: Roman Catholic 68%, Protestant 20%, indigenous beliefs 12%

ETHNIC MIX: Ovimbundu 37%, other 25%, Kimbundu 25%, Bakongo 13%

GOVERNMENT: Presidential system

CURRENCY: Readjusted kwanza = 100 lwei

Antarctica

The circumpolar continent of Antarctica is almost entirely covered by ice, some up to 1.2 miles (2 km) thick. It also contains 90% of the Earth's freshwater reserves.

GEOGRAPHY

The bulk of Antarctica's ice is contained in the Greater Antarctic Ice Sheet – a huge dome that rises steeply from the coast and flattens to a plateau in the interior.

CLIMATE

Powerful winds create a storm belt around the continent, which brings cloud, fog, and blizzards. Winter temperatures can fall to –112°F (–80°C).

PEOPLE & SOCIETY

No indigenous population. Scientists and logistical staff work at the 40 permanent, and as many as 100 temporary, research stations. A few Chilean settler families live on King George Island. Tourism is mostly by cruise ship to the Antarctic Peninsula. Annual tourist numbers are around 35,000.

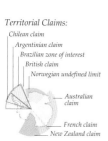

Territorial Claims:

Chilean claim
Argentinian claim
Brazilian zone of interest
British claim
Norwegian undefined limit
Australian claim
French claim
New Zealand claim

The Antarctic Treaty of 1959 holds all territorial claims in abeyance in the interest of international cooperation

FACTFILE

DATE OF FORMATION: 1961
TOTAL AREA: 5,405,000 sq. miles (14,000,000 sq. km)

◆ **INSIGHT:** *If the ice sheets of Antarctica were to melt, the world's oceans would rise by as much as 200–210 ft (60–65 m)*

Antigua & Barbuda

A former colony of Spain, France, and the UK, Antigua and Barbuda lies at the outer edge of the Leeward Islands group in the Caribbean, and includes the uninhabited islet of Redonda.

GEOGRAPHY
Mainly low-lying limestone and coral islands with some higher volcanic areas. Antigua's coast is indented with bays and harbors.

CLIMATE
Tropical, moderated by trade winds and sea breezes. Humidity and rainfall are low for the region.

PEOPLE & SOCIETY
Population almost entirely of African origin, with small communities of Europeans and South Asians. Women's status has risen as a result of greater access to education. Wealth disparities are small. The Bird family dominated politics from 1960, but lost power to the United Progressive Party (UPP) from 2004.

◆ **INSIGHT:** *In 1865, Redonda was "claimed" by an eccentric Englishman as a kingdom for his son*

THE ECONOMY
Tourism is the main source of revenue and the biggest provider of jobs. Financial services and Internet gambling are expanding. High debt.

ATLANTIC OCEAN

0   5 km
0   5 miles

17°40′

Codrington
Codrington
Lagoon

200m/656ft
Sea Level

Barbuda

17°35′

Palmetto
Point

61°50′

The Highlands

Spanish
Point

61°45′

Islands 30 miles (50 km) apart

V.C. Bird
Intl. Airport

1710′
Long I.

ST. JOHN'S

Antigua

Guiana I.

1705′

Bolans

Freetown

Green I.

61°55′

6140′

Falmouth

Guadeloupe Passage

61°50′   61°45′

1700′

Caribbean Sea

FACTFILE

OFFICIAL NAME: Antigua and Barbuda
DATE OF FORMATION: 1981
CAPITAL: St. John's
POPULATION: 90,156
TOTAL AREA: 170 sq. miles
(442 sq. km)
DENSITY: 530 people per sq. mile

LANGUAGES: English*, English patois
RELIGIONS: Anglican 45%, other Protestant 42%, Roman Catholic 10%, other 2%, Rastafarian 1%
ETHNIC MIX: Black African 95%, other 5%
GOVERNMENT: Parliamentary system
CURRENCY: E. Caribbean $ = 100 cents

Argentina

Argentina occupies most of southern South America.
After 30 years of intermittent military rule, democracy returned
in 1983. Economic crash in 2001 led to largest-ever debt default.

GEOGRAPHY

The Andes form a natural border
with Chile in the west. East are the
heavily wooded plains (Gran Chaco) and
treeless but fertile Pampas plains. Bleak
and arid Patagonia lies in the south.

CLIMATE

The Andes are semiarid in the north
and snowy in the south. Pampas have a
mild climate with summer rains.

PEOPLE & SOCIETY

People are largely of European
descent; over one-third are of Italian
origin. Indigenous peoples are now a tiny
minority, living mainly in Andean regions
or in the Gran Chaco. The middle classes
were worst hit by the economic
meltdown of 2001–2002.

◆ **INSIGHT:** *The Tango originated
in the poorer quarters of Buenos
Aires at the end of the 19th century*

THE ECONOMY

Agricultural exports led recovery.
Drought and global downturn in 2008.
Recession again in 2014, another default.

■	4000m/13124ft
■	3000m/9843ft
■	2000m/6562ft
■	1000m/3281ft
■	200m/656ft
	Sea Level

0   400 km
0   400 miles

FACTFILE

OFFICIAL NAME: Republic of Argentina

DATE OF FORMATION: 1816

CAPITAL: Buenos Aires

POPULATION: 41.4 million

TOTAL AREA: 1,068,296 sq. miles
(2,766,890 sq. km)

DENSITY: 39 people per sq. mile

LANGUAGES: Spanish*, Italian, Amerindian
languages

RELIGIONS: Roman Catholic 70%, other 18%,
Protestant 9%, Muslim 2%, Jewish 1%

ETHNIC MIX: Indo-European 97%, *Mestizo*
(European–Amerindian) 2%, Amerindian 1%

GOVERNMENT: Presidential system

CURRENCY: Argentine peso = 100 centavos

Armenia

The smallest of the former USSR's republics, Armenia lies landlocked in the Lesser Caucasus Mountains. After 1988, a confrontation with Azerbaijan dominated national life.

GEOGRAPHY
Rugged and mountainous, with expanses of semidesert and a large lake in the east: Sevana Lich.

CLIMATE
Continental climate, with little rainfall in the lowlands. The winters are often bitterly cold.

PEOPLE & SOCIETY
Christianity is the dominant religion, but minority groups are well integrated. War with Azerbaijan over the enclave of Nagorno Karabakh forced 350,000 Armenians living in Azerbaijan to return home, many to live in poverty. There are close and important ties to the 11-million-strong Armenian diaspora.

◆ INSIGHT: *In the 4th century, Armenia became the first country to adopt Christianity as its state religion*

THE ECONOMY
Overseas remittances and agriculture each account for a sixth of GDP. Main products are wine, tobacco, potatoes, and fruit. Well-developed machine-building and manufacturing — includes textiles and bottling of mineral water.

3000m/9843ft
2000m/6562ft
1000m/3281ft
500m/1640ft

GEORGIA
Alaverdi
Vanadzor
Gyumri
Sevan
AZERBAIJAN
TURKEY
Hrazdan
Ashtarak
Sevana Lich
Vagharshapat
Armavir
✦ YEREVAN
Ararat
AZERBAIJAN
Kapan
Aras
IRAN

0   50 km
0   50 miles

FACTFILE

OFFICIAL NAME: Republic of Armenia
DATE OF FORMATION: 1991
CAPITAL: Yerevan
POPULATION: 3 million
TOTAL AREA: 11,506 sq. miles (29,800 sq. km)
DENSITY: 261 people per sq. mile

LANGUAGES: Armenian*, Azeri, Russian
RELIGIONS: Armenian Apostolic Church (Orthodox) 88%, Armenian Catholic Church 6%, other 6%
ETHNIC MIX: Armenian 98%, Yezidi 1%, other 1%
GOVERNMENT: Parliamentary system
CURRENCY: Dram = 100 luma

Australia

An island continent in its own right, Australia is the world's sixth-largest country. European settlement began over 200 years ago. Most Australians now live in cities along the coast.

GEOGRAPHY

Located between the Indian and Pacific oceans, Australia has a variety of landscapes, including tropical rainforests, the arid plateaus, ridges, and vast deserts of the "red center," the lowlands and river systems draining into Lake Eyre, rolling tracts of pastoral land, and magnificent beaches around much of the coastline. In the far east are the mountains of the Great Dividing Range. Famous natural features include Uluru (Ayers Rock) and the Great Barrier Reef.

CLIMATE

The west and south are semi-arid with hot summers. The arid interior can reach 120°F (50°C) in the central desert areas. The north is hot throughout the year, and humid during the summer monsoon. East, southeast, and southwest coastal areas are temperate.

PEOPLE & SOCIETY

The first settlers arrived in Australia at least 100,000 years ago. Today, the Aborigines make up around 2% of the population. European colonization began in 1788, and was dominated by British and Irish immigrants, some of whom were convicts. White-only immigration drives brought many Europeans to Australia, but since the 1960s multiculturalism has been encouraged and most new settlers are Asian; Cantonese has overtaken Italian as the second most widely spoken language. Wealth disparities are small, but Aborigines, the exception in an otherwise integrated society, are marginalized: their average life expectancy is around ten years less than other Australians. Illegal immigration is a key political divide; Liberal–National government policies aim to turn back asylum seekers or process and resettle them offshore.

FACTFILE

OFFICIAL NAME: Commonwealth of Australia
DATE OF FORMATION: 1901
CAPITAL: Canberra
POPULATION: 23.3 million
TOTAL AREA: 2,967,893 sq. miles (7,686,850 sq. km)

DENSITY: 8 people per sq. mile
LANGUAGES: English*, Cantonese, other
RELIGIONS: Various Protestant 38%, Roman Catholic 26%, nonreligious 19%, other 17%
ETHNIC MIX: European 90%, Asian 7%, Aboriginal 2%, other 1%
GOVERNMENT: Parliamentary system
CURRENCY: Australian dollar = 100 cents

$ THE ECONOMY

Efficient mining and agriculture: particular success in viticulture. Large resource base: coal, iron ore, bauxite, and most other minerals. Protectionism abandoned to open up Australian markets. Concentration on trade with Asia: China's rapidly expanding demand for minerals means it has now surpassed Japan as Australia's major trading partner.

Upward trend in Asian visitor arrivals has strengthened tourism. The effects of droughts, floods, and cyclones have dented economic growth in recent years.

◆ **INSIGHT:** *Australia has the most endemic mammals and reptiles in the world. Species include marsupials such as the kangaroo and wombat, the egg-laying platypus, and the freshwater crocodile*

1000m/3281ft
500m/1640ft
200m/656ft
Sea Level
Below Sea Level

0   400 km

0   400 miles

Austria

Bordering eight countries in the heart of Europe, Austria was created in 1918 after the collapse of the Habsburg Empire. Neutral after World War II, it joined the EU in 1995.

GEOGRAPHY

Mainly mountainous. Alps and foothills cover the west and south. Lowlands in the east are part of the Danube River basin.

CLIMATE

Temperate continental climate. The western Alpine regions have colder winters and more rainfall.

PEOPLE & SOCIETY

Though Austrians speak German, they like to stress their distinctive identity in relation to Germany. Vienna is a major cultural center. Minorities are few; there are some ethnic Croats, Slovenes, and Hungarians, plus refugees from conflict in former Yugoslavia. Though strongly Roman Catholic, Austrian society is less conservative than some southern German *Länder*. Class divisions remain strong.

THE ECONOMY

Large manufacturing base, despite lack of energy resources. The skilled labor force is key to high-tech exports. Eurozone member. Limited GDP growth has returned since 2009 recession.

INSIGHT: *Many of the world's great composers were Austrian, including Mozart, Haydn, Schubert, and Strauss*

FACTFILE

OFFICIAL NAME: Republic of Austria
DATE OF FORMATION: 1918
CAPITAL: Vienna
POPULATION: 8.5 million
TOTAL AREA: 32,378 sq. miles (83,858 sq. km)
DENSITY: 266 people per sq. mile

LANGUAGES: German*, Croatian, Slovenian, Hungarian (Magyar)
RELIGIONS: Roman Catholic 78%, nonreligious 9%, other 8%, Protestant 5%
ETHNIC MIX: Austrian 93%, Croat, Slovene, and Hungarian 6%, other 1%
GOVERNMENT: Parliamentary system
CURRENCY: Euro = 100 cents

Azerbaijan

Situated on the western coast of the Caspian Sea, it was the first Soviet republic to declare independence in 1991. Territorial disputes with Armenia have dominated politics since.

GEOGRAPHY

Caucasus Mountains in west, including Naxçivan exclave south of Armenia. Flat, low-lying terrain on the coast of the Caspian Sea.

CLIMATE
Low rainfall. Continental, with bitter winters, inland. Subtropical in coastal regions.

PEOPLE & SOCIETY
Azeris, a Muslim people with ethnic links to Turks, form a large majority. Thousands of Armenians, Russians, and Jews have left since independence. Influx of half a million Azeri refugees fleeing war with Armenia over the disputed enclave of Nagorno Karabakh. Armenians there operate with de facto independence. The status of women deteriorated after the fall of communism but they are slowly regaining their position.

THE ECONOMY
Oil and natural gas exports drive economic growth. Pipeline to Ceyhan, Turkey, has opened up European market. Severe pollution in Baku.

 INSIGHT: *The fire-worshipping Zoroastrian faith originated in Azerbaijan in the 6th century BCE*

4000m/13124ft
3000m/9843ft
2000m/6562ft
1000m/3281ft
500m/1640ft
200m/656ft
Sea Level
Below Sea Level

0   100 km
0   100 miles

FACTFILE

OFFICIAL NAME: Republic of Azerbaijan
DATE OF FORMATION: 1991
CAPITAL: Baku
POPULATION: 9.4 million
TOTAL AREA: 33,436 sq. miles (86,600 sq. km)
DENSITY: 281 people per sq. mile

LANGUAGES: Azeri*, Russian
RELIGIONS: Shi'a Muslim 68%, Sunni Muslim 26%, Russian Orthodox 3%, Armenian Apostolic Church (Orthodox) 2%, other 1%
ETHNIC MIX: Azeri 91%, other 3%, Lazs 2%, Russian 2%, Armenian 2%
GOVERNMENT: Presidential system
CURRENCY: New manat = 100 gopik

NORTH & CENTRAL AMERICA
The Bahamas

Located off the Florida coast in the western Atlantic, the Bahamas comprises an archipelago of some 700 islands and 2400 cays, only around 30 of which are inhabited.

GEOGRAPHY
Long, mainly flat coral formations with a few low hills. Some islands have pine forests, lagoons, and mangrove swamps.

CLIMATE
Subtropical. Hot summers and mild winters. Heavy rainfall, especially in summer. Hurricanes can strike in July–December.

PEOPLE & SOCIETY
Over 60% of the population live on New Providence. Tourism employs over half of the labor force. There are marked wealth disparities, from urban professionals in the banking sector to traditional fishermen on outlying islands and illegal Haitian and Cuban immigrants. More women are now entering the professions. Government priorities are tackling narcotics trafficking and combating money laundering.

THE ECONOMY
Major tourist destination, especially for US visitors. Financial services: banking and insurance.

INSIGHT: *The country's extensive merchant fleet consists mainly of "flag-of-convenience" vessels registered by foreign owners*

FACTFILE
OFFICIAL NAME: Commonwealth of the Bahamas
DATE OF FORMATION: 1973
CAPITAL: Nassau
POPULATION: 400,000
TOTAL AREA: 5382 sq. miles (13,940 sq. km)

DENSITY: 103 people per sq. mile
LANGUAGES: English*, English Creole, French Creole
RELIGIONS: Baptist 32%, other 29%, Anglican 20%, Roman Catholic 19%
ETHNIC MIX: Black African 85%, other 15%
GOVERNMENT: Parliamentary system
CURRENCY: Bahamian dollar = 100 cents

Bahrain

Bahrain is an archipelago of 49 islands between the Qatar peninsula and the Saudi Arabian mainland. Only three of the islands are inhabited. It was the first Gulf emirate to export oil.

 GEOGRAPHY
All islands are low-lying. The largest, Bahrain Island, is mainly sandy plains and salt marshes.

 CLIMATE
Summers are hot and humid. Winters are mild. Low rainfall.

 PEOPLE & SOCIETY
The key social division is between the Shi'a majority and Sunni minority. Sunnis hold the best jobs in bureaucracy and business while Shi'as tend to do menial work. Bahrain is socially liberal. The al-Khalifa family has ruled since 1783, but transformed Bahrain into a constitutional monarchy in 2002. Protests calling for greater democracy rocked the country since the 2011 "Arab Spring".

◆ **INSIGHT:** *The 16 Hawar Islands were awarded to Bahrain in 2001 after a lengthy dispute with Qatar*

THE ECONOMY
Main exports are refined petroleum and aluminum products. As oil reserves run out, natural gas is of increasing importance. Major Middle East offshore banking center, hit by global banking crisis in 2008–2009.

FACTFILE

OFFICIAL NAME: Kingdom of Bahrain

DATE OF FORMATION: 1971

CAPITAL: Manama

POPULATION: 1.3 million

TOTAL AREA: 239 sq. miles (620 sq. km)

DENSITY: 4762 people per sq. mile

LANGUAGES: Arabic*

RELIGIONS: Muslim (mainly Shi'a) 99%, other 1%

ETHNIC MIX: Bahraini 63%, Asian 19%, other Arab 10%, Iranian 8%

GOVERNMENT: Mixed monarchical-parliamentary system

CURRENCY: Bahraini dinar = 1000 fils

Bangladesh

Bangladesh lies at the north end of the Bay of Bengal and frequently suffers devastating flood, cyclones, and famine. It seceded from Pakistan in 1971.

GEOGRAPHY
Mostly flat alluvial plains and deltas of the Brahmaputra and Ganges rivers. Southeast coasts are fringed with mangrove forests.

CLIMATE
Hot and humid. During the monsoon, water levels can rise 20 ft (6 m) above sea level.

PEOPLE & SOCIETY
After a period of military rule, Bangladesh returned to democracy in 1991; political instability has continued, however, and corruption is a major problem. A third of the population live in poverty, but living standards are improving. Women are prominent in politics, but their rights are neglected.

INSIGHT: *Torrential monsoon rains flood two-thirds of the country every year*

THE ECONOMY
Agriculture is vulnerable to unpredictable climate. Bangladesh accounts for 80% of world jute fiber exports. Poor infrastructure deters investment. Growing textile industry.

FACTFILE
OFFICIAL NAME: People's Republic of Bangladesh
DATE OF FORMATION: 1971
CAPITAL: Dhaka
POPULATION: 157 million
TOTAL AREA: 55,598 sq. miles (144,000 sq. km)

DENSITY: 3029 people per sq. mile
LANGUAGES: Bengali*, Urdu, Chakma, Marma, Garo, Khasi, Santhali, Tripuri, Mro
RELIGIONS: Muslim (mainly Sunni) 88%, Hindu 11%, other 1%
ETHNIC MIX: Bengali 98%, other 2%
GOVERNMENT: Parliamentary system
CURRENCY: Taka = 100 poisha

Barbados

Barbados is the most easterly of the Caribbean islands. Once solely inhabited by the native Arawak, Barbados was first colonized by British settlers in the 1620s.

GEOGRAPHY

Encircled by coral reefs. Fertile and predominantly flat, with a few gentle hills to the north.

CLIMATE

Moderate tropical climate. Sunnier and drier than its more mountainous neighbors.

PEOPLE & SOCIETY

Independent from the UK since 1966. Some latent tension between the economically dominant white community and the majority black population, but violence is rare. Increasing social mobility has enabled black Barbadians to enter the professions. Despite political stability, and good welfare and education services, pockets of abject poverty remain.

 INSIGHT: *Barbados retains a strong British influence and is referred to by its neighbors as "Little England"*

THE ECONOMY

Well-developed tourism sector based on climate and accessibility. Financial services, offshore banking, and information processing are key industries. Sugar production has dwindled. High cost of living.

FACTFILE

OFFICIAL NAME: Barbados

DATE OF FORMATION: 1966

CAPITAL: Bridgetown

POPULATION: 300,000

TOTAL AREA: 166 sq. miles (430 sq. km)

DENSITY: 1807 people per sq. mile

LANGUAGES: Bajan (Barbadian English), English*

RELIGIONS: Anglican 40%, other 24%, nonreligious 17%, Pentecostal 8%, Methodist 7%, Roman Catholic 4%

ETHNIC MIX: Black African 92%, White 3%, other 3%, mixed race 2%

GOVERNMENT: Parliamentary system

CURRENCY: Barbados dollar = 100 cents

Belarus

Literally "White Russia," Belarus lies landlocked in eastern Europe. It reluctantly became independent when the USSR broke up in 1991. It has few resources other than agriculture.

GEOGRAPHY

Mainly plains and low hills. The Dnieper and Dvina rivers drain the eastern lowlands. Vast Pripet Marshes in the southwest.

CLIMATE

Extreme continental climate. Winters are long, sub-freezing, but mainly dry; summers are hot.

PEOPLE & SOCIETY
Only 2% of people are non-Slav, so ethnic tension is minimal. Russian culture dominates. Belarus was the slowest ex-Soviet state to implement political reform; President Lukashenka has been labeled as Europe's last dictator. Enthusiasm for a merger with Russia has waned. Wealth is held by a small ex-Communist elite. Fallout from the 1986 Chernobyl nuclear disaster in Ukraine still seriously affects health and the environment.

THE ECONOMY
Industry outmoded and mainly state-owned. Depends on Russia for energy and raw materials: tensions over natural gas prices.

INSIGHT: *The number of cancer and leukemia cases soared after the 1986 Chernobyl disaster*

FACTFILE

OFFICIAL NAME: Republic of Belarus
DATE OF FORMATION: 1991
CAPITAL: Minsk
POPULATION: 9.4 million
TOTAL AREA: 80,154 sq. miles (207,600 sq. km)
DENSITY: 117 people per sq. mile

LANGUAGES: Belarussian*, Russian*
RELIGIONS: Orthodox Christian 80%, Roman Catholic 14%, other 4%, Protestant 2%
ETHNIC MIX: Belarussian 81%, Russian 11%, Polish 4%, Ukrainian 2%, other 2%
GOVERNMENT: Presidential system
CURRENCY: Belarussian rouble = 100 kopeks

Belgium

Belgium lies in northwestern Europe. Its history has been marked by tensions between the majority Dutch-speaking (Flemish) and minority French-speaking (Walloon) communities.

 GEOGRAPHY

Low-lying coastal plain covers two-thirds of the country. Land becomes hilly and forested in the southeast (Ardennes).

CLIMATE

Maritime climate with Gulf Stream influences. Mild temperatures, with heavy cloud cover and rain. More rainfall and weather fluctuations at the coast.

PEOPLE & SOCIETY

Since 1970, Flemish regions have become more prosperous than those of the minority Walloons, overturning traditional roles and increasing friction. Belgium moved to a federal system from 1980 in order to contain tensions, but recent fractious politics have raised doubts over the union's survival. The Flemish separatist N-VA heads the Flanders government and since 2010 has been the largest party at federal level. Brussels hosts key EU institutions.

THE ECONOMY

Variety of industrial exports, including steel, glassware, cut diamonds, and textiles. High levels of public debt. Bureaucracy larger than European average.

INSIGHT: *Belgium holds the world record for the country with the longest period without a government*

FACTFILE

OFFICIAL NAME: Kingdom of Belgium

DATE OF FORMATION: 1830

CAPITAL: Brussels

POPULATION: 11.1 million

TOTAL AREA: 11,780 sq. miles (30,510 sq. km)

DENSITY: 876 people per sq. mile

LANGUAGES: Dutch*, French*, German*

RELIGIONS: Roman Catholic 88%, other 10%, Muslim 2%

ETHNIC MIX: Fleming 58%, Walloon 33%, other 6%, Italian 2%, Moroccan 1%

GOVERNMENT: Parliamentary system

CURRENCY: Euro = 100 cents

Belize

Belize lies on the eastern shore of the Yucatan Peninsula. Formerly called British Honduras, Belize was the last Central American country to gain its independence, in 1981.

 GEOGRAPHY

Almost half the land area is forested. Low mountains in southeast. Flat swampy coastal plains.

 CLIMATE

Tropical. Very hot and humid, with May–December rainy season.

 PEOPLE & SOCIETY

English-speaking black Creoles are outnumbered by Spanish speakers, including native *mestizos* (European–Amerindian) and immigrants from neighboring states. The Creoles have traditionally dominated society, but high levels of emigration to the US have weakened their influence. The Afro-Carib *garifuna* have their own language. Corruption, and trafficking of people and narcotics, are major problems.

 INSIGHT: *Belize's barrier reef is the second-largest in the world*

THE ECONOMY

Tourism, agriculture, and offshore banking. Oil extraction began in 2005. Sugar, textiles, lobsters, and shrimp are exported. Serious hurricane damage is a recurring problem.

FACTFILE

OFFICIAL NAME: Belize

DATE OF FORMATION: 1981

CAPITAL: Belmopan

POPULATION: 300,000

TOTAL AREA: 8867 sq. miles (22,966 sq. km)

DENSITY: 34 people per sq. mile

LANGUAGES: English Creole, Spanish, English*, Mayan, Garifuna (Carib)

RELIGIONS: Roman Catholic 62%, other 20%, Anglican 12%, Methodist 6%

ETHNIC MIX: *Mestizo* 49%, Creole 25%, Maya 11%, other 9%, Garifuna 6%

GOVERNMENT: Parliamentary system

CURRENCY: Belizean dollar = 100 cents

Benin

Benin stretches north from the west African coast. In 1990, Benin became one of the pioneers of African democratization, ending 17 years of one-party Marxist-Leninist rule.

GEOGRAPHY

Sandy coastal region. Numerous lagoons lie just behind the shoreline. Forested plateaus inland. Mountains in the northwest.

CLIMATE

Hot and humid in the south. Two rainy seasons. Hot, dusty *harmattan* winds blow during the December–February dry season.

PEOPLE & SOCIETY

There are 42 different ethnic groups. The southern Fon have tended to dominate politics. Other major groups are the Adja and Yoruba. The northern Fulani follow a nomadic lifestyle. North–south tension is mainly due to the south being more developed. French culture, centered on Cotonou, is highly prized. Substantial differences in wealth reflect a strongly hierarchical society.

THE ECONOMY
Strong agricultural sector: cash crops include cotton, oil palm, and cashew nuts. Large-scale smuggling is a serious problem. France is the main aid donor. Recent floods.

INSIGHT:

Voodoo is thought to have originated in Benin, and was taken to Haiti by slaves

NIGER

BURKINA FASO

Malanville

Kandi

Natitingou

Ndali

Djougou

Parakou

TOGO

NIGERIA

Savè

Ouémé

500m/1640ft
200m/656ft
Sea Level

Lokossa

PORTO-NOVO

Ouidah — Cotonou

ATLANTIC OCEAN

0   100 km
0   100 miles

FACTFILE

OFFICIAL NAME: Republic of Benin

DATE OF FORMATION: 1960

CAPITAL: Porto-Novo

POPULATION: 10.3 million

TOTAL AREA: 43,483 sq. miles (112,620 sq. km)

DENSITY: 241 people per sq. mile

LANGUAGES: Fon, Bariba, Yoruba, Adja, Houeda, Somba, French*

RELIGIONS: Indigenous beliefs and Voodoo 50%, Christian 30%, Muslim 20%

ETHNIC MIX: Fon 41%, other 21%, Adja 16%, Yoruba 12%, Bariba 10%

GOVERNMENT: Presidential system

CURRENCY: CFA franc = 100 centimes

Bhutan

Perched in the eastern Himalayas between India and China lies the landlocked Kingdom of Bhutan. It is largely closed to the outside world to protect its culture; TV was banned until 1999.

GEOGRAPHY

Low, tropical southern strip rising through fertile central valleys to high Himalayas in the north. Around 70% of the land is forested.

CLIMATE

South is tropical, north is alpine, cold, and harsh. Central valleys warmer in east than west.

PEOPLE & SOCIETY

The king was absolute monarch until 1998, and the first democratic elections were held a decade later. Most people are devoutly Buddhist and originate from Tibet. The Hindu Nepalese settled in the south. Bhutan has 20 languages. In 1988, Dzongkha (a Tibetan dialect native to just 16% of the people) was made the official language. The Nepalese community regard this as "cultural imperialism," causing considerable ethnic tensions.

THE ECONOMY

Reliant on India for trade. Most people farm their own plots of land and herd cattle and yaks. Steep land unsuited for cultivation. Development of cash crops for Asian markets.

INSIGHT: *In 2004 Bhutan became the first country in the world to ban smoking and the sale of tobacco*

FACTFILE

OFFICIAL NAME: Kingdom of Bhutan
DATE OF FORMATION: 1656
CAPITAL: Thimphu
POPULATION: 800,000
TOTAL AREA: 18,147 sq. miles (47,000 sq. km)
DENSITY: 44 people per sq. mile

LANGUAGES: Dzongkha*, Nepali, Assamese
RELIGIONS: Mahayana Buddhist 75%, Hindu 25%
ETHNIC MIX: Drukpa 50%, Nepalese 35%, other 15%
GOVERNMENT: Mixed monarchical–parliamentary system
CURRENCY: Ngultrum = 100 chetrum

Bolivia

Landlocked high in central South America, Bolivia is one of the region's poorest countries. La Paz is the world's highest capital city: 13,385 feet (3631 m) above sea level.

GEOGRAPHY

A high windswept plateau, the *altiplano*, lies between two Andean mountain ranges. Semiarid grasslands to the east; dense tropical forests to the north.

CLIMATE

Altiplano has extreme tropical climate, with night-frost in winter. North and east are hot and humid.

PEOPLE & SOCIETY

Wealthy Spanish-descended families have traditionally controlled the economy. The indigenous majority faces widespread discrimination. Amerindian Evo Morales, president from 2005, is cutting poverty, redistributing land, and pushing for inter-national recognition of legal coca use.

 INSIGHT: *Between 1825 and 1982 Bolivia averaged more than one armed coup a year*

THE ECONOMY

Gold, silver, zinc, tin, oil, natural gas: all vulnerable to world price fluctuations. Social issues and nationalization of natural gas sector deter investors. Major coca producer. Lack of manufacturing. Rich eastern provinces want autonomy.

3000m/9843ft	
2000m/6562ft	
1000m/3281ft	
500m/1640ft	
200m/656ft	
Sea Level	

FACTFILE

OFFICIAL NAME: Plurinational State of Bolivia

DATE OF FORMATION: 1825

CAPITALS: La Paz (administrative); Sucre (judicial)

POPULATION: 10.7 million

TOTAL AREA: 424,162 sq. miles (1,098,580 sq. km)

DENSITY: 26 people per sq. mile

LANGUAGES: Aymara*, Quechua*, Spanish*

RELIGIONS: Roman Catholic 93%, other 7%

ETHNIC MIX: Quechua 37%, Aymara 32%, *Mestizo* (mixed European–Amerindian) 13%, European 10%, other 8%

GOVERNMENT: Presidential system

CURRENCY: Boliviano = 100 centavos

Bosnia & Herzegovina

Perched in the highlands of southeast Europe, Bosnia and Herzegovina was the focus of the bitter ethnic conflict that accompanied the early 1990s dissolution of the Yugoslav state.

GEOGRAPHY

Hills and mountains, with narrow river valleys. Lowlands in the north. Mainly deciduous forest covers about half of the total area.

CLIMATE

Continental. Hot summers and cold, often snowy winters.

PEOPLE & SOCIETY

Despite sharing the same origin and spoken language, Bosnians have been divided by history between Orthodox Serbs, Roman Catholic Croats, and Muslim Bosniaks. Ethnic cleansing was practiced by all sides in the civil war, displacing about 60% of the population. Hopes for EU integration will require further ethnic reconciliation.

INSIGHT: *The murder of Archduke Ferdinand of Austria in Sarajevo in 1914 triggered the First World War*

THE ECONOMY

Potential to recover status as a thriving market economy with a strong manufacturing base, but still struggles with resettling refugees and the legacy of war. Little foreign investment.

FACTFILE

OFFICIAL NAME: Bosnia and Herzegovina
DATE OF FORMATION: 1992
CAPITAL: Sarajevo
POPULATION: 3.8 million
TOTAL AREA: 19,741 sq. miles (51,129 sq. km)
DENSITY: 192 people per sq. mile

LANGUAGES: Bosnian*, Serbian*, Croatian*
RELIGIONS: Muslim (mainly Sunni) 40%, Orthodox Christian 31%, Roman Catholic 15%, other 10%, Protestant 4%
ETHNIC MIX: Bosniak 48%, Serb 34%, Croat 16%, other 2%
GOVERNMENT: Parliamentary system
CURRENCY: Marka = 100 pfeninga

Botswana

Landlocked in the heart of southern Africa, Botswana boasts the world's largest inland river delta. Diamonds provide potential wealth, but the country is crippled by HIV/AIDS.

GEOGRAPHY

Lies on vast plateau, high above sea level. Hills in the east. Kalahari Desert in center and southwest. Swamps and salt pans elsewhere and in Okavango Basin.

CLIMATE

Dry and prone to drought. Summer wet season, April–October. Winters are warm, with cold nights.

PEOPLE & SOCIETY

The nomadic San bushmen, the first inhabitants, are marginalized. One in five adults are living with HIV/AIDS: only Swaziland and Lesotho are worse affected. Life expectancy is around 64 years. Diamond revenue has widened wealth inequalities.

INSIGHT: *Water, Botswana's most precious resource, is honored in the name of the currency – pula*

THE ECONOMY

Overreliance on diamonds: vulnerable to world price fluctuations. Beef is exported to Europe. Tourism aimed at wealthy wildlife enthusiasts. AIDS is devastating the population.

FACTFILE

OFFICIAL NAME: Republic of Botswana

DATE OF FORMATION: 1966

CAPITAL: Gaborone

POPULATION: 2 million

TOTAL AREA: 231,803 sq. miles (600,370 sq. km)

DENSITY: 9 people per sq. mile

LANGUAGES: Setswana, English*, Shona, San, Khoikhoi, isiNdebele

RELIGIONS: Christian 70%, nonreligious 20%, traditional beliefs 6%, other 4%

ETHNIC MIX: Tswana 79%, Kalanga 11%, other 10%

GOVERNMENT: Presidential system

CURRENCY: Pula = 100 thebe

Brazil

Covering almost half of South America, Brazil is the site of the world's largest and ecologically most important rainforest. The country has immense natural and economic resources.

GEOGRAPHY

Rainforest grows around the massive Amazon River and its delta, covering almost half of Brazil's total land area. Apart from the basin of the River Plate to the south, the rest of the country consists of highlands. The mountainous east is part-forested and part-desert. The coastal plain in the southeast has swampy areas. The Atlantic coastline is 1240 miles (2000 km) long.

CLIMATE

Brazil's share of the Amazon Basin has a model tropical equatorial climate, with high temperatures and rainfall all year round. The Brazilian plateau has far greater seasonal variation. The dry northeast suffers frequent droughts, though coastal regions are occasionally flooded by bouts of torrential rain. The south has hot summers and cool winters.

PEOPLE & SOCIETY

Diverse population includes Amerindians, black people of African descent, European immigrants, and those of mixed race. Amerindians suffer prejudice from most other groups. Shanty towns in the cities attract poor migrants from the northeast. Urban crime, violent land disputes, and unchecked development in Amazonia tarnish Brazil's image as a modern nation. Catholicism and the family unit remain strong.

THE ECONOMY

Dominant regional economy. Huge potential for growth based on abundant natural resources. A leading exporter of coffee, sugar, soybeans, and orange juice. Social tension threatens stability. Infrastructure needs investment. Downturn in 2014.

Equator

COLOMBE

PERU

FACTFILE

OFFICIAL NAME: Federative Rep. of Brazil
DATE OF FORMATION: 1822
CAPITAL: Brasília
POPULATION: 200 million
TOTAL AREA: 3,286,470 sq. miles (8,511,965 sq. km)
DENSITY: 61 people per sq. mile

LANGUAGES: Portuguese*, German, Japanese, Italian, Spanish, Polish, Amerindian languages
RELIGIONS: Roman Catholic 74%, Protestant 15%, atheist 7%, other 4%
ETHNIC MIX: White 54%, mixed race 38%, Black 6%, other 2%
GOVERNMENT: Presidential system
CURRENCY: Real = 100 centavos

INSIGHT: *Since 1900, a third of Brazil's indigenous Amerindian groups have become extinct due to disease, starvation, or the forceful taking of land by miners, loggers, and settlers*

VENEZUELA

SURINAME

French Guiana (France)

GUYANA

Guiana Highlands

Boa Vista

Rio Branco

Macapá

Ilha de Marajó

Belém

Equator

ATLANTIC

OCEAN

apurá

Rio Negro

A m a z o n

Manaus

Amazon

São Luís

Santarém

Fortaleza

San Fernando de Noronha

Madeira

Juruá

Purus

B a s i n

Imperatriz

Parnaíba

Teresina

Tapajós

Xingu

Iriri

Tocantins

Porto Velho

Rio Branco

Juazeiro do Norte

Natal

João Pessoa

Campina Grande

Olinda

Recife

Chapada dos Parecis

Guaporé

São Manuel

Araguaia

Represa de Sobradinho

Maceió

Aracaju

BOLIVIA

Planalto de Mato Grosso

Taguatinga

Brazilian Highlands

Feira de Santana

Salvador

Cuiabá

BRASÍLIA

Goiânia

Itabuna

Vitória da Conquista

São Francisco

Montes Claros

Pantanal

Uberlândia

Uberaba

Governador Valadares

PARAGUAY

Campo Grande

Belo Horizonte

Paraná

Bauru

Ribeirão Preto

Vitória

Campinas

Nova Iguaçu

Campos

Londrina

São Paulo

Duque de Caxias

Santos

Rio de Janeiro

Curitiba

Joinville

Florianópolis

ARGENTINA

Caxias do Sul

Porto Alegre

Lagoa dos Patos

Pelotas

ATLANTIC

OCEAN

Rio Grande

URUGUAY

Mirim Lagoon

2000m/6562ft
1000m/3281ft
500m/1640ft
200m/656ft
Sea Level

0   500 km

0   500 miles

Brunei

Lying on the northern coast of the island of Borneo, Brunei is surrounded and divided in two by the Malaysian state of Sarawak. It has been independent since 1984.

GEOGRAPHY
Mostly dense lowland rainforest and mangrove swamps, with some mountains in the southeast.

CLIMATE
Tropical. Six-month rainy season with very high humidity.

PEOPLE & SOCIETY
Malays benefit from positive discrimination. Many in the Chinese community are stateless. Since a failed rebellion in 1962, Brunei has been ruled by decree of the sultan. In 1990, "Malay Muslim Monarchy" was introduced, promoting Islamic values as state ideology. Women, less restricted than in some Muslim states, usually wear headscarves but not the veil.

INSIGHT: *The sultan spent US$350 million building the world's largest palace at Bandar Seri Begawan*

THE ECONOMY
Oil and natural gas production has brought one of the world's highest standards of living. Massive overseas investments. Major consumer of high-tech hi-fi, video equipment, and Western designer clothes.

FACTFILE

OFFICIAL NAME: Brunei Darussalam

DATE OF FORMATION: 1984

CAPITAL: Bandar Seri Begawan

POPULATION: 400,000

TOTAL AREA: 2228 sq. miles (5770 sq. km)

DENSITY: 197 people per sq. mile

LANGUAGES: Malay*, English, Chinese

RELIGIONS: Muslim (mainly Sunni) 66%, Buddhist 14%, Christian 10%, other 10%

ETHNIC MIX: Malay 67%, Chinese 16%, other 11%, indigenous 6%

GOVERNMENT: Monarchy

CURRENCY: Brunei dollar = 100 cents

Bulgaria

Located in southeastern Europe, Bulgaria was under communist rule from 1947 to 1989. Significant political and economic reform since then enabled it to join the EU in 2007.

GEOGRAPHY

Mountains run east–west across center and along southern border. Danube plain in north, Thracian plain in southeast. Black Sea to the east.

CLIMATE

Hot summers, cooler at the coast. Snowy winters, especially in mountains. East winds bring seasonal extremes.

PEOPLE & SOCIETY

The communists tried forcibly to suppress cultural identities; once free movement was allowed in 1989, there was a large exodus of Bulgarian Turks. Privatizations in the 1990s left many Turks landless, prompting further emigration. Roma suffer discrimination at all levels of society. Women have equal rights in theory, but society remains patriarchal. EU accession included caveats demanding further action against organized crime, human trafficking, and corruption.

THE ECONOMY

Good agricultural production, including grapes, for well-developed wine industry, and tobacco. Expertise in software development. Industry and infrastructure are outdated.

INSIGHT: *Archaeologists have found evidence of wine-making in Bulgaria dating back over 5000 years*

FACTFILE

OFFICIAL NAME: Republic of Bulgaria

DATE OF FORMATION: 1908

CAPITAL: Sofia

POPULATION: 7.2 million

TOTAL AREA: 42,822 sq. miles (110,910 sq. km)

DENSITY: 169 people per sq. mile

LANGUAGES: Bulgarian*, Turkish, Romani

RELIGIONS: Bulgarian Orthodox 83%, Muslim 12%, other 4%, Roman Catholic 1%

ETHNIC MIX: Bulgarian 84%, Turkish 9%, Roma 5%, other 2%

GOVERNMENT: Parliamentary system

CURRENCY: Lev = 100 stotinki

Burkina Faso

The west African state of Burkina Faso was known as Upper Volta until 1984. It became a multiparty state in 1991, though former military ruler Blaise Compaoré remains in power.

 GEOGRAPHY
The Sahara covers the north of the country. The south is largely savanna. The three main rivers are the Black, White, and Red Voltas.

 CLIMATE
Tropical. Dry, cool weather November–February. Erratic rain March–April, mostly in southeast.

PEOPLE & SOCIETY
No single ethnic group is dominant, but the Mossi, from around Ouagadougou, have always played an important part in government. The people from the west are much more ethnically mixed. Extreme poverty has led to a strong sense of egalitarianism. Most women are still denied access to education, though their absence from public life belies their real power and social influence.

THE ECONOMY
Cotton is the major cash crop, but the encroaching Sahara Desert is restricting agriculture. Beneficiary of foreign debt cancellation plans.

INSIGHT: *Droughts and poor soils mean that many Burkinabés seek work southward in Ghana and Côte d'Ivoire*

FACTFILE

OFFICIAL NAME: Burkina Faso
DATE OF FORMATION: 1960
CAPITAL: Ouagadougou
POPULATION: 16.9 million
TOTAL AREA: 105,869 sq. miles (274,200 sq. km)
DENSITY: 160 people per sq. mile

LANGUAGES: Mossi, Fulani, French*, Tuareg, Dyula, Songhai
RELIGIONS: Muslim 55%, Christian 25%, traditional beliefs 20%
ETHNIC MIX: Mossi 48%, other 21%, Peul 10%, Lobi 7%, Bobo 7%, Mandé 7%
GOVERNMENT: Presidential system
CURRENCY: CFA franc = 100 centimes

Burundi

Small, densely populated and landlocked, Burundi lies just south of the equator, on the Nile–Congo watershed in central Africa. More than two-thirds of people live below the poverty line.

GEOGRAPHY
Hilly with high plateaus in center and savanna in the east. Great Rift Valley on western side.

CLIMATE
Temperate, with high humidity. Heavy and frequent rainfall, mostly October–May. Highlands have frost.

PEOPLE & SOCIETY
Burundi has been riven by ethnic conflict between majority Hutu and the Tutsi, who controlled the army – with repeated large-scale massacres: hundreds of thousands of people died between 1993 and 2004. The constitution now guarantees an ethnic balance in the government and army. Twa pygmies were not involved in the conflict.

 INSIGHT: *Burundi's fertility rate is one of the highest in Africa. On average, women have six children*

THE ECONOMY
Overwhelmingly agricultural economy, mostly subsistence. Small quantities of gold and tungsten. Potential of oil in Lake Tanganyika. Ongoing political fragility.

	2000m/6562ft
	1000m/3281ft
	500m/1640ft

FACTFILE

OFFICIAL NAME: Republic of Burundi
DATE OF FORMATION: 1962
CAPITAL: Bujumbura
POPULATION: 10.2 million
TOTAL AREA: 10,745 sq. miles (27,830 sq. km)
DENSITY: 1030 people per sq. mile

LANGUAGES: Kirundi*, French*, Kiswahili
RELIGIONS: Roman Catholic 62%, traditional beliefs 23%, Muslim 10%, Protestant 5%
ETHNIC MIX: Hutu 85%, Tutsi 14%, Twa 1%
GOVERNMENT: Presidential system
CURRENCY: Burundi franc = 100 centimes

Cambodia

Located on the Indochinese peninsula in southeast Asia, Cambodia has emerged from genocide, civil war, and invasion from Vietnam. Tourism has rebounded, and is a key income earner.

GEOGRAPHY

Mostly low-lying basin. Tônlé Sap (Great Lake) drains into the Mekong River. Forested mountains and plateau east of the Mekong.

CLIMATE

Tropical. High temperatures throughout the year. Heavy rainfall during May–October monsoon.

PEOPLE & SOCIETY

Devastated by US bombing, then by the Khmer Rouge regime, whose extreme Marxist program killed over a million between 1975 and 1979, Cambodia then endured further civil conflict and Vietnamese occupation. The effects are still felt, reflected in the high rates of orphans, widows, and land-mine victims. A fragile stability has lasted since elections in 1993. King Norodom Sihanouk, a key figure in politics, abdicated in 2004.

THE ECONOMY

Economy is heavily aid-reliant, still recovering from civil war. Rubber and timber are exported. Self-sufficient in rice. Garment industry is growing. Land disputes and corruption issues.

INSIGHT: *Cambodia has many impressive temples (including Angkor Wat), which date from when the country was the center of the Khmer Empire*

FACTFILE

OFFICIAL NAME: Kingdom of Cambodia
DATE OF FORMATION: 1953
CAPITAL: Phnom Penh
POPULATION: 15.1 million
TOTAL AREA: 69,900 sq. miles (181,040 sq. km)
DENSITY: 222 people per sq. mile

LANGUAGES: Khmer*, French, Chinese, Vietnamese, Cham
RELIGIONS: Buddhist 93%, Muslim 6%, Christian 1%
ETHNIC MIX: Khmer 90%, Vietnamese 5%, other 4%, Chinese 1%
GOVERNMENT: Parliamentary system
CURRENCY: Riel = 100 sen

Cameroon

Situated in the corner of the Gulf of Guinea, Cameroon was effectively a one-party state for 30 years. Multiparty elections, since 1992, regularly return that same party to power.

GEOGRAPHY

Over half the land is forested: equatorial rainforest in north, evergreen forest and wooded savanna in south. Mountains in the west.

CLIMATE

South is equatorial, with plentiful rainfall, declining inland. Far north is beset by drought.

PEOPLE & SOCIETY

Around 230 ethnic groups; no single group is dominant. The Bamileke is the largest, though it has never held political power. North–south tensions are diminished by the ethnic diversity. There is more rivalry between majority French- and minority English-speakers.

◆ **INSIGHT:** *Cameroon's name derives from the Portuguese word* camarões, *after the shrimp fished by the early European explorers*

THE ECONOMY
Oil reserves. Very diversified agricultural economy – timber, cocoa, bananas, coffee. Fuel smuggling from Nigeria undermines refinery profits. Corruption. Port for Chad and CAR.

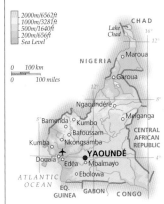

FACTFILE

OFFICIAL NAME: Republic of Cameroon
DATE OF FORMATION: 1960
CAPITAL: Yaoundé
POPULATION: 22.3 million
TOTAL AREA: 183,567 sq. miles (475,400 sq. km)
DENSITY: 124 people per sq. mile

LANGUAGES: Bamileke, Fang, Fulani, French*, English*
RELIGIONS: Roman Catholic 35%, traditional beliefs 25%, Muslim 22%, Protestant 18%
ETHNIC MIX: Cameroon highlanders 31%, Bantu 27%, other 21%, Kirdi 11%, Fulani 10%
GOVERNMENT: Presidential system
CURRENCY: CFA franc = 100 centimes

Canada

Canada extends from the Arctic to its US border along the 49th parallel. Unified under British rule from 1763, its development and expansion attracted large-scale immigration.

GEOGRAPHY

The world's second-largest country, stretching north to Cape Colombia on Ellesmere Island, south to Lake Erie, and across five time zones from the Pacific seaboard to Newfoundland. Arctic tundra and islands in the far north give way southward to forests, interspersed with lakes and rivers, and then the vast Canadian Shield, which covers over half the area of Canada. Rocky Mountains in west, beyond which are the Coast Mountains, islands, and fjords. Fertile lowlands in the east.

CLIMATE

Ranges from polar and subpolar in the north, to continental in the south. Winters in the interior are colder and longer than on the coast, with temperatures well below freezing and deep snow; summers are hotter. Pacific coast has the mildest winters.

PEOPLE & SOCIETY

Two-thirds of the population live in the Great Lakes–St. Lawrence lowlands, fostering some shared cultural values with the neighboring US. Important differences, however, include wider welfare provision and Commonwealth membership. The French-speaking Québécois wish to preserve their culture and language from further Anglicization, and demand to be recognized as a "distinct society." The government welcomes ethnic diversity among immigrants, promoting a policy that encourages each group to maintain its own culture. Other sizable immigrant groups include Chinese, Italians, Germans. Ukrainians, and Portuguese. Land claims made by the indigenous peoples are being redressed. Nunavut, an Inuit-governed territory that covers nearly a quarter of Canada's land area, was created from a portion of the Northwest Territories in 1999. Women are well represented at most levels of business and government.

FACTFILE

OFFICIAL NAME: Canada
DATE OF FORMATION: 1867
CAPITAL: Ottawa
POPULATION: 35.2 million
TOTAL AREA: 3,855,171 sq. miles (9,984,670 sq. km)
DENSITY: 10 people per sq. mile

LANGUAGES: English*, French*, Chinese, other
RELIGIONS: Roman Catholic 44%, Protestant 29%, other and nonreligious 27%
ETHNIC ORIGIN: British, French, and other European 87%, Asian 9%, Amerindian, Métis, and Inuit 4%
GOVERNMENT: Parliamentary system
CURRENCY: Canadian dollar = 100 cents

THE ECONOMY

Wide-ranging resources, providing exports, cheap energy, and raw materials for manufacturing, underpin a high standard of living, with smaller wealth disparities than in the US. Prices for primary exports fluctuate, but the high oil price has encouraged development of Alberta's vast oil fields. Manufactured exports have flourished under growing global competition, especially since the creation in 1994 of the NAFTA free trade area, but reliance on the US market makes the Canadian economy vulnerable to US slowdowns. Unemployment rose during the 2009 recession, but the economy rebounded quickly.

INSIGHT: *The Magnetic North Pole, where the dipping needle of a compass stands still, migrates across northern Canada*

3000m/9843ft
2000m/6562ft
1000m/3281ft
500m/1640ft
200m/656ft
Sea Level

0   400 km
0   400 miles

Cape Verde

Off the west coast of Africa, in the Atlantic Ocean, lies the group of islands that make up Cape Verde, a Portuguese colony until it gained independence in 1975.

 GEOGRAPHY
Ten main islands and eight smaller islets, all of volcanic origin. Mostly mountainous, with steep cliffs and rocky headlands.

 CLIMATE
Warm, and very dry. Subject to droughts that can sometimes last for years at a time.

 PEOPLE & SOCIETY
Most people are of mixed Portuguese–African origin (*Mestiço*); the rest are descendants of African slaves or more recent immigrants. Creolization of the culture negates ethnic tensions. Over half of the population live on Santiago. Around 700,000 Cape Verdeans live abroad, mostly in the US.

◆ **INSIGHT:** *Poor soils and lack of surface water mean that Cape Verde is dependent on food aid*

THE ECONOMY
Most people are subsistence farmers. Clothing is the main export. No natural resources. Mid-Atlantic location ensures work maintaining ships and planes.

FACTFILE

OFFICIAL NAME: Republic of Cape Verde
DATE OF FORMATION: 1975
CAPITAL: Praia
POPULATION: 500,000
TOTAL AREA: 1557 sq. miles (4033 sq. km)
DENSITY: 321 people per sq. mile

LANGUAGES: Portuguese Creole, Portuguese*
RELIGIONS: Roman Catholic 97%, other 2%, Protestant (Church of the Nazarene) 1%
ETHNIC MIX: *Mestiço* 71%, African 28%, European 1%
GOVERNMENT: Mixed presidential-parliamentary system
CURRENCY: Escudo = 100 centavos

Central African Republic

The Central African Republic (CAR) is a landlocked
country lying between the basins of the Chad and Congo Rivers.
Politics suffers frequent interruption by coups and rebellions.

GEOGRAPHY

Comprises a low plateau,
covered by scrub or savanna. North
is arid. Equatorial rainforests in the
south. The Ubangi River forms the
border with the Democratic Republic
of the Congo.

CLIMATE

The south is equatorial; the
north is hot and dry. Rain occurs all
year round, with heaviest falls
between July and October.

PEOPLE & SOCIETY

The Baya and Banda are the
largest ethnic groups, but the lingua
franca is Sango, a trading creole spoken
by the minorities in the south who
have traditionally provided most
political leaders. Less than 2% of the
population live in the north. Recent
rebellions by northern militias have
displaced thousands of people.

THE ECONOMY

Dominated by subsistence farming.
Exports include diamonds, cotton,
timber, and coffee. Aid needed to
support refugees. Instability and
poor infrastructure hinder progress.

INSIGHT: *"Emperor" Bokassa's
eccentric rule from 1965 to 1979 was
followed by military dictatorship until
democracy was restored in 1993*

FACTFILE

OFFICIAL NAME: Central African Republic

DATE OF FORMATION: 1960

CAPITAL: Bangui

POPULATION: 4.6 million

TOTAL AREA: 240,534 sq. miles
(622,984 sq. km)

DENSITY: 19 people per sq. mile

LANGUAGES: Sango, Banda, Gbaya,
French*

RELIGIONS: Traditional beliefs 35%, Roman
Catholic 25%, Protestant 25%, Muslim 15%

ETHNIC MIX: Baya 33%, Banda 27%,
other 17%, Mandjia 13%, Sara 10%,

GOVERNMENT: Transitional regime

CURRENCY: CFA franc = 100 centimes

Chad

Landlocked in north-central Africa, Chad has had a turbulent history since independence from France in 1960. Intermittent periods of civil war followed a military coup in 1975.

GEOGRAPHY

Mostly plateaus sloping westward to Lake Chad. Northern third is Sahara. Tibesti Mountains in north rise to 10,826 ft (3300 m).

CLIMATE

Three distinct zones: desert in north, semiarid region in center, and tropics in south.

PEOPLE & SOCIETY

Half the population live in Chad's southern fifth. The northern third has only 100,000 people, mainly Muslim Toubou nomads. Democracy was restored in 1996 by ex-coup leader Idriss Déby, who has won all elections since. Instability has continued, first with tension between Muslims and southern Christians and, more recently, with rebellions in the east.

INSIGHT: *Lake Chad is slowly drying up – it is now estimated to be just 3% of the size it was in 1963*

THE ECONOMY

The discovery of oil, and the opening of a pipeline to the coast via Cameroon, are transforming Chad's economy, though the new wealth is unlikely to reach most people.

3000m/9843ft
2000m/6562ft
1000m/3281ft
500m/1640ft
200m/656ft
Sea Level

0   200 km
0   200 miles

FACTFILE

OFFICIAL NAME: Republic of Chad
DATE OF FORMATION: 1960
CAPITAL: N'Djaména
POPULATION: 12.8 million
TOTAL AREA: 495,752 sq. miles
(1,284,000 sq. km)
DENSITY: 26 people per sq. mile

LANGUAGES: French*, Sara, Arabic*, Maba
RELIGIONS: Muslim 51%, Christian 35%,
traditional beliefs 7%, animist 7%
ETHNIC MIX: Other 30%, Sara 28%,
Mayo-Kebbi 12%, Arab 12%,
Ouaddai 9%, Kanem-Bornou 9%
GOVERNMENT: Presidential system
CURRENCY: CFA franc = 100 centimes

Chile

Chile extends in a ribbon down the west coast of South America. It returned to elected civilian rule in 1989 after a referendum forced out military dictator General Pinochet.

GEOGRAPHY
Fertile valleys in the center between the coast and the Andes. Atacama Desert in north. Deep-sea channels, lakes, and fjords in south.

CLIMATE
Arid in the north. Hot, dry summers and mild winters in the center. Higher Andean peaks have glaciers and year-round snow. Very wet and stormy in the south.

PEOPLE & SOCIETY
Most people are *mestizo* (mixed Spanish–Amerindian descent), and are highly urbanized. General Pinochet's dictatorship was brutally repressive, but the business and middle classes prospered. Over a third of the population live in Santiago, many in large slums. There are three main indigenous groups, including the Rapa Nui of Easter Island.

THE ECONOMY
World's biggest copper producer. Growth in foreign investment due to political stability. Exports include wine, fishmeal, fruits, and salmon. Serious earthquake damage in 2010.

INSIGHT:
Chile's Atacama Desert is the driest place on Earth, making it the perfect location for hi-tech space observatories

- 4000m/13124ft
- 3000m/9843ft
- 2000m/6562ft
- 1000m/3281ft
- Sea Level

0   300 km
0   300 miles

PERU
BOLIVIA
Arica
Iquique
Atacama Desert
Antofagasta
PACIFIC OCEAN
Viña del Mar **SANTIAGO**
Valparaíso
Rancagua
Talcahuano Talca
Concepción Chillán
Valdivia
Temuco
Puerto Montt
Isla de Chiloé
Andes
ARGENTINA
Punta Arenas
Strait of Magellan
Cape Horn

FACTFILE

OFFICIAL NAME: Republic of Chile
DATE OF FORMATION: 1818
CAPITAL: Santiago
POPULATION: 17.6 million
TOTAL AREA: 292,258 sq. miles (756,950 sq. km)
DENSITY: 61 people per sq. mile

LANGUAGES: Spanish*, Amerindian languages
RELIGIONS: Roman Catholic 89%, other and nonreligious 11%
ETHNIC MIX: *Mestizo* and European 90%, other Amerindian 9%, Mapuche 1%
GOVERNMENT: Presidential system
CURRENCY: Chilean peso = 100 centavos

China

Covering a vast area of eastern Asia, China is bordered by 14 countries. A one-party Communist state since 1949, it has recently become a dominant force in global manufacturing.

GEOGRAPHY

A land of huge physical diversity, China has a long Pacific coastline to the east. Two-thirds of the country is uplands. The southwestern mountains include Tibet, the world's highest plateau; in the northwest, the Tien Shan Mountains separate the arid Tarim and Dzungarian basins. The rolling hills and plains of the low-lying east are home to two-thirds of the population.

CLIMATE

China is divided into two main climatic regions. The north and west are semiarid or arid, with extreme temperature variations. The south and east are warmer and more humid, with year-round rainfall. Winter temperatures vary with latitude, but are warmest on the subtropical southeast coast. Summer temperatures are more uniform, rising above 70°F (21°C).

PEOPLE & SOCIETY

Most people are Han Chinese. The rest of the population belong to one of 55 minority nationalities, or recognized ethnic groups. Many of these groups have a disproportionate political significance as they live in strategic border areas. A policy of resettling Han Chinese in remote regions is deeply resented and has led to uprisings in Xinjiang and Tibet. The government has relaxed the one-child family policy, particularly for minorities, after some small groups were brought close to extinction. Chinese society is patriarchal in practice, and generations tend to live together. However, economic change is breaking down the social controls of the Mao Zedong era. Divorce and unemployment are rising. A resurgence of religious belief has occurred in recent years. Materialism has replaced the puritanism of the past; there are now more cell phones in China than in the US.

FACTFILE

OFFICIAL NAME: People's Republic of China
DATE OF FORMATION: 960
CAPITAL: Beijing
POPULATION: 1.39 billion
TOTAL AREA: 3,705,386 sq. miles (9,596,960 sq. km)
DENSITY: 385 people per sq. mile

LANGUAGES: Mandarin*, Cantonese, other
RELIGIONS: Nonreligious 59%, traditional beliefs 20%, other 13%, Buddhist 6%, Muslim 2%
ETHNIC MIX: Han 92%, other 4%, Hui 1%, Miao 1%, Manchu 1%, Zhuang 1%
GOVERNMENT: One-party state
CURRENCY: Yuan = 10 jiao = 100 fen

THE ECONOMY

China has shifted from a centrally planned to a market-oriented economy; liberalization has gone furthest in the south where the emerging business class is based. Exports led annual GDP growth of over 10% in 2003–2007. Faced with a global downturn from 2008, Chinese stimulus packages boosted domestic spending. The buying power of China's huge market for raw materials and consumer goods helped drive global recovery. China is now the world's largest exporter and second-largest economy. The Twelfth Five-Year Plan (2011–2015) seeks to limit population growth and improve social infrastructure.

INSIGHT: *China has the world's oldest continuous civilization. Its recorded history began 4000 years ago, with the Shang dynasty*

4000m/13124ft
3000m/9843ft
2000m/6562ft
1000m/3281ft
500m/1640ft
200m/656ft
Sea Level

0   400 km
0   400 miles

Colombia

Lying in northwest South America, Colombia has coastlines on both the Caribbean and the Pacific. It is primarily noted for its coffee, emeralds, gold, and cocaine trafficking.

GEOGRAPHY
The densely forested and almost uninhabited east is separated from the western coastal plains by the Andes, which divide into three ranges (cordilleras) with intervening valleys.

CLIMATE
Coastal plains are hot and wet. The highlands are much cooler. The equatorial east has two wet seasons.

PEOPLE & SOCIETY
Most Colombians are of mixed blood. Blacks and Amerindians have the least political representation. Civil conflict since the 1960s has killed over 220,000 people and displaced more than five million. The fighting is deeply entwined with the narcotics trade. Violent crime is common.

 INSIGHT: *Colombia is the world's main source of emeralds*

THE ECONOMY
Healthy and diversified export sector – includes coffee and coal. Considerable growth potential, but narcotics-related violence and corruption deter foreign investors.

FACTFILE

OFFICIAL NAME: Republic of Colombia
DATE OF FORMATION: 1819
CAPITAL: Bogotá
POPULATION: 48.3 million
TOTAL AREA: 439,733 sq. miles (1,138,910 sq. km)
DENSITY: 120 people per sq. mile
LANGUAGES: Spanish*, Wayuu, Páez, other Amerindian languages
RELIGIONS: Roman Catholic 95%, other 5%
ETHNIC MIX: *Mestizo* (European–Amerindian) 58%, White 20%, European–African 14%, African 4%, African–Amerindian 3%, Amerindian 1%
GOVERNMENT: Presidential system
CURRENCY: Colombian peso = 100 centavos

Comoros

Off the east African coast, between Mozambique and Madagascar, lies the archipelago republic of the Comoros, comprising three main islands and a number of smaller islets.

GEOGRAPHY
Main islands are of volcanic origin and are heavily forested. The remainder are coral atolls.

CLIMATE
Hot and humid all year round, especially on the coasts. November to May is hottest and wettest period.

PEOPLE & SOCIETY
The Comoros has absorbed a diversity of people over the years, including Africans, Arabs, Polynesians,and Persians. There have also been Portuguese, Dutch, French, and Indian immigrants. Ethnic discord is rare, but regional tensions between islands are marked. The country is politically unstable and there have been frequent coups. A fragile new federal system was introduced in 2002, though in 2009 the island presidents were reduced to governors. A political and business elite controls most of the wealth.

THE ECONOMY
One of the world's poorest countries. Subsistence-level farming. Vanilla and cloves are main cash crops. Lack of basic infrastructure.

◆ **INSIGHT:** *The Comoros is the world's largest producer of ylang-ylang – an extract from tree blossom used in manufacturing perfumes*

FACTFILE

OFFICIAL NAME: Union of the Comoros

DATE OF FORMATION: 1975

CAPITAL: Moroni

POPULATION: 700,000

TOTAL AREA: 838 sq. miles (2170 sq. km)

DENSITY: 813 people per sq. mile

LANGUAGES: Arabic*, Comoran*, French*

RELIGIONS: Muslim (mainly Sunni) 98%, Roman Catholic 1%, other 1%

ETHNIC MIX: Comoran 97%, other 3%

GOVERNMENT: Presidential system

CURRENCY: Comoros franc = 100 centimes

Congo

Astride the equator in west-central Africa, this former French colony emerged from 20 years of Marxist-Leninist rule in 1990. Democracy was soon overshadowed by years of violence.

GEOGRAPHY

Mostly forest- or savanna-covered plateaus, drained by the Ubangi and Congo river systems. Narrow coastal plain is lined with sand dunes and lagoons.

CLIMATE

Hot, tropical. Temperatures rarely fall below 86°F (30°C). Two wet and two dry seasons. Rainfall is heaviest south of the equator.

PEOPLE & SOCIETY

One of the most tribally conscious and heavily urbanized countries in Africa, with most people living in the Brazzaville–Pointe-Noire region. Main tensions are between the Bakongo in the north and the Mbochi in the south. Relative peace was secured in 1999, and "ninja" rebels in the Pool region, around Brazzaville, signed a peace deal in 2003.

THE ECONOMY

Oil provides over 85% of export revenue. Timber is extracted. Foreign debt cut by two-thirds in 2010. Industrial base around Brazzaville and Pointe-Noire.

INSIGHT: *In 1970, Congo became the first African country to declare itself a communist state*

FACTFILE

OFFICIAL NAME: Republic of the Congo

DATE OF FORMATION: 1960

CAPITAL: Brazzaville

POPULATION: 4.4 million

TOTAL AREA: 132,046 sq. miles (342,000 sq. km)

DENSITY: 33 people per sq. mile

LANGUAGES: Kongo, Teke, Lingala, French*

RELIGIONS: Traditional beliefs 50%, Roman Catholic 35%, Protestant 13%, Muslim 2%

ETHNIC MIX: Bakongo 51%, Teke 17%, other 16%, Mbochi 11%, Mbédé 5%

GOVERNMENT: Presidential system

CURRENCY: CFA franc = 100 centimes

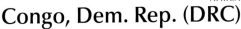

Congo, Dem. Rep. (DRC)

A former Belgian colony in east-central Africa, the Democratic Republic of the Congo (DRC) is Africa's second-largest country and the scene of one of its worst regional wars.

GEOGRAPHY
Rainforested basin of Congo River occupies 60% of the land area. High mountain ranges and lakes stretch down the eastern border.

CLIMATE
Tropical and humid. Distinct wet and dry seasons south of the equator. The north is mainly wet.

PEOPLE & SOCIETY
There are 12 main ethnic groups and around 190 smaller ones. Civil war from 1996 drew neighboring countries into a bloody conflict. The indigenous forest pygmies, victimized in the war, are now a marginalized group. A tentative peace deal in 2003 has been undermined by intercommunal violence in the east.

INSIGHT: *The DRC's rainforests comprise 6% of the world's, and 50% of Africa's, remaining woodlands*

THE ECONOMY
Rich resource base: minerals (copper, coltan, cobalt, diamonds) dominate export earnings. War and decades of corruption have caused economic collapse. Food aid is needed to ease humanitarian crisis.

2000m/6562ft	
1000m/3281ft	
500m/1640ft	
200m/656ft	
Sea Level	

0   200 km
0   200 miles

FACTFILE

OFFICIAL NAME: Democratic Republic of the Congo
DATE OF FORMATION: 1960
CAPITAL: Kinshasa
POPULATION: 67.5 million
TOTAL AREA: 905,563 sq. miles (2,345,410 sq. km)

DENSITY: 77 people per sq. mile
LANGUAGES: Kiswahili, Tshiluba, French*
RELIGIONS: Christian 70%, Kimbanguist 10%, Muslim 10%, traditional beliefs and other 10%
ETHNIC MIX: Other 55%, Mongo, Luba, Kongo, and Mangbetu-Azande 45%
GOVERNMENT: Presidential system
CURRENCY: Congolese franc = 100 centimes

Costa Rica

Costa Rica, Central America's most stable country, is rich in pristine scenery and exotic wildlife. Its neutrality in foreign affairs is long-standing, but it has strong ties with the US.

GEOGRAPHY
Coastal plains of swamp and savanna rise to a fertile central plateau, which leads to a mountain range with active volcanic peaks.

CLIMATE
Hot and humid in coastal regions. Temperate central uplands. High annual rainfall.

PEOPLE & SOCIETY
Most people are *mestizo*, of partly Spanish–partly Amerindian origin. There is a black, English-speaking minority and around 35,000 indigenous Amerindians. Plantation owners are the wealthiest group, while one in five people live in poverty. Nonetheless, living standards are high for the region, and education and healthcare provision is good.

 INSIGHT: *Costa Rica's 1949 constitution bans a national army*

THE ECONOMY
Main exports are bananas, coffee, pineapples, and beef, but all vulnerable to fluctuating world prices. Stability has attracted multinationals. History of high inflation. Pioneer of eco-tourism. Plans to be the world's first carbon neutral country (by 2025).

3000m/9843ft	
2000m/6562ft	
1000m/3281ft	
500m/1640ft	
200m/656ft	
Sea Level	

FACTFILE

OFFICIAL NAME: Republic of Costa Rica

DATE OF FORMATION: 1838

CAPITAL: San José

POPULATION: 4.9 million

TOTAL AREA: 19,730 sq. miles (51,100 sq. km)

DENSITY: 249 people per sq. mile

LANGUAGES: Spanish*, English Creole, Bribri, Cabecar

RELIGIONS: Roman Catholic 71%, Evangelical 14%, nonreligious 11%, other 4%

ETHNIC MIX: *Mestizo* and European 94%, Black 3%, Chinese 1%, Amerindian 1%, other 1%

GOVERNMENT: Presidential system

CURRENCY: C.R. colón = 100 céntimos

Côte d'Ivoire (Ivory Coast)

One of the larger nations along the coast of west Africa, Côte d'Ivoire is the world's biggest cocoa producer. Since 2002 its image of stability has been rocked by civil war and electoral chaos.

GEOGRAPHY
Sandy coastal strip and rainforested interior, with savanna plateau in north.

CLIMATE
Hot all year. Two wet seasons in south; north has one, with lower rainfall.

PEOPLE & SOCIETY
Over 60 tribes; largest is the Baoulé (an Akan group). Southern Christians harbor resentment against non-Ivorian Muslims in the north. Plantations employ millions of migrant workers (including children), though thousands fled back to Burkina during the 2002–2005 civil war. Rebels joined a transitional government in 2007. President Gbagbo delayed elections until 2010 and then refused to step down; civil conflict led to his ouster.

INSIGHT: *The Basilica of Our Lady of Peace in Yamoussoukro is the largest church in the world*

THE ECONOMY
Main crops are cocoa and coffee. Oil is now major export. Good infrastructure. Lack of professional training. Instability deters investment.

1000m/3281ft	
500m/1640ft	
200m/656ft	
Sea Level	

0   100 km
0   100 miles

FACTFILE

OFFICIAL NAME: Republic of Côte d'Ivoire

DATE OF FORMATION: 1960

CAPITAL: Yamoussoukro

POPULATION: 20.3 million

TOTAL AREA: 124,502 sq. miles (322,460 sq. km)

DENSITY: 165 people per sq. mile

LANGUAGES: Akan, French*, Krou, Voltaïque

RELIGIONS: Muslim 38%, Roman Catholic 25%, traditional beliefs 25% , Protestant 6%, other 6%

ETHNIC MIX: Akan 42%, Voltaïque 18%, Mandé du Nord 17%, Krou 11%, Mandé du Sud 10% other 2%

GOVERNMENT: Presidential system

CURRENCY: CFA franc = 100 centimes

Croatia

Though it was controlled by Hungary from medieval times and was a part of the Yugoslav state for much of the 20th century, Croatia has a very strong national identity.

GEOGRAPHY

Rocky, mountainous Adriatic coastline is dotted with islands. Interior is a mixture of wooded mountains and broad valleys.

CLIMATE

The interior has a temperate continental climate. Mediterranean climate along the Adriatic coast.

PEOPLE & SOCIETY

Croats are distinguished from Bosniaks and Serbs by their Roman Catholic faith and use of the Latin alphabet. Many Serbs fled Croatia during the early 1990s conflict that accompanied Yugoslavia's breakup. Croatia's entry into the EU, delayed by border disputes with Slovenia, finally occurred in 2013.

◆ **INSIGHT:** *Croatia only regained control of Serb-occupied Eastern Slavonia, around Vukovar, in 1998*

THE ECONOMY

The war cost the economy an estimated $50 billion. Unemployment has been persistently high. Corruption deters foreign investment. Tourism is mainly on the Dalmatian coast. EU membership.

FACTFILE

OFFICIAL NAME: Republic of Croatia

DATE OF FORMATION: 1991

CAPITAL: Zagreb

POPULATION: 4.3 million

TOTAL AREA: 21,831 sq. miles (56,542 sq. km)

DENSITY: 197 people per sq. mile

LANGUAGES: Croatian*

RELIGIONS: Roman Catholic 88%, other 7%, Orthodox Christian 4%, Muslim 1%

ETHNIC MIX: Croat 90%, Serb 5%, other 5%

GOVERNMENT: Parliamentary system

CURRENCY: Kuna = 100 lipa

Cuba

A former Spanish colony, Cuba is the largest island in the Caribbean. It became the only communist country in the Americas after Fidel Castro seized power in 1959.

GEOGRAPHY

Mostly fertile plains and basins. Three mountainous areas. Forests of pine and mahogany cover one-quarter of the country.

CLIMATE

Subtropical. Hot all year round, and very hot in summer. Heaviest rainfall in the mountains. Hurricanes can strike in the fall.

PEOPLE & SOCIETY

The Castro regime has reduced formerly extreme wealth disparities, given education a high priority, and established an efficient health service. Political dissent, however, is not tolerated. A dramatic fall in living standards since the late 1980s has led thousands of Cubans to flee to the US, to seek asylum. About 70% of Cubans are of Spanish descent. There is little ethnic tension.

THE ECONOMY

Sugar industry now superseded by tourism and nickel. US trade embargo, since 1961. Shortages drive black market. Parallel use of US dollar (1993–2004), and then convertible peso, boosted investment but created a "dollarized" elite: dual peso system to be scrapped.

INSIGHT: *Fidel Castro had become the world's longest-serving non-hereditary ruler before handing power to his brother Raúl in 2006*

FACTFILE

OFFICIAL NAME: Republic of Cuba

DATE OF FORMATION: 1902

CAPITAL: Havana

POPULATION: 11.3 million

TOTAL AREA: 42,803 sq. miles (110,860 sq. km)

DENSITY: 264 people per sq. mile

LANGUAGES: Spanish*

RELIGIONS: Nonreligious 49%, Roman Catholic 40%, atheist 6%, other 4%, Protestant 1%

ETHNIC MIX: Mulatto (mixed race) 51%, White 37%, Black 11%, Chinese 1%

GOVERNMENT: One-party state

CURRENCY: Cuban peso = 100 centavos

Cyprus

 Cyprus lies south of Turkey in the eastern Mediterranean. Since 1974, it has been partitioned between the Turkish-occupied north and the Greek-Cypriot south.

GEOGRAPHY
Mountains in the center-west give way to a fertile plain in the east, flanked by hills to the northeast.

CLIMATE
Mediterranean. Summers are hot and dry. Winters are mild, with snow in the mountains.

PEOPLE & SOCIETY
The Greek majority practice Orthodox Christianity. Since the 16th century, a minority community of Turkish Muslims has lived in the north of the island. In 1974 Turkish troops occupied the north and proclaimed the Turkish Republic of Northern Cyprus (TRNC), but it is recognized only by Turkey. Over 100,000 mainland Turks have settled there since. UN-led mediation failed to reunite the island ahead of EU accession in 2004, so the north was left out of membership; peace talks continue.

THE ECONOMY
Tourism. Eurozone member. Weathered 2009 downturn, but banks crashed in 2013: Cypriots lost savings under IMF/EU bailout terms. North lacks investment and wages are lower.

INSIGHT: *The Green Line, which separates north from south, was opened for the first time in 2003*

Mediterranean Sea
Dipkarpaz (Rizokarpason)
Turkish Republic of Northern Cyprus
(only recognized by Turkey)
Lapta (Lápithos)
Girne (Kerýnia)
Famagusta Bay
Güzelyurt (Mórfou)
Değirmenlik (Kythréa)
NICOSIA
Gazimağusa (Famagusta)
Troodos Mts.
Lárnaka
Agia Nápa
Páfos
Sovereign Base Area (to UK)
Lemesós (Limassol)
Sovereign Base Area (to UK)
Akrotiri

≈≈≈ *Cease-fire line*

0   25 km
0   25 miles

1000m/3281ft
500m/1640ft
200m/656ft
Sea Level

FACTFILE

OFFICIAL NAME: Republic of Cyprus
DATE OF FORMATION: 1960
CAPITAL: Nicosia
POPULATION: 1.1 million
TOTAL AREA: 3571 sq. miles (9250 sq. km)
DENSITY: 308 people per sq. mile

LANGUAGES: Greek*, Turkish*
RELIGIONS: Orthodox Christian 78%, Muslim 18%, other 4%
ETHNIC MIX: Greek 81%, Turkish 11%, other 8%
GOVERNMENT: Presidential systems
CURRENCY: Euro = 100 cents (new Turkish lira in TRNC = 100 kurus)

Czech Republic

Once part of Czechoslovakia, a central European communist state in 1948–1989, the Czech Republic peacefully dissolved its union with Slovakia in 1993. It joined the EU in 2004.

 GEOGRAPHY

Landlocked in central Europe. Bohemia, the western territory, is a plateau surrounded by mountains. Moravia, in the east, is characterized by hills and lowlands.

 CLIMATE

Cool, sometimes cold winters and warm summer months, which bring most of the annual rainfall.

 PEOPLE & SOCIETY

Secular and urban society, with high divorce rates. Czechs make up the vast majority of the population, while the next largest group are Moravians. The 300,000 Slovaks left after partition are now permitted dual citizenship. Ethnic tensions are few, but there is widespread hostility toward the Roma minority. A new commercial elite is emerging alongside postcommunist entrepreneurs.

$ THE ECONOMY

Traditional heavy industries (machinery, iron, car-making) have been successfully privatized. Prague attracts tourists. Skilled workforce. Will join euro in 2017 at earliest.

◆ INSIGHT: *Charles University in Prague was founded in the 13th century*

1000m/3281ft
500m/1640ft
200m/656ft
Sea Level

0   50 km
0   50 miles

FACTFILE

OFFICIAL NAME: Czech Republic

DATE OF FORMATION: 1993

CAPITAL: Prague

POPULATION: 10.7 million

TOTAL AREA: 30,450 sq. miles (78,866 sq. km)

DENSITY: 351 people per sq. mile

LANGUAGES: Czech*, Slovak, Hungarian (Magyar)

RELIGIONS: Roman Catholic 39%, atheist 38%, other 18%, Protestant 3%, Hussite 2%

ETHNIC MIX: Czech 90%, other 4%, Moravian 4%, Slovak 2%

GOVERNMENT: Parliamentary system

CURRENCY: Czech koruna = 100 haleru

Denmark

Denmark occupies the Jutland peninsula and over 400 islands in southern Scandinavia. Greenland and the Faeroe Islands are self-governing associated territories.

GEOGRAPHY

Fertile farmland covers two-thirds of the terrain, which is among the flattest in the world. About 100 islands are inhabited.

CLIMATE

Damp, temperate climate with mild summers and cold, wet winters. Rainfall is moderate.

PEOPLE & SOCIETY

Income distribution is the most even in the West. Danish liberalism is challenged over immigration: cultural clashes have arisen with immigrant minorities. Almost all women now work; Denmark is a world leader in childcare provision. Marriage is becoming less common, even for couples with children.

◆ **INSIGHT:** *Denmark is Europe's oldest kingdom – the monarchy dates back to the 10th century*

THE ECONOMY

Natural gas and oil reserves. Skilled workforce key to high-tech industrial success. Pork, bacon, dairy products are exported. Opted not to join the euro, though its currency is pegged.

200m/656ft
Sea Level

Skagerrak

Aalborg

Jylland   Kattegat

Viborg   Randers
Silkeborg
Herning   Århus   Helsingør

Horsens   COPENHAGEN
                   Roskilde
Esbjerg   Vejle
Kolding   Odense
          Fyn   Sjælland
North   Svendborg   Næstved
Sea   Sønderborg   Baltic Sea

GERMANY   Lolland   Nykøbing
                    Falster

Bornholm
Rønne

0   50 km
0   50 miles

FACTFILE

OFFICIAL NAME: Kingdom of Denmark
DATE OF FORMATION: 950
CAPITAL: Copenhagen
POPULATION: 5.6 million
TOTAL AREA: 16,639 sq. miles (43,094 sq. km)
DENSITY: 342 people per sq. mile

LANGUAGES: Danish*
RELIGIONS: Evangelical Lutheran 95%, Roman Catholic 3%, Muslim 2%
ETHNIC MIX: Danish 96%, other (including Scandinavian and Turkish) 3%, Faeroese and Inuit 1%
GOVERNMENT: Parliamentary system
CURRENCY: Danish krone = 100 øre

Djibouti

A city-state with a desert hinterland, Djibouti lies in northeast Africa on the Red Sea. Once known as the French Territory of the Afars and Issas, independence came in 1977.

GEOGRAPHY
Mainly low-lying desert and semidesert, with a volcanic mountain range in the north.

CLIMATE
Almost no rain, though the monsoon is very humid. The 109°F (45°C) heat of summer is unbearable.

PEOPLE & SOCIETY
The main ethnic groups are the Issas in the south, and the nomadic Afars in the north. Tensions between them developed into a guerrilla war in 1991–1994. Smaller tribal groups make up the rest of the population, and the rural peoples are mostly nomadic. Wealth is concentrated in Djibouti city. France exerts considerable influence in Djibouti, supporting it financially and maintaining a naval base and a military garrison.

THE ECONOMY
Djibouti's major assets are its ports in a key Red Sea location.

INSIGHT: *Chewing the leaves of the mildly narcotic qat shrub is an age-old social ritual in Djibouti*

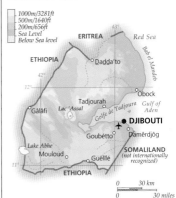

FACTFILE

OFFICIAL NAME: Republic of Djibouti
DATE OF FORMATION: 1977
CAPITAL: Djibouti
POPULATION: 900,000
TOTAL AREA: 8494 sq. miles (22,000 sq. km)
DENSITY: 101 people per sq. mile

LANGUAGES: Somali, Afar, French*, Arabic*
RELIGIONS: Muslim (mainly Sunni) 94%, Christian 6%
ETHNIC MIX: Issa 60%, Afar 35%, other 5%
GOVERNMENT: Presidential system
CURRENCY: Djibouti franc = 100 centimes

Dominica

Dominica is renowned as the Caribbean island that resisted European colonization until the 18th century. It achieved independence from the UK in 1978.

GEOGRAPHY

Mountainous and densely forested. Volcanic activity has given the land very fertile soils, hot springs, geysers, and black sand beaches.

CLIMATE

Tropical, cooled by constant trade winds. Heavy annual rainfall. Tropical depressions and hurricanes are likely June–November.

PEOPLE & SOCIETY

The majority of Dominicans are descendants of African slaves brought over to work on banana plantations. The Carib Territory on the northeast of the island is home to the only surviving indigenous community in the Caribbean. Wealth disparities are not as marked as elsewhere in the region, but the alleviation of poverty has become a major plank of government policy.

THE ECONOMY

Based on bananas, but has lost preferential access to EU market. Some diversification: flowers, coffee, fruit. Agriculture vulnerable to hurricanes. Eco-tourism. Some offshore banking.

INSIGHT: *Dominica is known as "Nature Island," due to its spectacular flora and fauna*

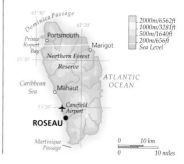

FACTFILE

OFFICIAL NAME: Commonwealth of Dominica

DATE OF FORMATION: 1978

CAPITAL: Roseau

POPULATION: 73,286

TOTAL AREA: 291 sq. miles (754 sq. km)

DENSITY: 253 people per sq. mile

LANGUAGES: French Creole, English*

RELIGIONS: Roman Catholic 77%, Protestant 15%, other 8%

ETHNIC MIX: Black 87%, mixed race 9%, Carib 3%, other 1%

GOVERNMENT: Parliamentary system

CURRENCY: East Caribbean dollar = 100 cents

Dominican Republic

The Dominican Republic occupies the eastern two-thirds of the island of Hispaniola in the Caribbean. Spanish-speaking, it seeks closer ties to the anglophone West Indies.

GEOGRAPHY

Highlands and rainforested mountains – including the highest peak in the Caribbean, Pico Duarte – interspersed with fertile valleys. Extensive coastal plain in the east.

CLIMATE
Hot and humid close to sea level, cooler at altitude. Heavy rainfall, especially in the northeast.

PEOPLE & SOCIETY

White landowners – especially those descended from the original Spanish settlers – form the wealthy elite. The mixed-race majority controls commerce and forms the bulk of the professional middle classes. White and mixed-race women are entering the professions. Great disparities of wealth exist; the black and Haitian-immigrant populations occupy the bottom of the social ladder.

THE ECONOMY
Mining (nickel and gold), sugar, and textiles. Tourism, remittances, and exports all rely heavily on US market. Hidden economy based on trans-shipment of narcotics to the US.

◆ **INSIGHT:** *Santo Domingo is the oldest city in the Americas. It was founded in 1496 by the brother of Christopher Columbus*

FACTFILE

OFFICIAL NAME: Dominican Republic

DATE OF FORMATION: 1865

CAPITAL: Santo Domingo

POPULATION: 10.4 million

TOTAL AREA: 18,679 sq. miles (48,380 sq. km)

DENSITY: 557 people per sq. mile

LANGUAGES: Spanish*, French Creole

RELIGIONS: Roman Catholic 95%, other and nonreligious 5%

ETHNIC MIX: Mixed race 73%, European 16%, Black African 11%

GOVERNMENT: Presidential system

CURRENCY: Dominican Republic peso = 100 centavos

East Timor

East Timor occupies the once Portuguese-owned eastern half of the island of Timor. Invaded by Indonesia in 1975, it became independent in 2002 following a long struggle.

GEOGRAPHY

A narrow coastal plain gives way to forested highlands. The mountain backbone rises to 9715 ft (2963 m).

CLIMATE

Tropical. Heavy rain in wet season (December–March), then dry and hot, particularly in the north.

PEOPLE & SOCIETY

The population is almost entirely Roman Catholic. The Timorese are a mix of Malay and Papuan peoples, and many indigenous Papuan tribes survive. There is an urban Chinese minority, and ethnic Indonesian settlers became numerous after annexation in 1975. Preindependence violence in 1999 was politically rather than ethnically motivated. Women do not have access to the professions and levels of domestic violence are notably high. Living standards are low.

THE ECONOMY

Widespread poverty. Violence in 1999 damaged infrastructure. Riots in 2006 undermined stability, further deterring foreign investment. Agreement with Australia on division of oil revenue from the Timor Sea.

◆ **INSIGHT:** *Once dependent on sandalwood, the economy is being transformed by oil under the Timor Sea*

FACTFILE

OFFICIAL NAME: Democratic Republic of Timor-Leste

DATE OF FORMATION: 2002

CAPITAL: Dili

POPULATION: 1.1 million

TOTAL AREA: 5756 sq. miles (14,874 sq. km)

DENSITY: 195 people per sq. mile

LANGUAGES: Tetum* (Portuguese/Austronesian), Bahasa Indonesia, Portuguese*

RELIGIONS: Roman Catholic 95%, other (including Muslim and Protestant) 5%

ETHNIC MIX: Malay/Papuan groups c. 85%, Indonesian c. 13%, Chinese 2%

GOVERNMENT: Parliamentary system

CURRENCY: US dollar = 100 cents

Ecuador

Once part of the Inca heartland, Ecuador lies on the western coast of South America. Its territory includes the fascinating Galápagos Islands, 610 miles (970 km) to the west.

GEOGRAPHY

Broad coastal plain, inter-Andean central highlands, dense jungle in upper Amazon basin.

CLIMATE
The climate is hot and moist on the coast, cool in the Andes, and hot equatorial in the Amazon basin.

PEOPLE & SOCIETY
Most people are of Amerindian–Spanish extraction (*mestizo*). Black communities exist on the coast. The strong and largely unified Amerindian movement leads the pressure for social reform. Recent left-wing policies have given greater rights to women, the poor, and Amerindians. Extreme poverty has fallen from 17% in 2006 to 8.6% in 2013.

◆ **INSIGHT:** *Darwin's study on the Galápagos Islands in 1856 played a major part in his theory of evolution*

THE ECONOMY
Oil provides around half of export earnings. World's biggest banana exporter. Use of US dollar offers stability, but less control. Defaulted on debt in 2008, prioritizing social spending.

FACTFILE

OFFICIAL NAME: Republic of Ecuador

DATE OF FORMATION: 1830

CAPITAL: Quito

POPULATION: 15.7 million

TOTAL AREA: 109,483 sq. miles (283,560 sq. km)

DENSITY: 147 people per sq. mile

LANGUAGES: Spanish*, Quechua, other Amerindian languages

RELIGIONS: Roman Catholic 95%; Protestant, Jewish, and other 5%

ETHNIC MIX: *Mestizo* 77%, White 11%, Amerindian 7%, Black African 5%

GOVERNMENT: Presidential system

CURRENCY: US dollar = 100 cents

Egypt

Occupying the northeast corner of Africa, Egypt is divided by the highly fertile Nile Valley. A long tradition of ethnic and religious tolerance has been shaken by the rise in Islamism.

GEOGRAPHY
Fertile Nile Valley separates arid Libyan Desert from smaller semiarid eastern desert. Sinai peninsula has mountains in south.

CLIMATE
Summers are very hot, but winters are cooler. Rainfall is negligible, except on the coast.

PEOPLE & SOCIETY
Mubarak's military-backed regime was ousted in a popular uprising in the "Arab Spring" of 2011, but the subsequent elected Muslim Brotherhood government was in turn ousted. Clashes between Muslims and Copts are rising. Women's access to education and economic status are threatened by Islamism. Rapidly growing population. Poverty in the south.

◆ **INSIGHT:** *In 450 BCE Herodotus visited the already-ancient pyramids*

THE ECONOMY
Oil and gas. Cotton. Tolls from the Suez Canal. Tourist industry and foreign investment affected by terrorist attacks and ongoing political instability.

2000m/6562ft
1000m/3281ft
500m/1640ft
200m/656ft
Sea Level
Below Sea Level

0 — 200 km
0 — 200 miles

FACTFILE

OFFICIAL NAME: Arab Republic of Egypt
DATE OF FORMATION: 1936
CAPITAL: Cairo
POPULATION: 82.1 million
TOTAL AREA: 386,660 sq. miles (1,001,450 sq. km)
DENSITY: 214 people per sq. mile

LANGUAGES: Arabic*, French, English, Berber
RELIGIONS: Muslim (mainly Sunni) 90%, Coptic Christian and other 10%
ETHNIC MIX: Egyptian 99%, other (Nubian, Armenian, Greek, Berber) 1%
GOVERNMENT: Transitional regime
CURRENCY: Egyptian pound = 100 piastres

El Salvador

El Salvador is Central America's smallest and most densely populated country. Already struggling to recover from a civil war in the 1980s, it was badly struck by earthquakes in 2001.

GEOGRAPHY
El Salvador is a narrow coastal belt backed by two mountain ranges. There is a central plateau. The country is located within a seismic zone, and there are more than 20 volcanic peaks.

CLIMATE
Tropical coastal belt is very hot, with seasonal rains. Cooler, temperate climate in highlands.

PEOPLE & SOCIETY
Ethnic tensions are few. Economic disparities sparked the 1981–1991 civil war between the US-backed government and left-wing FMLN guerrillas; 75,000 people died, many of them unarmed civilians, and human rights abuses were widespread. In 2009 the FMLN won the presidency, but wealth disparities still exist despite some reform. Gangs now control much of daily life; the murder rate is rising again despite a 2012 truce.

THE ECONOMY
Coffee, sugar. Garment industry. Overseas remittances. Frequent natural disasters damage infrastructure and deepen country's reliance on aid. Most businesses suffer extortion by gangs. Violence deters investors and tourism.

INSIGHT: *Independent since 1841, El Salvador is named after Jesus Christ, "the savior" of Christians*

2000m/6562ft
1000m/3281ft
500m/1640ft
200m/656ft
Sea Level

0   25 km
0   25 miles

FACTFILE

OFFICIAL NAME: Republic of El Salvador
DATE OF FORMATION: 1841
CAPITAL: San Salvador
POPULATION: 6.3 million
TOTAL AREA: 8124 sq. miles (21,040 sq. km)
DENSITY: 788 people per sq. mile

LANGUAGES: Spanish*
RELIGIONS: Roman Catholic 80%, Evangelical 18%, other 2%
ETHNIC MIX: *Mestizo* (European–Amerindian) 90%, White 9%, Amerindian 1%
GOVERNMENT: Presidential system
CURRENCY: Salvadorean colón = 100 centavos; US dollar = 100 cents

Equatorial Guinea

Comprising the mainland territory of Río Muni and five islands on the west coast of central Africa, Equatorial Guinea, despite its name, lies just north of the equator.

GEOGRAPHY
The islands are mountainous and volcanic. The mainland is lower, with mangrove swamps along the coast.

CLIMATE
The island of Bioco is extremely wet and humid. The mainland is only marginally drier and cooler.

PEOPLE & SOCIETY
Equatorial Guinea is the only Spanish-speaking country in Africa. Río Muni is sparsely populated and most people there are Fang, an ethnic group also found in Cameroon and northern Gabon. Bioco is populated by Bubi and a minority of Creoles known as Fernandinos. Tensions between the two territories have been reignited by the discovery of oil off Bioco. Wealth is concentrated in the ruling clan; oil revenue since 1995 has made little impact on most people.

THE ECONOMY
Oil and gas now account for almost all of exports; the government has promised to reinvest oil funds in development. Timber, cocoa, coffee.

INSIGHT: *In 2003, state radio declared President Obiang Nguema to be "like God in Heaven"*

FACTFILE

OFFICIAL NAME: Republic of Equatorial Guinea

DATE OF FORMATION: 1968

CAPITAL: Malabo

POPULATION: 800,000

TOTAL AREA: 10,830 sq. miles (28,051 sq. km)

DENSITY: 74 people per sq. mile

LANGUAGES: Spanish*, Fang, Bubi, French*

RELIGIONS: Roman Catholic 90%, other 10%

ETHNIC MIX: Fang 85%, other 11%, Bubi 4%

GOVERNMENT: Presidential system

CURRENCY: CFA franc = 100 centimes

Eritrea

Lying along the southwest shore of the Red Sea, Eritrea won a long war for independence from Ethiopia in 1993. The two neighbors fought a bitter border war in 1998–2000.

GEOGRAPHY
Mostly consists of rugged mountains, bush, and the Danakil Desert, which falls below sea level.

CLIMATE
Warm in the mountains; desert areas are hot. Droughts from July onward are common.

PEOPLE & SOCIETY
Tigrinya-speakers, mainly Orthodox Christians, are the most numerous of nine main ethnic groups. A strong sense of nationhood has been forged by war. Women played a vital role in combat. Around two-thirds of people are subsistence farmers. Multiparty elections, due under the 1997 constitution, are yet to be held.

◆ **INSIGHT:** *Eritrea was modern Italy's first African colony. It's named for the ancient Greek for Red Sea: Erythra Thalassa*

THE ECONOMY
Legacy of disruption and destruction from wars; resettlement of refugees. Susceptible to drought and famine; dependent on food aid. Most of the population live at subsistence level. Potential for extraction of gold, copper, and oil. Red Sea location: port at Massawa.

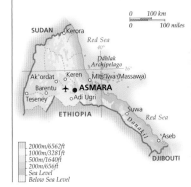

FACTFILE
OFFICIAL NAME: State of Eritrea
DATE OF FORMATION: 1993
CAPITAL: Asmara
POPULATION: 6.3 million
TOTAL AREA: 46,842 sq. miles (121,320 sq. km)
DENSITY: 139 people per sq. mile
LANGUAGES: Tigrinya*, English*, Tigre, Afar, Arabic*, Saho, Bilen, Kunama, Nara, Hadareb
RELIGIONS: Christian 50%, Muslim 48%, other 2%
ETHNIC MIX: Tigray 50%, Tigre 31%, other 9%, Saho 5%, Afar 5%
GOVERNMENT: Mixed presidential–parliamentary system
CURRENCY: Nakfa = 100 cents

Estonia

The smallest and most Western-oriented of the former Soviet-ruled Baltic states, Estonia is also the most developed, but its standard of living is well below the EU average.

GEOGRAPHY
Estonia's terrain is flat, boggy, and partly forested, with over 1500 islands. Lake Peipus forms much of the eastern border with Russia.

CLIMATE
Maritime, with some continental extremes. Harsh winters, with cool summers and damp springs.

PEOPLE & SOCIETY
Estonians are related ethnically and linguistically to the Finns. Friction between ethnic Estonians and the large Russian minority led to a reassertion of Estonian culture and language. Outright discrimination against the Russian language was only ended in 2000. Estonians are predominantly Lutheran. Families are small. The divorce rate has reduced since the 1980s peak. Market reforms have increased prosperity; a few people have become very rich.

THE ECONOMY
Timber and oil shale. Good productivity. Strong growth accompanied EU accession in 2004, but first EU country to enter recession in 2008. Drastic spending cuts aided quick revival. Joined eurozone in 2011. Low debt burden.

INSIGHT: *Estonia pioneered online voting in 2007, and voting by cell phone in 2011*

FACTFILE

OFFICIAL NAME: Republic of Estonia
DATE OF FORMATION: 1991
CAPITAL: Tallinn
POPULATION: 1.3 million
TOTAL AREA: 17,462 sq. miles (45,226 sq. km)
DENSITY: 75 people per sq. mile

LANGUAGES: Estonian*, Russian
RELIGIONS: Evangelical Lutheran 56%, Orthodox Christian 25%, other 19%
ETHNIC MIX: Estonian 69%, Russian 25%, other 4%, Ukrainian 2%
GOVERNMENT: Parliamentary system
CURRENCY: Euro = 100 cents

Ethiopia

The former empire of Ethiopia once dominated northeast Africa. A Marxist regime in 1974–1991, now a free-market democracy, it has suffered economic, civil, and natural crises.

GEOGRAPHY
Great Rift Valley divides mountainous northwest region from desert lowlands in northeast and southeast. Ethiopian Plateau is drained mainly by the Blue Nile.

CLIMATE
Moderate, with summer rains. Highlands are warm, with night frost and snowfalls on the mountains.

PEOPLE & SOCIETY
76 Ethiopian nationalities speak 286 languages. Oromo (or Gallas) are the largest group. Ethnic representation is a major political issue. Orthodox Christianity has a very ancient history in Ethiopia. Former emperor Haile Selassie inspired Rastafarianism.

 INSIGHT: *King Solomon and the Queen of Sheba are said to have founded the Kingdom of Abyssinia (Ethiopia) c. 1000 BCE*

THE ECONOMY
Overwhelmingly dependent on agriculture; coffee is main export crop. War-damaged infrastructure and periodic serious droughts and famines undermine growth. There is a heavy reliance on food aid. Landlocked since secession of Eritrea.

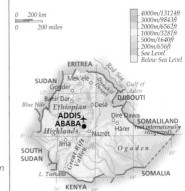

0	200 km
0	200 miles

4000m/13124ft
3000m/9843ft
2000m/6562ft
1000m/3281ft
500m/1640ft
200m/656ft
Sea Level
Below Sea Level

ERITREA
SUDAN
Mek'elē
Gonder
Bahir Dar
Blue Nile
Ethiopian
ADDIS
ABABA
Highlands
Jima
SOUTH
SUDAN
L. Turkana
KENYA
Red Sea
Gulf of Aden
DJIBOUTI
Desē
Dire Dawa
Hārer
SOMALILAND
(not internationally recognized)
Nazrēt
Ogaden
SOMALIA
Great Rift Valley

FACTFILE

OFFICIAL NAME: Federal Democratic Republic of Ethiopia

DATE OF FORMATION: 1896

CAPITAL: Addis Ababa

POPULATION: 94.1 million

TOTAL AREA: 435,184 sq. miles (1,127,127 sq. km)

DENSITY: 220 people per sq. mile

LANGUAGES: Amharic*, Tigrinya, other

RELIGIONS: Orthodox Christian 40%, Muslim 40%, traditional beliefs 15%, other 5%

ETHNIC MIX: Oromo 40%, Amhara 25%, other 13%, Sidama 9%, Tigray 7%, Somali 6%

GOVERNMENT: Parliamentary system

CURRENCY: Birr = 100 cents

Fiji

A volcanic archipelago in the South Pacific, with two large islands and 880 islets. Tensions between native Fijians and the Indian minority have sparked a succession of coups.

GEOGRAPHY

Main islands are mountainous, fringed by coral reefs. Remainder are limestone and coral formations.

CLIMATE

Tropical. High temperatures all year round. Cyclones are a hazard.

PEOPLE & SOCIETY

The British introduced workers from India in the late 19th century, and by 1946 their descendants outnumbered the ethnic Fijians. Ethnic-Fijian nationalism is strong. Many Indo-Fijians left after the 1987 coup, restoring ethnic Fijians to a majority. The first Indo-Fijian-dominated government was ousted in 2000. The army led another coup in 2006: elections were held in 2014. Women are lobbying for more rights.

◆ **INSIGHT:** *Both Fijians and Indians practice fire-walking; Indians walk on hot embers, Fijians on heated stones*

THE ECONOMY

Tourism was main sector, though damaged by instability. Coups have also caused international isolation. All sectors struggling: sugar production, gold mining, textiles, timber, and commercial fishing.

1000m/3281ft
500m/1640ft
Sea Level

PACIFIC OCEAN

Yasawa Group · Nabavatu · Labasa · Vanua Levu
Nabouwalu
Bligh Water · Koro · Taveuni
Lautoka · Rakiraki · Ovalau
Viti Levu · Koro Sea · Gau · Lakeba Passage
Sigatoka · Lau Group
Kadavu Passage · Moala
Kadavu

PACIFIC OCEAN

178°E   180°

0   100 km
0   100 miles

FACTFILE

OFFICIAL NAME: Republic of Fiji
DATE OF FORMATION: 1970
CAPITAL: Suva
POPULATION: 900,000
TOTAL AREA: 7054 sq. miles (18,270 sq. km)
DENSITY: 128 people per sq. mile

LANGUAGES: Fijian, English*, Hindi, Urdu, Tamil, Telugu
RELIGIONS: Hindu 38%, Methodist 37%, Roman Catholic 9%, Muslim 8%, other 8%
ETHNIC MIX: Melanesian (Fijian) 51%, Indian 44%, other 5%
GOVERNMENT: Parliamentary system
CURRENCY: Fiji dollar = 100 cents

Finland

Finland's language and national identity have been influenced by both its Scandinavian and Russian neighbors. Once aligned with the USSR, Finland is now a member of the EU.

GEOGRAPHY
South and center are flat, with low hills and many lakes. Uplands and low mountains in the north. 60% of the land area is forested.

CLIMATE
Long, harsh winters with frequent snowfalls. Short, warmer summers. Rainfall is low, and decreases northward.

PEOPLE & SOCIETY
One in four of the population lives in the Greater Helsinki region. Swedish-speakers live mainly in the Åland Islands in the southwest. The Sámi (Lapps) lead a seminomadic existence inside the Arctic Circle. Women make up 48% of the labor force, continuing a long tradition of equality between the sexes. Finnish women were the first in Europe to get the vote, in 1906, and the first in the world able to stand for parliament. Families tend to be close-knit.

THE ECONOMY
Strong engineering and electronics sectors: home of Nokia. Wood, pulp, and paper production.

◆ **INSIGHT:** *Finland has Europe's largest inland waterway system*

FACTFILE

OFFICIAL NAME: Republic of Finland
DATE OF FORMATION: 1917
CAPITAL: Helsinki
POPULATION: 5.4 million
TOTAL AREA: 130,127 sq. miles (337,030 sq. km)
DENSITY: 46 people per sq. mile

LANGUAGES: Finnish*, Swedish*, Sámi
RELIGIONS: Evangelical Lutheran 83%, other 15%, Orthodox Christian 1%, Roman Catholic 1%
ETHNIC MIX: Finnish 93%, other (including Sámi) 7%
GOVERNMENT: Parliamentary system
CURRENCY: Euro = 100 cents

France

Stretching across western Europe, from the English Channel (la Manche) to the Mediterranean Sea, France was Europe's first modern republic, and is still a leading industrial power.

GEOGRAPHY

Broad plain covers northern half of the country. High mountain ranges in the east and southwest, with a mountainous plateau in the center.

CLIMATE

Three main climates: temperate and damp northwest; continental east; and Mediterranean south.

PEOPLE & SOCIETY

Strong national identity coexists with pronounced regional differences, including local languages. Immigration laws have been tightened since the 1970s, but ethnic minorities growing up in city suburbs feel increasingly alienated. Wearing the veil is banned in public. New equality laws are under debate.

INSIGHT: *France is the most popular tourist destination in the world, with over 80 million visitors a year*

THE ECONOMY

Chemicals, electronics, heavy engineering, cars, and aircraft typify a strong and diversified export sector. World leader in cosmetics, perfumes, and quality wines. Modernized agriculture.

FACTFILE

OFFICIAL NAME: French Republic
DATE OF FORMATION: 987
CAPITAL: Paris
POPULATION: 64.3 million
TOTAL AREA: 211,208 sq. miles (547,030 sq. km)
DENSITY: 303 people per sq. mile
LANGUAGES: French*, Provençal, German, Breton, Catalan, Basque

RELIGIONS: Roman Catholic 88%, Muslim 8%, Protestant 2%, Jewish 1%, Buddhist 1%
ETHNIC MIX: French 90%, North African 6%, German (Alsace) 2%, Breton 1%, other 1%
GOVERNMENT: Mixed presidential–parliamentary system
CURRENCY: Euro = 100 cents

Gabon

Gabon is a former French colony straddling the equator on Africa's west coast. Independent since 1960, it returned to multiparty politics in 1990, after 22 years of one-party rule.

GEOGRAPHY
Low plateaus and mountains lie beyond the coastal strip. Two-thirds of the land is covered by rainforest.

CLIMATE
Hot and tropical, with little distinction between seasons. Cold Benguela current cools the coast.

PEOPLE & SOCIETY
Some 40 different languages are spoken. The Fang, who live mainly in the north, are the largest ethnic group, but have yet to gain control of the government. Oil wealth has led to the growth of an affluent middle class, but one in three people still live in poverty. Menial jobs are done by immigrant workers. Education follows the French system. With 87% of people living in towns, Gabon is one of Africa's most urbanized countries. The government is encouraging population growth.

THE ECONOMY
Oil accounts for 75% of exports, but reserves are dwindling: not much post-oil planning. High debt problem. Tropical hardwoods and manganese.

INSIGHT: *Libreville was founded as a settlement for freed French slaves in 1849*

FACTFILE

OFFICIAL NAME: Gabonese Republic
DATE OF FORMATION: 1960
CAPITAL: Libreville
POPULATION: 1.7 million
TOTAL AREA: 103,346 sq. miles (267,667 sq. km)
DENSITY: 17 people per sq. mile
LANGUAGES: Fang, French*, Punu, Sira, Nzebi, Mpongwe
RELIGIONS: Christian (mainly Roman Catholic) 55%, traditional beliefs 40%, other 4%, Muslim 1%
ETHNIC MIX: Fang 26%, Shira-punu 24%, other 24%, foreign residents 15%, Nzabi-duma 11%
GOVERNMENT: Presidential system
CURRENCY: CFA franc = 100 centimes

Gambia

Gambia is a riverbank state on the west coast of Africa, almost entirely surrounded by Senegal. It was renowned for its stability until its government was overthrown in a coup in 1994.

GEOGRAPHY
Located on the narrow strip of land bordering the Gambia River. Long, sandy beaches are backed by mangrove swamps along the river. Savanna and tropical forests higher up.

CLIMATE
Subtropical, with wet, humid months July–October, and warm, dry season November–May.

PEOPLE & SOCIETY
Little tension between various ethnic groups. The largest group, the Mandinka, has traditionally held power. Islam is a strong social influence, though there is no official state religion. A small expatriate community from the UK lives on the coast. Seasonal migrants come from neighboring states to harvest groundnuts each year. Women are active as traders. Yahya Jammeh, who led the 1994 coup, is still the elected president.

THE ECONOMY
Around 75% of the labor force is involved in agriculture. Groundnuts are the principal crop. Fish stocks are declining. Eco-tourism is promoted, though most visitors come for the beaches. Banjul is one of west Africa's finest deepwater ports: significant re-export trade. Smuggling problems.

INSIGHT: *Overfishing in the waters off Gambia and Senegal, mainly by foreign vessels, is a growing problem*

FACTFILE

OFFICIAL NAME: Republic of the Gambia
DATE OF FORMATION: 1965
CAPITAL: Banjul
POPULATION: 1.8 million
TOTAL AREA: 4363 sq. miles (11,300 sq. km)
DENSITY: 466 people per sq. mile

LANGUAGES: Mandinka, Fulani, Wolof, Jola, Soninke, English*
RELIGIONS: Sunni Muslim 90%, Christian 8%, traditional beliefs 2%
ETHNIC MIX: Mandinka 42%, Fulani 18%, Wolof 16%, Jola 10%, Serahuli 9%, other 5%
GOVERNMENT: Presidential system
CURRENCY: Dalasi = 100 butut

Georgia

Located on the eastern shore of the Black Sea, Georgia has been torn by civil war and ethnic disputes since achieving independence from the Soviet Union in 1991.

GEOGRAPHY
Kura Valley lies between Caucasus Mountains in the north and Lesser Caucasus range in south. Lowlands along the Black Sea coast.

CLIMATE
Subtropical along the coast, changing to continental extremes at high altitudes. Rainfall is moderate.

PEOPLE & SOCIETY
Paternalistic society, with strong family, cultural, and literary traditions. Georgia was converted to Christianity in 326 CE. Armenians in the south are the poorest group. Civil conflicts in the early 1990s against Abkhaz and Osset separatists displaced 300,000 people. Abkhazia and South Ossetia now effectively operate as separate states, backed up by Russian forces since the 2008 war. Russia opposes Georgian hopes of joining the EU and NATO.

THE ECONOMY
Transit revenues from pipelines taking oil to the West. Long-established and booming wine industry. Political instability. Fast pace of reforms in late 2000s, at cost of high unemployment.

◆ **INSIGHT:** *Western Georgia was the land of the legendary Golden Fleece of Greek mythology*

FACTFILE

OFFICIAL NAME: Georgia

DATE OF FORMATION: 1991

CAPITAL: Tbilisi

POPULATION: 4.3 million

TOTAL AREA: 26,911 sq. miles (69,700 sq. km)

DENSITY: 160 people per sq. mile

LANGUAGES: Georgian*, Russian, Azeri, Armenian, Mingrelian, Ossetian, Abkhazian

RELIGIONS: Georgian Orthodox 74%, Muslim 10%, Russian Orthodox 10%, Armenian Apostolic Church (Orthodox) 4%, other 2%

ETHNIC MIX: Georgian 84%, Armenian 6%, Azeri 6%, Russian 2%, Ossetian 1%, other 1%

GOVERNMENT: Presidential system

CURRENCY: Lari = 100 tetri

Germany

Europe's strongest industrial power and its most populous nation, Germany was divided after military defeat in 1945 into a free-market west and a communist east, but reunified in 1990.

GEOGRAPHY
Central European coastal plains in the north, rising to rolling hills of central region and Alps in far south.

CLIMATE
Damp, temperate in northern and central regions. Continental extremes in mountainous south.

PEOPLE & SOCIETY
Regionalism is strong. The north is mainly Protestant, while the south is staunchly Roman Catholic. Social and economic differences still exist between east and west. Turks are the largest single ethnic minority; many came as guest workers in the 1950s–1970s. Immigration rules now favor skilled workers. Feminism is strong.

◆ **INSIGHT:** *Germany's rivers and canals carry as much freight as its busy highways*

THE ECONOMY
Major exporter of electronics, heavy engineering, chemicals, and cars. Worst recession for 60 years in 2008–2009. Aging population.

2000m/6562ft
1000m/3281ft
500m/1640ft
200m/656ft
Sea Level

0   100 km
0   100 miles

FACTFILE

OFFICIAL NAME: Federal Republic of Germany
DATE OF FORMATION: 1871
CAPITAL: Berlin
POPULATION: 82.7 million
TOTAL AREA: 137,846 sq. miles (357,021 sq. km)

DENSITY: 613 people per sq. mile
LANGUAGES: German*, Turkish
RELIGIONS: Protestant 34%, Roman Catholic 33%, other 30%, Muslim 3%
ETHNIC MIX: German 92%, other 3%, other European 3%, Turkish 2%
GOVERNMENT: Parliamentary system
CURRENCY: Euro = 100 cents

Ghana

The heartland of the ancient Ashanti kingdom, Ghana in west Africa was once known as the Gold Coast. It has experienced intermittent periods of military rule since independence in 1957.

GEOGRAPHY

Mostly low-lying. The west is covered by rainforest. One of the world's largest artificial lakes – Lake Volta – was created by damming the White Volta River.

CLIMATE

Tropical. There are two wet seasons in the south, but the north is drier, and has just one.

PEOPLE & SOCIETY

Around 75 cultural-linguistic groups. The largest is the Akan, who include the Ashanti and Fanti peoples. Southern peoples are richer and more urban than those of the north. There are few tribal tensions. Family ties are strong. Women play a major role in market trading. The 2000 election saw Ghana's first peaceful handover of power. Poverty levels have been significantly reduced.

THE ECONOMY

World's second-largest cocoa producer. Oil discovered in 2007: on stream from 2010. Hardwood trees such as maple and sapele. Gold mining.

INSIGHT: *Ghana was the first colony in west Africa to gain independence*

FACTFILE

OFFICIAL NAME: Republic of Ghana

DATE OF FORMATION: 1957

CAPITAL: Accra

POPULATION: 25.9 million

TOTAL AREA: 92,100 sq. miles (238,540 sq. km)

DENSITY: 292 people per sq. mile

LANGUAGES: Twi, Fanti, Ewe, Ga, Adangbe, Gurma, Dagomba (Dagbani), English*

RELIGIONS: Christian 69%, Muslim 16%, traditional beliefs 9%, other 6%

ETHNIC MIX: Akan 49%, Mole-Dagbani 17%, Ewe 13%, other 13%, Ga and Ga-Adangbe 8%

GOVERNMENT: Presidential system

CURRENCY: Cedi = 100 pesewas

Greece

The Balkan state of Greece is bounded on three sides by the Mediterranean, Aegean, and Ionian seas. It has a strong seafaring tradition, with some of the world's richest shipowners.

GEOGRAPHY

Mountainous peninsula and over 2000 islands. Large plain along the mainland's Aegean coast.

CLIMATE

Mainly Mediterranean, with dry, hot summers. Alpine climate in northern mountain areas.

PEOPLE & SOCIETY

Postwar industrial development altered the dominance of agriculture and seafaring. Rural exodus to cities has been stemmed but a third of the population lives in Athens. Age-old culture and Greek Orthodox Church balance social mobility. Civil marriage and divorce only legalized in 1982. There has been much recent civil unrest against severe austerity measures.

◆ **INSIGHT:** *The modern Olympics, first held in Athens in 1896, evolved from Olympia's ancient Greek games*

THE ECONOMY

Public debt and budget deficit very high: EU bailouts to avoid bankruptcy. World's largest shipping fleet. One of Europe's top tourist destinations. Fruit, vegetables, olives. Large black economy.

FACTFILE

OFFICIAL NAME: Hellenic Republic
DATE OF FORMATION: 1829
CAPITAL: Athens
POPULATION: 11.1 million
TOTAL AREA: 50,942 sq. miles (131,940 sq. km)
DENSITY: 220 people per sq. mile

LANGUAGES: Greek*, Turkish, Macedonian, Albanian
RELIGIONS: Orthodox Christian 98%, Muslim 1%, other 1%
ETHNIC MIX: Greek 98%, other 2%
GOVERNMENT: Parliamentary system
CURRENCY: Euro = 100 cents

Grenada

The southernmost of the Windward Islands, Grenada made world headlines in 1983 when the US and Caribbean allies mounted an invasion to sever links with Castro's Cuba.

GEOGRAPHY
Volcanic in origin, with densely forested central mountains. Its territory also includes the islands of Carriacou and Petite Martinique.

CLIMATE
Tropical, tempered by trade winds. Hurricanes are a hazard in the July–November wet season.

PEOPLE & SOCIETY
Grenadians are mainly of African origin; their traditions remain strong, especially on Carriacou. Inter-ethnic marriage has reduced tensions between the groups. Extended families, often headed by women, are the norm. Wealth disparities are not marked, but levels of poverty are growing.

◆ **INSIGHT:** *Known as "the spice island of the Caribbean," it is the world's second-largest nutmeg producer*

THE ECONOMY
Severe damage from Hurricane Ivan in 2004 to crops and 90% of buildings; reconstruction taking years. Nutmeg, cocoa, bananas, and mace. Smuggling is a serious problem.

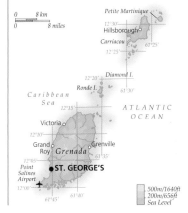

FACTFILE

OFFICIAL NAME: Grenada

DATE OF FORMATION: 1974

CAPITAL: St. George's

POPULATION: 109,590

TOTAL AREA: 131 sq. miles (340 sq. km)

DENSITY: 837 people per sq. mile

LANGUAGES: English*, English Creole

RELIGIONS: Roman Catholic 68%, Anglican 17%, other 15%

ETHNIC MIX: Black African 82%, *Mulatto* (mixed race) 13%, East Indian 3%, other 2%

GOVERNMENT: Parliamentary system

CURRENCY: East Caribbean dollar = 100 cents

Guatemala

The largest and most populous nation on the Central American isthmus, Guatemala returned to civilian rule in 1986 after 32 years of violent and repressive military rule.

GEOGRAPHY
Narrow Pacific coastal plain. Central highlands with volcanoes. Short coast on the Caribbean Sea. Tropical rainforests in the north.

CLIMATE
Tropical: hot and humid in coastal regions and north. More temperate in central highlands.

PEOPLE & SOCIETY
Amerindians, concentrated in the highlands, form a majority. Power, wealth, and land are controlled by *ladinos* (Westernized Amerindians and *mestizos*). Catholicism is predominant, mixed with Amerindian beliefs. Literacy is low. A quarter of the population live on less than $2 a day. Violent crime is a problem.

INSIGHT: *Guatemala, which means "land of trees," was the center of the ancient Mayan civilization*

THE ECONOMY
Coffee, sugar, and bananas are top exports. Tourism. Damage from natural disasters. Marked wealth inequalities inhibit domestic market.

FACTFILE

OFFICIAL NAME: Republic of Guatemala
DATE OF FORMATION: 1838
CAPITAL: Guatemala City
POPULATION: 15.5 million
TOTAL AREA: 42,042 sq. miles (108,890 sq. km)
DENSITY: 370 people per sq. mile

LANGUAGES: Quiché, Mam, Cakchiquel, Kekchí, Spanish
RELIGIONS: Roman Catholic 65%, Protestant 33%, other and nonreligious 2%
ETHNIC MIX: Amerindian 60%, *Mestizo* (European–Amerindian) 30%, other 10%
GOVERNMENT: Presidential system
CURRENCY: Quetzal = 100 centavos

Guinea

Located on the west coast of Africa, Guinea was the first French colony in Africa to gain independence, in 1958. The country was under military rule in 1984–1995 and 2008–2010.

GEOGRAPHY
Coastal plains and mangrove swamps in west rise to forested or savanna highlands in the south. Semidesert in the north.

CLIMATE
Tropical, with a wet season April–October. Conakry is especially rainy. Hot, dry *harmattan* wind blows from Sahara during dry season.

PEOPLE & SOCIETY
Peul and Malinké make up most of the population, but rivalries between them have allowed coastal peoples such as the Soussou to dominate politics. Daily life revolves around the extended family. Women acquired influence under Marxist party rule between 1958 and 1984, but the Muslim revival since then has reversed the trend. Private enterprise has created a business class. A deadly Ebola outbreak hit the country in 2014.

THE ECONOMY
Substantial gold, diamond, and especially bauxite reserves. Cash crops: bananas, coffee, pineapples, palm oil. Poor infrastructure. Instability.

INSIGHT: *The colors of Guinea's flag represent the three words of the country's motto: work (red), justice (yellow), and solidarity (green)*

FACTFILE

OFFICIAL NAME: Republic of Guinea
DATE OF FORMATION: 1958
CAPITAL: Conakry
POPULATION: 11.7 million
TOTAL AREA: 94,925 sq. miles (245,857 sq. km)
DENSITY: 123 people per sq. mile

LANGUAGES: Pulaar, Malinké, Soussou, French*
RELIGIONS: Muslim 85%, Christian 8%, traditional beliefs 7%
ETHNIC MIX: Peul 40%, Malinké 30%, Soussou 20%, other 10%
GOVERNMENT: Presidential system
CURRENCY: Guinea franc = 100 centimes

Guinea-Bissau

Known as Portuguese Guinea while a colony, Guinea-Bissau lies on Africa's west coast. Since 1994, its nascent democracy has been plagued by coups and rebellions.

 GEOGRAPHY
Low-lying, apart from savanna highlands in northeast. Rainforests and swamps are found along coastal areas.

 CLIMATE
Tropical, with wet season May–November and dry season December–April. Hot, dry *harmattan* desert wind blows during dry season.

PEOPLE & SOCIETY
The largest ethnic group is the Balante, who live in the south. Though only around 1% of the population, the mixed race Portuguese–African *mestiços* dominate the top ranks of government and bureaucracy. Most people live and work on small family farms, grouped in self-contained villages. The bulk of the urban population live in Bissau, where they face economic hardship. Narcotics traffickers are taking advantage of the ongoing instability.

THE ECONOMY
Mostly subsistence farming. Lack of sufficiency in rice staple. Main cash crop is cashew nuts. Major cocaine transit route from South America to Europe. Offshore oil as yet untapped. Fisheries and timber potential.

INSIGHT: *In 1974, Guinea-Bissau became the first Portuguese colony to gain independence*

FACTFILE

OFFICIAL NAME: Republic of Guinea-Bissau
DATE OF FORMATION: 1974
CAPITAL: Bissau
POPULATION: 1.7 million
TOTAL AREA: 13,946 sq. miles (36,120 sq. km)
DENSITY: 157 people per sq. mile

LANGUAGES: Portuguese Creole, Balante, Fulani, Malinké, Portuguese*
RELIGIONS: Traditional beliefs 50%, Muslim 40%, Christian 10%
ETHNIC MIX: Balante 30%, Fulani 20%, other 16%, Mandyako 14%, Mandinka 13%, Papel 7%
GOVERNMENT: Presidential system
CURRENCY: CFA franc = 100 centimes

Guyana

On the northeast coast of South America, Guyana is the continent's only English-speaking country. Independent since 1966, it has close ties with the anglophone Caribbean.

GEOGRAPHY
Mainly artificial coast, reclaimed by dikes and dams from swamps and tidal marshes. Forests cover 85% of the interior, rising to savanna uplands and mountains.

CLIMATE
Tropical. Coast cooled by sea breezes. Lowlands are hot, wet, and humid. Highlands are a little cooler.

PEOPLE & SOCIETY
Guyana is a complex multiracial society. Tension exists between the Afro-Guyanese, descended from slaves, and the Indo-Guyanese, descendants of laborers brought over after slavery was abolished. Politics is highly polarized around this split and has often spilled over into violence on the streets. Amerindian subsistence farmers are the poorest people in society and have little representation.

THE ECONOMY
Diverse exports: gold, sugar, fish, bauxite, rice, timber, diamonds. Debt relief granted. Narcotics transit zone.

INSIGHT: *Guyana means "land of many waters," reflecting its dense network of rivers*

FACTFILE

OFFICIAL NAME: Cooperative Republic of Guyana

DATE OF FORMATION: 1966

CAPITAL: Georgetown

POPULATION: 800,000

TOTAL AREA: 83,000 sq. miles (214,970 sq. km)

DENSITY: 11 people per sq. mile

LANGUAGES: English Creole, Hindi, Tamil, Amerindian languages, English*

RELIGIONS: Christian 57%, Hindu 28%, Muslim 10%, other 5%

ETHNIC MIX: East Indian 43%, Black African 30%, mixed race 17%, Amerindian 9%, other 1%

GOVERNMENT: Presidential system

CURRENCY: Guyanese dollar = 100 cents

Haiti

Formerly a French colony, Haiti shares the Caribbean island of Hispaniola with the Dominican Republic. At independence in 1804, it became the world's first black republic.

GEOGRAPHY
Predominantly mountainous, with forests and fertile plains.

CLIMATE
Tropical, with rain throughout the year. Humid in coastal areas, much cooler in the mountains.

PEOPLE & SOCIETY
Most Haitians are of African descent. A few have European roots, primarily French. The rigid class structure maintains vast disparities of wealth. The majority of the population live in extreme poverty; Haiti is one of the poorest countries in the Americas. A combination of political oppression and a collapsing economy led thousands to seek asylum in the US or the Dominican Republic. Though most are Christians, many Haitians practice Voodoo, which was recognized as an official religion in 2003.

THE ECONOMY
Fragile economy completely shattered by 2010 earthquake. Ongoing problems of instability, hurricane damage, high unemployment, narcotics trafficking.

INSIGHT: *A slave rebellion headed by Toussaint Louverture in 1791 led to Haiti's independence*

FACTFILE

OFFICIAL NAME: Republic of Haiti
DATE OF FORMATION: 1804
CAPITAL: Port-au-Prince
POPULATION: 10.3 million
TOTAL AREA: 10,714 sq. miles (27,750 sq. km)
DENSITY: 968 people per sq. mile

LANGUAGES: French Creole*, French*
RELIGIONS: Roman Catholic 55%, Protestant 28%, other (including Voodoo) 16%, nonreligious 1%
ETHNIC MIX: Black African 95%, *Mulatto* (mixed race) and European 5%
GOVERNMENT: Presidential system
CURRENCY: Gourde = 100 centimes

Honduras

Straddling the Central American isthmus, Honduras returned to democratic rule in 1984, after a period of military government. Hurricane Mitch devastated the country in 1998.

GEOGRAPHY

Narrow plains along both coasts, with a mountainous interior, cut by river valleys. Tropical forests, swamps, and lagoons in the east.

CLIMATE

Tropical coastal lowlands are hot and humid, with May–October rains. Interior is cooler and drier.

PEOPLE & SOCIETY

The majority of the population is *mestizo* (mixed European–Amerindian). An English-speaking *garífuna* (black) community and Miskito Amerindians struggle to preserve their rights to land along the remote Caribbean coast. Women's status remains low. Wealth inequalities are large and poverty is at the root of social tension. Two-thirds of the population live in poverty. The army ousted the president in 2009. Violent crime is a major issue.

THE ECONOMY

Garments, coffee, bananas, and shellfish are exported. Remittances account for a fifth of GDP. Debt relief from 2005. Mineral potential. High underemployment and corruption.

INSIGHT: *The Honduran currency is named after a Lenca Indian chief who was the main leader of resistance to the Spanish conquest in the 16th century*

FACTFILE

OFFICIAL NAME: Republic of Honduras

DATE OF FORMATION: 1838

CAPITAL: Tegucigalpa

POPULATION: 8.1 million

TOTAL AREA: 43,278 sq. miles (112,090 sq. km)

DENSITY: 187 people per sq. mile

LANGUAGES: Spanish*, Garífuna (Carib), English Creole

RELIGIONS: Roman Catholic 97%, Protestant 3%

ETHNIC MIX: *Mestizo* 90%, Black African 5%, Amerindian 4%, White 1%

GOVERNMENT: Presidential system

CURRENCY: Lempira = 100 centavos

Hungary

Landlocked in central Europe, Hungary was one of the twin centers of the once-great Habsburg Empire. It lost two-thirds of its historical territory for supporting Germany in WW I.

GEOGRAPHY

Landlocked. Fertile plains in east and northwest; west and north are hilly. The Danube River cuts through the country and the capital.

CLIMATE

Continental, with wet springs, late but very hot summers, and cold, cloudy winters. The transition between seasons tends to be sudden.

PEOPLE & SOCIETY

Hungary's population has been shrinking since the 1980s. Mostly ethnic Hungarian (Magyar), there are small minorities of Germans, Jews, and neighboring peoples. Roma face particular discrimination. The government is greatly concerned about the fate of ethnic Hungarians in Romania, Serbia, and Slovakia. Hungary joined the EU in 2004. Working hours are longer than in western Europe.

THE ECONOMY

Strong industrial base. Hard-hit by 2007–2009 global downturn: currency plummeted. IMF bailout to avoid meltdown. Spending cuts. Fast growth in 2014. No date set for joining euro.

INSIGHT: *The Hungarian language is Asian in origin and is most closely related to Finnish*

0   50 km
0   50 miles

500m/1640ft
200m/656ft
Sea Level

FACTFILE

OFFICIAL NAME: Hungary
DATE OF FORMATION: 1918
CAPITAL: Budapest
POPULATION: 10 million
TOTAL AREA: 35,919 sq. miles (93,030 sq. km)
DENSITY: 280 people per sq. mile

LANGUAGES: Hungarian (Magyar)*
RELIGIONS: Roman Catholic 52%, Calvinist 16%, other 15%, nonreligious 14%, Lutheran 3%
ETHNIC MIX: Magyar 90%, Roma 4%, German 3%, Serb 2%, other 1%
GOVERNMENT: Parliamentary system
CURRENCY: Forint = 100 fillér

Iceland

Europe's westernmost country, Iceland's strategic ocean location straddles the Mid-Atlantic Ridge. Its spectacular landscape is largely uninhabited, aside from coastal towns.

 GEOGRAPHY

Grassy coastal lowlands, with fjords in the north. Central plateau of cold lava desert, geothermal springs, and glaciers. Around 200 volcanoes, with numerous geysers and solfataras.

 CLIMATE

Its location in the middle of the Gulf Stream moderates the climate. Mild winters and brief, cool summers.

PEOPLE & SOCIETY

Icelanders share a strong national identity, with few foreign residents. Their language has changed little in 700 years, in part due to the country's isolation. There is high social mobility, free health care, and low-cost heating (geothermal and hydropower). Iceland's recent banking collapse and near financial ruin has swung the long-running debate over EU membership in favor of joining.

THE ECONOMY

Once reliant on fish. Aluminum smelting. Tourism. Banks overexposed in 2007–2009 global downturn. Nation bankrupt, króna depreciated 90%.

INSIGHT: *The word geyser is taken from Geysir (the "gusher") in southwest Iceland*

FACTFILE

OFFICIAL NAME: Republic of Iceland

DATE OF FORMATION: 1944

CAPITAL: Reykjavík

POPULATION: 300,000

TOTAL AREA: 39,768 sq. miles (103,000 sq. km)

DENSITY: 8 people per sq. mile

LANGUAGES: Icelandic*

RELIGIONS: Evangelical Lutheran 84%, nonreligious 3%, Roman Catholic 3%, other (mostly Christian) 10%

ETHNIC MIX: Icelandic 94%, other 5%, Danish 1%

GOVERNMENT: Parliamentary system

CURRENCY: Icelandic króna = 100 aurar

India

India is the world's second most populous country and largest democracy. Despite some success in reducing the birth rate, its population will probably overtake China's by 2028.

GEOGRAPHY

Separated from northern Asia by the Himalaya mountain range, India forms a subcontinent. As well as the Himalayas, there are two other main geographical regions, the Indo-Gangetic plain, which lies between the foothills of the Himalayas and the Vindhya Mountains, and the central-southern Deccan plateau. The Ghats are smaller mountain ranges located on the east and west coasts.

CLIMATE

Varies greatly according to latitude, altitude, and season. Most of India has three seasons: hot, wet, and cool. Summer temperatures in the north can reach 104°F (40°C). Monsoon rains normally break in June, petering out in September to October. In the cool season, the weather is mainly dry. The climate in the warmer south is less variable than in the north.

PEOPLE & SOCIETY

India's planners, overseeing an economic revolution, see its growing population rather than environmental constraints as the main brake on development. Nationwide awareness campaigns promote birth control but cultural and religious pressures encourage large families. Rural deprivation spurs urban migration, to live in sprawling slums. Over 70% of people survive on less than $2 a day. The majority of Indians are Hindu. Various attempts to reform the Hindu caste system, which determines social standing and even marriage, have met with violent opposition. Severe tensions exist between Hindus and the Muslim minority, especially in Kashmir and Gujarat. Smaller ethnic groups exist in the northeast, and many struggle for greater autonomy. Over two million people are living with HIV/AIDS.

FACTFILE

OFFICIAL NAME: Republic of India
DATE OF FORMATION: 1947
CAPITAL: New Delhi
POPULATION: 1.25 billion
TOTAL AREA: 1,269,338 sq. miles (3,287,590 sq. km)
DENSITY: 1091 people per sq. mile

LANGUAGES: Hindi*, English*, Urdu, Bengali, Marathi, Telugu, Tamil, Bihari, Gujarati, Kanarese
RELIGIONS: Hindu 81%, Muslim 13%, Christian 2%, Sikh 2%, Buddhist 1%, other 1%
ETHNIC MIX: Indo-Aryan 72%, Dravidian 25%, Mongoloid and other 3%
GOVERNMENT: Parliamentary system
CURRENCY: Indian rupee = 100 paise

$ THE ECONOMY

One of the world's fastest-growing economies. Protectionism has given way to free-market economics. Tea, gems, textiles exported. High-tech industries, outsourcing center. Success of "Bollywood" films. Cheap labor. Huge market, held back by poverty.

◆ **INSIGHT:** *India's national animal, the tiger, was depicted as early as 4000 years ago by the Mohenjo-Daro civilization*

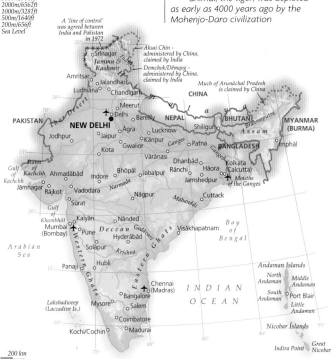

5000m/16405ft
4000m/13124ft
3000m/9843ft
2000m/6562ft
1000m/3281ft
500m/1640ft
200m/656ft
Sea Level

A 'line of control' was agreed between India and Pakistan in 1972

Srinagar
Jammu & Kashmir

Aksai Chin - administered by China, claimed by India
Demchok/Dêmqog - administered by China, claimed by India

Amritsar
Jalandhar
Ludhiana
Chandigarh

Much of Arunāchal Pradesh is claimed by China

CHINA

Thar Desert

PAKISTAN

Meerut
Delhi
Bareilly

Himalaya

NEPAL

BHUTAN

Brahmaputra

MYANMAR (BURMA)

NEW DELHI
Jodhpur
Jaipur
Agra
Kānpur
Lucknow
Ganges
Patna
Shiligun
Assam

Imphāl

Gwalior
Kota

Vārānasi
Dhanbād
Ranchi
Jamshedpur

Kolkata (Calcutta)
Hāora

BANGLADESH

Rann of Kachchh
Gulf of Kachchh

Ahmadābād
Indore
Bhopāl
Jabalpur

Narmada

Jāmnagar
Rājkot

Vadodara
Sūrat
Nāgpur

Mouths of the Ganges

Mahānadi

Cuttack

Gulf of Khambhāt

Kalyān
Mumbai (Bombay)
Pune

Nānded
Godāvari

Hyderābād

Deccan

Visākhapatnam

Western Ghats

Solāpur
Krishna

Eastern Ghats

Bay of Bengal

Arabian Sea

Panaji
Hubli

Andaman Islands
North Andaman
Middle Andaman

Lakshadweep (Laccadive Is.)

Chennai (Madras)

Bangalore
Mysore
Salem

South Andaman
Port Blair
Little Andaman

INDIAN OCEAN

Coimbatore
Kochi/Cochin
Madurai

Nicobar Islands

Indira Point
Great Nicobar

0   200 km
0   200 miles

Indonesia

Formerly called the Dutch East Indies, Indonesia is the world's largest archipelago, with 18,108 islands scattered across 3000 miles (5000 km). It is the world's fourth most populous nation.

GEOGRAPHY

Indonesia is highly mountainous, with numerous tropical swamps. The land is covered with dense rainforest, especially on New Guinea, where it remains largely unexplored. There are more than 200 volcanoes, many of which are still active. Earthquakes, eruptions, and tsunamis are hazards. The islands of Java, Bali, Lombok, Sumatra, and Borneo were once joined together by dry land, which has since been submerged by rising sea levels. Coastal lowland development distinguishes some of the large islands.

CLIMATE

The climate is predominantly tropical monsoon. Variations relate mainly to differences in latitude and altitude; hilly areas are cooler overall. Rain falls throughout the year, often in thunderstorms, but there is a relatively dry season from June to September.

PEOPLE & SOCIETY

The basic Melanesian–Malay ethnic division disguises a diverse society. Bahasa Indonesia, the national language, coexists with at least 250 other spoken languages or dialects. Attempts by the Javanese

FACTFILE

OFFICIAL NAME: Republic of Indonesia
DATE OF FORMATION: 1949
CAPITAL: Jakarta
POPULATION: 250 million
TOTAL AREA: 741,096 sq. miles (1,919,440 sq. km)
DENSITY: 360 people per sq. mile

LANGUAGES: Javanese, Sundanese, Madurese, Bahasa Indonesia*, Dutch
RELIGIONS: Sunni Muslim 86%, Christian 9%, Hindu 2%, other 2%, Buddhist 1%
ETHNIC MIX: Javanese 41%, other 32%, Sundanese 15%, coastal Malays 12%
GOVERNMENT: Presidential system
CURRENCY: Rupiah = 100 sen

political elite to suppress local cultures have been vigorously opposed, especially by the Aceh of northern Sumatra, and the Papuans. Religious and interethnic hostility is a problem, with clashes between Christians and Muslims in many areas, and discrimination against ethnic Chinese leading to mob attacks on their businesses. Gender equality is enshrined in law; women are active in public life.

THE ECONOMY

Varied resources, especially natural gas. Cheap and plentiful labor pool. Sizable state-owned sector, and state control of prices of basic goods. Large foreign debt rescheduled. The 2004 tsunami, which killed over 130,000 people, devastated northern Sumatra. Bureaucracy and corruption damage business confidence. Regional conflicts and terrorist attacks deter tourists and investors. Piracy is rife.

4000m/13124ft
3000m/9843ft
2000m/6562ft
1000m/3281ft
500m/1640ft
Sea Level

INSIGHT: Indonesia has a very youthful population: almost 30% of its people are under 15 years of age

0 _____ 500 km
0 _____ 500 miles

Iran

Since the 1979 Islamic fundamentalist revolution led by Ayatollah Khomeini, the Middle Eastern country of Iran has been the world's largest theocracy.

GEOGRAPHY

High desert plateau with large salt pans in the east. West and north are mountainous. Coastal land bordering Caspian Sea is rainy and forested.

CLIMATE

Desert climate. Hot summers, and bitterly cold winters. Area around the Caspian Sea is more temperate.

PEOPLE & SOCIETY

Many ethnic groups, including Persians, Azaris (ethnically related to Azeris), and Kurds. Militant Shi'a Islamism has dominated since the 1979 revolution. The mullahs' belief that adherence to religious values is more important than economic welfare has led to fall in living standards. Female emancipation has been reversed. Student-backed demonstrations favoring greater liberalism have been suppressed. International sanctions press for end of uranium enrichment program.

THE ECONOMY

A leading oil producer, though sanctions limit exports. Government restricts contact with the West, blocking acquisition of vital technology. High unemployment, inflation. Black market.

◆ **INSIGHT:** *More than a hundred offenses carry the death penalty*

3000m/9843ft
2000m/6562ft
1000m/3281ft
500m/1640ft
200m/656ft
Sea Level

0   200 km
0   200 miles

FACTFILE

OFFICIAL NAME: Islamic Republic of Iran
DATE OF FORMATION: 1502
CAPITAL: Tehran
POPULATION: 77.4 million
TOTAL AREA: 636,293 sq. miles (1,648,000 sq. km)
DENSITY: 123 people per sq. mile

LANGUAGES: Farsi*, Azeri, Luri, Gilaki, Arabic, Mazanderani, Kurdish, Turkmen, Baluchi
RELIGIONS: Shi'a Muslim 89%, Sunni Muslim 9%, other 2%
ETHNIC MIX: Persian 51%, Azari 24%, other 10%, Lur and Bakhtiari 8%, Kurdish 7%
GOVERNMENT: Islamic theocracy
CURRENCY: Iranian rial = 100 dinars

Iraq

Oil-rich Iraq is situated in the central Middle East. The last five decades have been dominated by dictatorship, war, and civil strife. A US-led Coalition ousted Saddam Hussein in 2003.

GEOGRAPHY

Mainly desert. The Tigris and Euphrates rivers water fertile regions and create the southern marshland. Mountains along northeast border.

CLIMATE

Southern deserts have hot, dry summers and mild winters. North has dry summers, but winters can be harsh in the mountains. Rainfall is low.

PEOPLE & SOCIETY

Carved out of remnants of the Ottoman Empire, Iraq is home to Arab Muslims (mainly Shi'a, some Sunni), northern Kurds (persecuted under Saddam), and smaller minorities. Since Saddam's removal, sectarian violence has overshadowed efforts to build democracy. US forces withdrew in 2011. By 2014 Islamic State jihadists controlled part of the country. After years of war and sanctions, poverty is widespread.

THE ECONOMY

Economy and infrastructure have been destroyed. Given stability and aid for reconstruction, hopes of recovery would rest on massive oil reserves.

◆ INSIGHT: *As Mesopotamia, Iraq was the site where the Sumerians established the world's first civilization*

FACTFILE

OFFICIAL NAME: Republic of Iraq
DATE OF FORMATION: 1932
CAPITAL: Baghdad
POPULATION: 33.8 million
TOTAL AREA: 168,753 sq. miles (437,072 sq. km)
DENSITY: 200 people per sq. mile

LANGUAGES: Arabic*, Kurdish*, Turkic languages, Armenian, Assyrian
RELIGIONS: Shi'a Muslim 60%, Sunni Muslim 35%, other (including Christian) 5%
ETHNIC MIX: Arab 80%, Kurdish 15%, Turkmen 3%, other 2%
GOVERNMENT: Parliamentary system
CURRENCY: New Iraqi dinar = 1000 fils

Ireland

In the Atlantic Ocean off the west coast of Britain, the Irish Republic governs about 85% of the island of Ireland, with the remainder (Northern Ireland) being part of the UK.

GEOGRAPHY
Low mountain ranges along an irregular coastline surround an inland plain punctuated by lakes, undulating hills, and peat bogs.

CLIMATE
The Gulf Stream accounts for the mild and wet climate. Snow is rare, except in the mountains.

PEOPLE & SOCIETY
Though homogeneous in ethnicity and Roman Catholic by religion, society has undergone a major generational change, liberalizing birth control, divorce, abortion, and general attitudes. Traditionally an emigrant nation, except for a decade of net immigration in the 2000s. Ireland and the UK signed a peace deal over Northern Ireland in 1998.

 INSIGHT: *About 40% of Irish people can speak Irish Gaelic*

THE ECONOMY
Efficient agriculture, electronics, and food-processing industries. Rapid growth until 2008: housing bubble burst, banks faltered. Large EU bailouts to avoid bankruptcy. Struggling with budget deficit.

1000m/3281ft
500m/1640ft
200m/656ft
Sea Level

UNITED KINGDOM (Northern Ireland)

Donegal
Donegal Bay
Sligo
Westport
Galway
Athlone
Shannon Airport
Tralee
Killarney
Tipperary
Clonmel
Cork
Limerick
Lough Derg
Mullingar
DUBLIN
Dún Laoghaire
Wicklow Mts.
Kilkenny
Wexford
Waterford
Dundalk
ATLANTIC OCEAN
Irish Sea
Celtic Sea

0   50 km
0   50 miles

FACTFILE

OFFICIAL NAME: Ireland
DATE OF FORMATION: 1922
CAPITAL: Dublin
POPULATION: 4.6 million
TOTAL AREA: 27,135 sq. miles (70,280 sq. km)
DENSITY: 173 people per sq. mile

LANGUAGES: English*, Irish Gaelic*
RELIGIONS: Roman Catholic 87%, other and nonreligious 10%, Anglican 3%
ETHNIC MIX: Irish 99%, other 1%
GOVERNMENT: Parliamentary system
CURRENCY: Euro = 100 cents

Israel

Created as a new state in 1948, Israel lies on the eastern shore of the Mediterranean. Palestinian resistance to Israeli occupation has led to years of fierce violence.

GEOGRAPHY
Coastal plain. Desert in the south. In the east lie the Great Rift Valley and the Dead Sea – the lowest point on the Earth's land surface.

CLIMATE
Summers are hot and dry. Wet season, March–November, is mild.

PEOPLE & SOCIETY
Large numbers of Jews settled in Palestine before Israel was founded in 1948. After World War II, there was a massive increase in immigration. Sephardi Jews from the Middle East and Mediterranean are now in the majority, but Ashkenazi Jews from central Europe still dominate business and politics. Palestinians in Gaza and Jericho gained limited autonomy in 1994 but Israeli–Palestinian talks on a two-state solution, backed by most of the world, have repeatedly foundered.

THE ECONOMY
High-tech industries, modern infrastructure, and educated workforce, but hampered by conflict and boycotts.

INSIGHT: *All Jews worldwide have the right to Israeli citizenship*

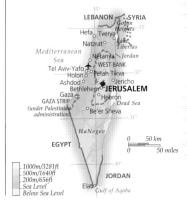

FACTFILE

OFFICIAL NAME: State of Israel
DATE OF FORMATION: 1948
CAPITAL: Jerusalem (not internationally recognized)
POPULATION: 7.7 million
TOTAL AREA: 8019 sq. miles (20,770 sq. km)

DENSITY: 981 people per sq. mile
LANGUAGES: Hebrew*, Arabic*, Yiddish, German, Russian, Polish, Romanian, Persian
RELIGIONS: Jewish 76%, Muslim (mainly Sunni) 16%, other 4%, Christian 2%, Druze 2%
ETHNIC MIX: Jewish 76%, Arab 20%, other 4%
GOVERNMENT: Parliamentary system
CURRENCY: Shekel = 100 agorot

Italy

The Italian peninsula was home to the Roman Empire, one of the greatest ancient civilizations. The south has two famous volcanoes, Vesuvius and Etna.

GEOGRAPHY

The Appennines form the backbone of a rugged peninsula, extending from the Alps into the Mediterranean Sea. Alluvial plain in the north.

CLIMATE

Mediterranean in the south. Seasonal extremes in the mountains and on the northern alluvial plain.

PEOPLE & SOCIETY

Ethnically homogeneous, but with a gulf between the prosperous, industrial north and the poorer, agricultural south. Strong regional identities persist, especially on Sicily and Sardinia. Family ties remain strong, though the influence of the Roman Catholic Church has lessened.

INSIGHT: *Italy was a collection of dukedoms, monarchies, and city-states before unification in the 1860s*

THE ECONOMY

World leader in industrial and product design, fashion, textiles. Strong tourism and agriculture. Large public sector debt: austerity packages. Reforms have failed to restore GDP growth. Lack of jobs.

3000m/9843ft
2000m/6562ft
1000m/3281ft
500m/1640ft
200m/656ft
Sea Level

SWITZERLAND AUSTRIA
Bolzano SLOVENIA
Milano Verona Trieste
Torino Venezia
FRANCE Parma Golfo di Venezia
Genova Bologna Rimini
Pisa Firenze SAN MARINO
Perugia Ancona
Adriatic Sea
ROME
VATICAN CITY
Bari
Sassari Napoli Taranto Lecce
Sardegna Salerno
(Sardinia) Tyrrhenian Ionian
Cagliari Sea Cosenza Sea
Mediterranean Palermo Messina
Sea Sicilia Siracusa
(Sicily)

0   100 km
0   100 miles

FACTFILE

OFFICIAL NAME: Italian Republic

DATE OF FORMATION: 1861

CAPITAL: Rome

POPULATION: 61 million

TOTAL AREA: 116,305 sq. miles (301,230 sq. km)

DENSITY: 537 people per sq. mile

LANGUAGES: Italian*, German, French, Rhaeto-Romanic, Sardinian

RELIGIONS: Roman Catholic 85%, other and nonreligious 13%, Muslim 2%

ETHNIC MIX: Italian 94%, other 4%, Sardinian 2%

GOVERNMENT: Parliamentary system

CURRENCY: Euro = 100 cents

Jamaica

First colonized by the Spanish and then by the English, the Caribbean island of Jamaica achieved independence in 1962. It remains an influential force in Caribbean politics.

GEOGRAPHY

Mainly mountainous, with lush tropical vegetation. Inaccessible limestone area in the northwest. Low, irregular coastal plains are broken by hills and plateaus.

CLIMATE

Tropical. Hot and humid at sea level, with temperate mountain areas. Hurricanes are likely June–November.

PEOPLE & SOCIETY

Social tensions result from vast disparities in wealth, rather than race. Economic and political life is dominated by a few wealthy, long-established families. Many women hold senior positions in public life. Armed crime, much of it narcotics-related, is a problem. Large areas of Kingston, which have their own patois, are ruled by violent gangs. Jamaican music styles are influential worldwide.

THE ECONOMY

Major bauxite producer, though sector vulnerable to changes in world prices. Tourism and light industry. Sugar, bananas, coffee, and rum are exported. Debt burden dominates budget. High underemployment.

INSIGHT: *Jamaica's Rastafarians revere the late emperor of Ethiopia, Haile Selassie, as their spiritual leader, and see Africa as their spiritual home*

FACTFILE

OFFICIAL NAME: Jamaica
DATE OF FORMATION: 1962
CAPITAL: Kingston
POPULATION: 2.8 million
TOTAL AREA: 4243 sq. miles (10,990 sq. km)
DENSITY: 670 people per sq. mile
LANGUAGES: English Creole, English*

RELIGIONS: Other and nonreligious 45%, other Protestant 20%, Church of God 18%, Baptist 10%, Anglican 7%
ETHNIC MIX: Black African 91%, *Mulatto* (mixed race) 7%, European and Chinese 1%, East Indian 1%
GOVERNMENT: Parliamentary system
CURRENCY: Jamaican dollar = 100 cents

Japan

Japan is located off the east Asian coast and comprises four principal islands and over 3000 smaller ones. A powerful economy, it has an emperor as ceremonial head of state.

GEOGRAPHY

The terrain is predominantly mountainous, with fertile coastal plains; over two-thirds is woodland. There is no single continuous mountain range; the mountains divide into many small land blocks separated by lowlands and dissected by numerous river valleys. The islands lie on the Pacific "Ring of Fire," and earthquakes and volcanic eruptions are frequent. The Pacific coast is vunerable to tsunamis. There are numerous hot springs.

CLIMATE

Generally temperate–oceanic. Spring is warm and sunny, while summer is hot and humid, with high rainfall. In western Hokkaido and northwest Honshu, winters are very cold, with heavy snowfall. Freak storms and damaging floods in recent years have raised concern over global climate changes.

PEOPLE & SOCIETY

One of the most racially homogeneous societies in the world. A sense of order and social structure was founded on a strongly ingrained respect for elders and social superiors. In business, this underpinned the now much-diluted "lifetime employer" concept, where company allegiance determined social life as well as career. There is little tradition of generational rebellion, but the youth market is powerful and current fashions focus on teenagers. The education system is highly pressurized. Nongraduates have difficulty reaching management-level jobs, so competition for university places is intense. Long-term jobs for women are now the norm. One of the world's best healthcare systems and increased longevity have led to an aging population, with one in four people already over 65. The cost of living is high, especially in Tokyo.

FACTFILE

OFFICIAL NAME: Japan
DATE OF FORMATION: 1590
CAPITAL: Tokyo
POPULATION: 127 million
TOTAL AREA: 145,882 sq. miles (377,835 sq. km)
DENSITY: 874 people per sq. mile

LANGUAGES: Japanese*, Korean, Chinese
RELIGIONS: Shinto and Buddhist 76%, Buddhist 16%, other (including Christian) 8%
ETHNIC MIX: Japanese 99%, other (mainly Korean) 1%
GOVERNMENT: Parliamentary system
CURRENCY: Yen = 100 sen

THE ECONOMY

World's third-largest economy. A market leader in high-tech electronics and cars. Global spread of business. Once-revolutionary management and production methods. Long-term research and development. Talent for developing ideas from abroad. Protectionism in domestic economy. Reform of financial sector delayed by traditional economic power brokers. Major coal importer. Retreat from nuclear power after massive damage caused by 2011 earthquake and tsunami: resulting energy imports bill ended 30 years of trade surpluses.

INSIGHT: *The Japanese are still among the world's most avid newspaper readers, with daily sales around 47 million copies*

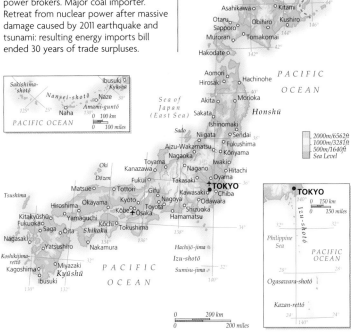

La Pérouse Strait

Kuril Islands
Administered by the Russ. Fed.,
claimed by Japan

Kuril Is.

Hokkaidō

Asahikawa ○ ✈ Kitami
Otaru ○ Obihiro ○ Kushiro
Sapporo ○ Muroran ○ Tomakomai

Hakodate ○

Aomori ○
Hirosaki ○ Hachinohe

PACIFIC

OCEAN

Akita ○ Morioka

Sakata ○ **Honshū**

Ishinomaki
Niigata ○ ○ Sendai
Aizu-Wakamatsu ○ Fukushima
Nagaoka ○ ○ Kōriyama
Iwaki ○ Hitachi
Nagano ○ Oyama
Takasaki ○ Chiba
Kawasaki ○ ★TOKYO
Gifu ○ ○ Odawara
Nagoya ○ Shizuoka
Toyota ○ Hamamatsu

Toyama ○
Kanazawa ○
Fukui ○

Oki
Dōzen
Matsue ○ ○ Tottori
Hiroshima ○ Okayama ○ Kyōto ○
Yamaguchi ○ Kōchi ✈ ○Kōbe ○ Osaka

Kitakyūshū ○
Fukuoka ○
Saga ○ Ōita
Nagasaki ○
Yatsushiro ○ Nakamura
Koshikijima-rettō
Kagoshima ○ Miyazaki ○ **Kyūshū**
Ibusuki ○

Shikoku
Tokushima ○

Hachijō-jima
Izu-shotō
Sumisu-jima

PACIFIC

OCEAN

Tsushima

Sea of
Japan
(East Sea)

Sado

Inset (upper left)
Sakishima-shotō
Ibusuki ○ Kyūshū
Nansei-shotō Naze
Amami-guntō
Naha

PACIFIC OCEAN

0   100 km
0   100 miles

2000m/6562ft
1000m/3281ft
500m/1640ft
Sea Level

Inset (Tokyo)
★TOKYO

0   150 km
0   150 miles

Izu-shotō

Philippine
Sea

PACIFIC
OCEAN

Ogasawara-shotō

Kazan-rettō

0   200 km
0   200 miles

Jordan

The Kingdom of Jordan lies east of Israel, and borders the Palestinian West Bank. Usually pro-Western in outlook, Jordan fears the rise of Islamists in Syria and Iraq.

GEOGRAPHY
Mostly desert plateaus, with occasional salt pans. Lowest parts lie along the eastern shores of the Dead Sea and the Jordan River.

CLIMATE
Hot, dry summers. Cool, wet winters. Areas below sea level very hot in summer, and warm in winter.

PEOPLE & SOCIETY
Jordanians are mainly Muslim with a strong national identity, but with Bedouin roots. The monarchy's power base lies among the rural tribes, which also provide the backbone of the army. Protests since 2011 have elicited gradual political reform, with greater powers for parliament. Jordan ceded its claim to the West Bank to the aspiring Palestinian state in 1988. Palestinian refugees make up over a third of the population. Recent influx of over 600,000 Syrian refugees.

THE ECONOMY
Lack of water. Exports garments, potash, fertilizers, and phosphates. Tourism hit by regional instability.

◆ **INSIGHT:** *The Nabataean ruins of the ancient city of Petra attract thousands of tourists every year*

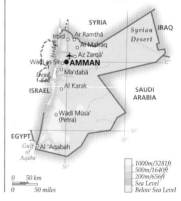

FACTFILE

OFFICIAL NAME: Hashemite Kingdom of Jordan
DATE OF FORMATION: 1946
CAPITAL: Amman
POPULATION: 7.3 million
TOTAL AREA: 35,637 sq. miles (92,300 sq. km)

DENSITY: 213 people per sq. mile
LANGUAGES: Arabic*
RELIGIONS: Sunni Muslim 92%, Christian 6%, other 2%
ETHNIC MIX: Arab 98%, Circassian 1%, Armenian 1%
GOVERNMENT: Monarchy
CURRENCY: Jordanian dinar = 1000 fils

Kazakhstan

Kazakhstan was the last of the former Soviet
republics to declare independence. Foreign investment in
the oil and natural gas sector is strengthening its regional power.

 GEOGRAPHY
Mainly steppe. Volga Delta and
Caspian Sea in the west. Central plateau.
Inhospitable Altai Mountains in the east.
Semidesert in the south.

 CLIMATE
Dry continental. Temperature
variations between desert south and
northern steppes are large. Winters are
mildest near the Caspian Sea.

 PEOPLE & SOCIETY
Kazakhstan's ethnic diversity
arose mainly from forced settlements
there during Soviet times. Since
independence, the proportion of ethnic
Russians has dropped. Many emigrated,
while ethnic Kazakhs arrived from
neighboring states. Very few Kazakhs
maintain a traditional nomadic lifestyle,
but Islam and loyalty to clans remain
strong. There are significant disparities
of wealth.

$ THE ECONOMY
Vast mineral resources: natural gas,
oil, bismuth, uranium, and cadmium. Oil
pipelines to China and Black Sea. Many
Western investors. Wheat exported. Sale
of farmland legal only since 2003.

◆ INSIGHT: *The Soviet-built Baykonyr
space center is still an important launch
site for international missions*

3000m/9843ft
2000m/6562ft
1000m/3281ft
500m/1640ft
200m/656ft
Sea Level
Below Sea level

0   400 km
0   400 miles

FACTFILE

OFFICIAL NAME: Republic of Kazakhstan
DATE OF FORMATION: 1991
CAPITAL: Astana
POPULATION: 16.4 million
TOTAL AREA: 1,049,150 sq. miles
(2,717,300 sq. km)
DENSITY: 16 people per sq. mile

LANGUAGES: Kazakh*, Russian, Ukrainian,
German, Uzbek, Tatar, Uighur
RELIGIONS: Muslim (mainly Sunni) 47%,
Orthodox Christian 44%, other 9%
ETHNIC MIX: Kazakh 57%, Russian 27%,
other 8%, Ukrainian 3%, Uzbek 3%, German 2%
GOVERNMENT: Presidential system
CURRENCY: Tenge = 100 tiyn

Kenya

Kenya straddles the equator on Africa's east coast. After nearly 40 years in power, the KANU party was soundly defeated in elections in 2002. Corruption is a serious issue.

GEOGRAPHY

A central plateau is divided by the Great Rift Valley. North of the equator is mainly semidesert. To the east lies a fertile coastal belt.

CLIMATE

The coast and the Great Rift Valley are hot and humid. The plateau interior is temperate. The northeastern desert is hot and dry. Rain usually falls April–May and October–November.

PEOPLE & SOCIETY

70 ethnic groups share about 40 languages. Strong clan and family links in rural areas are being weakened by urban migration. Poverty, severe drought, and years of high population growth exacerbate ethnic tensions.

INSIGHT: *Kenya has more than 60 game reserves, national parks, and marine reservations*

THE ECONOMY

Tourism, hurt by sporadic violence. Flowers, tea, and coffee. Sizable informal economy. Diversified manufacturing sector. Needs food aid, especially to cope with 2011 famine. Oil exploration.

FACTFILE

OFFICIAL NAME: Republic of Kenya
DATE OF FORMATION: 1963
CAPITAL: Nairobi
POPULATION: 44.4 million
TOTAL AREA: 224,961 sq. miles (582,650 sq. km)
DENSITY: 203 people per sq. mile

LANGUAGES: Kiswahili*, English*, Kikuyu, Luo, Kalenjin, Kamba
RELIGIONS: Christian 80%, Muslim 10% traditional beliefs 9%, other 1%
ETHNIC MIX: Other 28%, Kikuyu 22%, Luhya 14%, Luo 14%, Kalenjin 11%, Kamba 11%
GOVERNMENT: Presidential system
CURRENCY: Kenya shilling = 100 cents

Kiribati

Situated in the mid-Pacific, the islands adopted the name Kiribati (pronounced "Keer-ee-bus," a corruption of their former name "Gilberts") upon independence from Britain in 1979.

GEOGRAPHY

Kiribati consists of three groups of tiny, very low-lying coral atolls scattered across 1,930,000 sq. miles (5 million sq. km) of ocean. Most of the 33 atolls have central lagoons.

CLIMATE

Central islands have a maritime equatorial climate. Those to north and south are tropical, with constant high temperatures. There is little rainfall.

PEOPLE & SOCIETY

Officially I-Kiribati, many local people still refer to themselves as Gilbertese. Almost all are Micronesian, apart from the inhabitants of the island of Banaba, who employed anthropologists to establish their racial distinction. Most people are poor subsistence farmers and many travel abroad to work. The islands are effectively ruled by traditional chiefs.

THE ECONOMY

Since exhaustion of Banaba's phosphate deposits in 1980, copra (dried coconut) and fish have become the main exports. Foreign aid and remittances are vital to compensate for Kiribati's isolation and lack of resources.

INSIGHT: *In 1981, the UK paid A$10 million to Banabans to compensate for the destruction of their island by mining*

All land under 200m/656ft

PACIFIC OCEAN

Tungaru

Tarawa

Banaba

Equator

Kiritimati

Line Islands

170°

Phoenix Islands

180°

170°

160°

Millennium Island

10°

150°

Tarawa 173°

1°30'N

Betio   Bonriki

BAIRIKI

0   600 km

0   600 miles

FACTFILE

OFFICIAL NAME: Republic of Kiribati

DATE OF FORMATION: 1979

CAPITAL: Bairiki (Tarawa Atoll)

POPULATION: 103,248

TOTAL AREA: 277 sq. miles (717 sq. km)

DENSITY: 377 people per sq. mile

LANGUAGES: English*, Kiribati

RELIGIONS: Roman Catholic 55%, Kiribati Protestant Church 36%, other 9%

ETHNIC MIX: Micronesian 99%, other 1%

GOVERNMENT: Presidential system

CURRENCY: Australian dollar = 100 cents

North Korea

Separated from the democratic South by the world's most heavily defended border, the Stalinist North Korean state has been isolated from the outside world since 1948.

GEOGRAPHY

Mostly mountainous, with fertile plains in the southwest.

CLIMATE

Continental. Warm summers and cold winters, especially in the north, where snow is common.

PEOPLE & SOCIETY

Life is heavily regulated. Cult of personality is more powerful than the state-controlled religions, which include Korea's own Chondogyo. Women are expected to work and to run the home. Children are looked after in state-run crèches. The Korean Worker's Party is the sole party. Its elite have a privileged lifestyle. Globally condemned for its nuclear weapons tests, the regime's grip on power perpetuates its pariah status.

 INSIGHT: *Internet access is limited, and restricted to the political elite*

THE ECONOMY

Minerals are only resource. Vital aid streams lost with global collapse of communism after 1989. Decades of economic mismanagement have led to chronic food shortages. Lack of fuel. Disproportionate defense budget.

2000m/6562ft	
1000m/3281ft	
500m/1640ft	
200m/656ft	
Sea Level	

FACTFILE

OFFICIAL NAME: Democratic People's Republic of Korea
DATE OF FORMATION: 1948
CAPITAL: Pyongyang
POPULATION: 24.9 million
TOTAL AREA: 46,540 sq. miles (120,540 sq. km)

DENSITY: 536 people per sq. mile
LANGUAGES: Korean*
RELIGIONS: Government-controlled religions include Chondogyo, Buddhism, and Christianity
ETHNIC MIX: Korean 100%
GOVERNMENT: One-party state
CURRENCY: North Korean won = 100 chon

South Korea

South Korea occupies the southern half of the Korean peninsula. Under US sponsorship, it was separated from the communist North in 1948 and is now a capitalist economy.

GEOGRAPHY
Over 80% is mountainous and two-thirds is forested. The flattest and most populous parts lie along the west coast and in the extreme south.

CLIMATE
There are four distinct seasons. Winters are dry, and bitterly cold. Summers are hot and humid.

PEOPLE & SOCIETY
Inhabited for the last 2000 years by a single ethnic group. The nuclear family is replacing traditional extended households. Since the 1953 armistice, the Koreas have remained technically at war. Reunification is the ultimate goal, but the two sides fluctuate between harsh rhetoric or belligerence and conciliation, allowing cross-border family reunions.

INSIGHT: *Half of all Koreans are named Kim, Lee, Park, or Choi*

THE ECONOMY
World's biggest shipbuilder. High-tech goods and cars: rising demand from China. Strong regional competition. Aging population.

FACTFILE

OFFICIAL NAME: Republic of Korea
DATE OF FORMATION: 1948
CAPITALS: Seoul; Sejong City (administrative)
POPULATION: 49.3 million
TOTAL AREA: 38,023 sq. miles (98,480 sq. km)
DENSITY: 1293 people per sq. mile

LANGUAGES: Korean*
RELIGIONS: Mahayana Buddhist 47%, Protestant 38%, Roman Catholic 11%, Confucianist 3%, other 1%
ETHNIC MIX: Korean 100%
GOVERNMENT: Presidential system
CURRENCY: South Korean won = 100 chon

Kosovo

Once part of the former Yugoslav state, Kosovo seceded from Serbia in 2008. International recognition, mainly from Western countries, is strongly opposed by Serbia and Russia.

GEOGRAPHY

Landlocked and mountainous, with two plains in the east and west.

CLIMATE

Continental, with warm, sunny summers and cold, snowy winters.

PEOPLE & SOCIETY

The balance of Albanians to Serbs in Kosovo has changed dramatically over centuries, both groups suffering interethnic violence at various times. Attacks against Albanians in the late 1990s caused a million to flee. After NATO stepped in, many Serbs left: Albanians now form a 92% majority. Most Albanians are Muslim. Serbs dominate three northern provinces, which have threatened to secede.

INSIGHT: *The UN administered Kosovo in 1999–2008 after NATO intervention to stop Serb ethnic cleansing*

THE ECONOMY

One of the poorest countries in Europe. Aid and remittances cover a large trade deficit. Organized crime: smuggling of fuel, cigarettes, and cement. Uncertain status deters foreign investors. High unemployment. Use of euro has helped fight inflation. Lignite deposits. Inefficient agriculture.

FACTFILE

OFFICIAL NAME: Republic of Kosovo
DATE OF FORMATION: 2008
CAPITAL: Prishtinë
POPULATION: 1.8 million
TOTAL AREA: 4212 sq. miles (10,908 sq. km)
DENSITY: 427 people per sq. mile

LANGUAGES: Albanian*, Serbian*, Bosniak, Gorani, Roma, Turkish
RELIGIONS: Muslim 92%, Roman Catholic 4%, Orthodox Christian 4%
ETHNIC MIX: Albanian 92%, Serb 4%, Bosniak and Gorani 2%, Turkish 1%, Roma 1%
GOVERNMENT: Parliamentary system
CURRENCY: Euro = 100 cents

Kuwait

Kuwait lies at the northwest tip of the Gulf, dwarfed by its neighbors Iraq, Iran, and Saudi Arabia. It was a British protectorate until 1961, when full independence was granted.

GEOGRAPHY
Terrain is low-lying desert. The lowest land is in the north. Cultivation is only possible along the coast.

CLIMATE
Summers are very hot and dry. Winters are cooler, with some rain and occasional frost at night.

PEOPLE & SOCIETY
Oil-rich monarchy, ruled by the al-Sabah family. It is a conservative Sunni Muslim society, but women are relatively free. Nonetheless, a 1999 decree giving women the vote was blocked for six years in parliament by Islamic traditionalists. Immigrant workers, from other Arab states, India, and Pakistan, now outnumber native citizens. US-led forces rescued Kuwait after the 1990 Iraqi invasion, and later used it as a launchpad for the 2003 invasion to oust Saddam Hussein.

THE ECONOMY
Oil and natural gas dominate the economy. Skilled workforce, raw materials, and food are imported. High standard of living. Financial services: stock market lost 40% of value in 2008.

◆ **INSIGHT:** *During the 1991 Gulf War, Iraq deliberately set fire to 800 of Kuwait's 950 oil wells*

FACTFILE

OFFICIAL NAME: State of Kuwait
DATE OF FORMATION: 1961
CAPITAL: Kuwait City
POPULATION: 3.4 million
TOTAL AREA: 6880 sq. miles (17,820 sq. km)
DENSITY: 494 people per sq. mile

LANGUAGES: Arabic*, English
RELIGIONS: Sunni Muslim 45%, Shi'a Muslim 40%, Christian, Hindu, and other 15%
ETHNIC MIX: Kuwaiti 45%, other Arab 35%, South Asian 9%, other 7%, Iranian 4%
GOVERNMENT: Monarchy
CURRENCY: Kuwaiti dinar = 1000 fils

Kyrgyzstan

A small and mountainous landlocked state in central Asia, Kyrgyzstan is one of the least urbanized ex-Soviet republics, and was slow to develop its own sense of cultural identity.

 ## GEOGRAPHY

The mountainous spurs of the Tien Shan range contain glaciers, alpine meadows, forests, and narrow valleys. Semidesert in the west.

 ## CLIMATE

Varies from permanent snow and cold deserts at high altitudes, to hot deserts in low regions.

 ## PEOPLE & SOCIETY

Ethnic Kyrgyz have only been in the majority since the late 1980s – due to a high birth rate and the emigration of ethnic Russians. Wary of losing skills vital to the economy, the government has attempted to deter Russians from leaving; concessions include making Russian an official language. There are some tensions between Kyrgyz and Uzbeks, and a trend toward greater Islamization, particularly in the poorer south.

THE ECONOMY

Mainly still under state control; corruption issues. Agriculture employs a third of the labor force. Cotton, wool, meat, and tobacco exports. Mercury, gold, and antimony are mined. Great potential for hydroelectric power.

◆ **INSIGHT:** *Kyrgyz folklore is based around the 1000-year-old poem, Manas, which takes a week to recite*

FACTFILE

OFFICIAL NAME: Kyrgyz Republic

DATE OF FORMATION: 1991

CAPITAL: Bishkek

POPULATION: 5.5 million

TOTAL AREA: 76,641 sq. miles (198,500 sq. km)

DENSITY: 72 people per sq. mile

LANGUAGES: Kyrgyz*, Russian*, Uzbek, Tatar, Ukrainian

RELIGIONS: Muslim (mainly Sunni) 70%, Orthodox Christian 30%

ETHNIC MIX: Kyrgyz 69%, Uzbek 14%, Russian 9%, other 6%, Dungan 1%, Uighur 1%

GOVERNMENT: Presidential system

CURRENCY: Som = 100 tiyin

Laos

A French colony prior to 1953, Laos lies landlocked in southeast Asia. Heavily bombed during the Vietnam War, it fell in 1975 to communist insurgents, whose regime remains in power.

GEOGRAPHY
Largely forested mountains, broadening in the north to a plateau. Lowlands along the Mekong Valley.

CLIMATE
Monsoon rains September–May. The rest of the year is hot and dry.

PEOPLE & SOCIETY
There are over 60 ethnic groups. Lowland Laotians (Lao Loum) live along the Mekong River and are rice farmers. Upland and highland Laotians (Lao Theung and Lao Soung) traditionally employ environmentally damaging slash-and-burn farming, and grow illegal cash crops (notably opium). Government efforts to reform these practices are resisted.

◆ **INSIGHT:** *Three small Laotian kingdoms were unified under French control in 1899*

THE ECONOMY
One of world's least developed nations. Poor infrastructure. Gold, copper, electricity, timber, garments, and coffee are exported. Levels of foreign investment are rising.

0	100 km
0	100 miles

CHINA
MYANMAR (BURMA)
VIETNAM
Phongsali
Xam Nua
Houayxay
Mekong
Louangphabang
Xaignabouli
Xiangkhoang
THAILAND
Ban Naxon
Pakxan
VIENTIANE
THAILAND
Thakhek
Khanthabouli
Muang
Khôngxédôn
Salavan
Pakxé
CAMBODIA

2000m/6562ft
1000m/3281ft
500m/1640ft
200m/656ft
Sea Level

FACTFILE

OFFICIAL NAME: Lao People's Democratic Republic

DATE OF FORMATION: 1953

CAPITAL: Vientiane

POPULATION: 6.8 million

TOTAL AREA: 91,428 sq. miles (236,800 sq. km)

DENSITY: 76 people per sq. mile

LANGUAGES: Lao*, Mon-Khmer, Yao, Vietnamese, Chinese, French

RELIGIONS: Buddhist 65%, other (including animist) 34%, Christian 1%

ETHNIC MIX: Lao Loum 66%, Lao Theung 30%, Lao Soung 2%, other 2%

GOVERNMENT: One-party state

CURRENCY: New kip = 100 at

Latvia

Latvia lies on the east coast of the Baltic Sea. Like its Baltic neighbors, it regained independence from Moscow in 1991, and joined the EU and NATO in 2004.

GEOGRAPHY

A flat coastal plain which is deeply indented by the Gulf of Riga. Poor drainage creates many bogs and swamps in the forested interior.

CLIMATE

Temperate, with warm summers and cold winters. There is steady rainfall throughout the year.

PEOPLE & SOCIETY

Latvians make up just under two-thirds of the population and are mostly Lutheran. They have been officially favored by the state since 1991 over the largely Orthodox Christian Russian minority. Latvian was declared the only official language in 2000 and has been used exclusively in schools since 2004. This discrimination has strained relations with neighboring Russia. Women enjoy full equality. The divorce rate is high.

THE ECONOMY

Service-led economy. After fast growth, global credit crunch brought Latvia to verge of bankruptcy in 2008: banks were bailed out, stringent austerity measures imposed. Worst recession in EU ensued. Back to fastest growth in EU in 2012–2013. Adopted euro in 2014.

 INSIGHT: *In Latvia, life expectancy for men is ten years less than for women*

FACTFILE

OFFICIAL NAME: Republic of Latvia
DATE OF FORMATION: 1991
CAPITAL: Riga
POPULATION: 2.1 million
TOTAL AREA: 24,938 sq. miles (64,589 sq. km)
DENSITY: 84 people per sq. mile

LANGUAGES: Latvian*, Russian
RELIGIONS: Other 43%, Lutheran 24%, Roman Catholic 18%, Orthodox Christian 15%
ETHNIC MIX: Latvian 62%, Russian 27%, Belarussian 3%, other 4%, Ukrainian 2%, Polish 2%
GOVERNMENT: Parliamentary system
CURRENCY: Euro = 100 cents

Lebanon

Once a vibrant cultural hotspot, Lebanon suffered 14 years of civil war and occupation until a 1989 peace deal. It now fears spillover from neighboring Syria's own civil war.

GEOGRAPHY

Behind a narrow Mediterranean coastal plain, two parallel mountain ranges run the length of the country, separated by the fertile Beqaa Valley.

CLIMATE

Winters are mild and summers are hot, with high coastal humidity. Snow falls on high ground in winter.

PEOPLE & SOCIETY

Huge gulf exists between the poor and a small, rich elite. Politics reflects divisions between the traditional ruling Maronite Christians and Sunni and Shi'a Muslims. A 1989 power-sharing deal ended civil war. Syria acted as power broker until made to withdraw in 2005. Political crises add to instability. Israel attacked in 2006 in a botched bid to crush Iran-backed Hezbollah militants. Lebanon hosts over a million Syrian refugees and 450,000 from Palestine.

THE ECONOMY

Wine and fruit. Much infrastructure destroyed. Instability undermines Beirut's role as regional financial center. High public debt. Refugee influx.

◆ **INSIGHT:** *The Cedar of Lebanon has been the nation's symbol for more than 2000 years*

FACTFILE

OFFICIAL NAME: Lebanese Republic

DATE OF FORMATION: 1941

CAPITAL: Beirut

POPULATION: 4.8 million

TOTAL AREA: 4015 sq. miles (10,400 sq. km)

DENSITY: 1215 people per sq. mile

LANGUAGES: Arabic*, French, Armenian, Assyrian

RELIGIONS: Muslim 60%, Christian 39%, other 1%

ETHNIC MIX: Arab 95%, Armenian 4%, other 1%

GOVERNMENT: Parliamentary system

CURRENCY: Lebanese pound = 100 piastres

Lesotho

The landlocked Kingdom of Lesotho is entirely surrounded by – and economically dependent on – South Africa, which even sent in troops to restore calm after rioting in 1998.

GEOGRAPHY

A high mountainous plateau, cut by valleys and ravines. The Maluti Range runs through the center. The Drakensberg Range lies to the east.

CLIMATE

Temperate. Summers are hot with torrential rain storms. Snow is frequent in the mountains in winter.

PEOPLE & SOCIETY

The overwhelming majority of people are Sotho, though there are some South Asians, Europeans, and Chinese. A strong sense of national identity has tended to minimize ethnic tensions. Many men work as migrant laborers in South Africa, leaving women to run households.

INSIGHT: *Lesotho has one of the highest literacy rates in Africa – but one of the highest rates of HIV/AIDS too*

THE ECONOMY

Dependent on South Africa. Water and energy exported from Highlands Water Scheme. Subsistence farming. Garment exports struggle to compete. HIV/AIDS is depleting workforce.

3000m/9843ft
2000m/6562ft
1000m/3281ft

0   50 km
0   50 miles

FACTFILE

OFFICIAL NAME: Kingdom of Lesotho

DATE OF FORMATION: 1966

CAPITAL: Maseru

POPULATION: 2.1 million

TOTAL AREA: 11,720 sq. miles (30,355 sq. km)

DENSITY: 179 people per sq. mile

LANGUAGES: English*, Sesotho*, isiZulu

RELIGIONS: Christian 90%, traditional beliefs 10%

ETHNIC MIX: Sotho 99%, European and Asian 1%

GOVERNMENT: Parliamentary system

CURRENCY: Loti = 100 lisente

Liberia, on Africa's Atlantic coast, was founded as a republic of freed slaves. A brutal coup in 1980 and years of civil war have left a legacy of gang violence and looting.

GEOGRAPHY
A coastline of beaches and mangrove swamps rises to forested plateaus and highlands inland.

CLIMATE
High temperatures. There is only one wet season, from May to October, except in the extreme southeast.

PEOPLE & SOCIETY

The key social distinction used to be between Americo-Liberians – descendants of freed slaves – and the indigenous tribal peoples. However, political assimilation and intermarriage have eased tensions. Intertribal tension is now a much more serious problem, fueling the 1990–2003 civil war. A deadly Ebola outbreak hit the country in 2014.

INSIGHT: *Liberia is named after the people liberated from slavery who arrived from the US in the 1800s*

THE ECONOMY
War caused economic collapse. Rubber is key export. Bans now lifted on timber and diamond exports. Revenue from merchant shipping licenses. Debt burden. Vast iron ore reserves. Shutdown in 2014 due to Ebola.

FACTFILE

OFFICIAL NAME: Republic of Liberia

DATE OF FORMATION: 1847

CAPITAL: Monrovia

POPULATION: 4.3 million

TOTAL AREA: 43,000 sq. miles (111,370 sq. km)

DENSITY: 116 people per sq. mile

LANGUAGES: Kpelle, Vai, Bassa, Kru, Grebo, Kissi, Gola, Loma, English*

RELIGIONS: Christian 40%, traditional beliefs 40%, Muslim 20%

ETHNIC MIX: Indigenous tribes (12 groups) 49%, Kpellé 20%, Bassa 16%, Gio 8%, Krou 7%

GOVERNMENT: Presidential system

CURRENCY: Liberian dollar = 100 cents

Libya

Situated on north Africa's Mediterranean coast, Libya was declared a revolutionary state in 1969 by Colonel Gaddafi. Civil war, launched in the 2011 "Arab Spring," ousted his regime.

GEOGRAPHY

Apart from the coastal strip and a mountain range in the south, Libya is desert or semidesert.

CLIMATE

Hot and arid. The coastal area has a temperate climate, with mild, wet winters and hot, dry summers.

PEOPLE & SOCIETY

Once a nation of nomads and livestock herders, it is almost 80% urban. Gaddafi's revolution wiped out private enterprise and the middle classes, and promoted Islam and African unity. Sanctions were lifted after Libya offered compensation for terrorist bombings and ended its Weapons of Mass Destruction (WMD) program. In 2011, rebels from the east took power with international help, but failed to unite the country. Tripoli is in the sway of Islamist militias, while rival parliaments vie for political control.

THE ECONOMY

Oil is key export. Dates, olives, and fruit grow in oases, but most food is imported. Recent instability. Corruption and mismanagement.

◆ **INSIGHT:** *90% of Libya is still desert, despite grand irrigation projects*

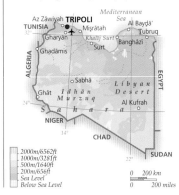

FACTFILE

OFFICIAL NAME: State of Libya
DATE OF FORMATION: 1951
CAPITAL: Tripoli
POPULATION: 6.2 million
TOTAL AREA: 679,358 sq. miles (1,759,540 sq. km)
DENSITY: 9 people per sq. mile

LANGUAGES: Arabic*, Tuareg
RELIGIONS: Muslim (mainly Sunni) 97%, other 3%
ETHNIC MIX: Arab and Berber 97%, other 3%
GOVERNMENT: Transitional regime
CURRENCY: Libyan dinar = 1000 dirhams

Liechtenstein

Perched in the Alps between Switzerland and Austria, the state of Liechtenstein became an independent principality of the Holy Roman Empire in 1719. It has close links with Switzerland.

 GEOGRAPHY
The upper Rhine Valley covers the western third of the country. The mountains and narrow valleys of the eastern Alps make up the remainder.

 CLIMATE
Warm, dry summers. Winters are cold, with heavy snow in the mountains from December to March.

 PEOPLE & SOCIETY
The principality's role as a financial center accounts for its many foreign residents (a third of the population). Half of the workforce are cross-border commuters. Living standards are high, with few social tensions. Linked by a customs union since 1924, Switzerland handles Liechtenstein's foreign affairs and defense issues.

 INSIGHT: *Women in Liechtenstein obtained the vote only in 1984*

THE ECONOMY
Banking secrecy (now modified) and low taxes help attract foreign investment. Anti-money-laundering rules are recent. Diversified exports include precision instruments, dental products, and chemicals.

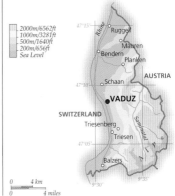

2000m/6562ft
1000m/3281ft
500m/1640ft
200m/656ft
Sea Level

47°15'
Ruggell
Mauren
Bendern
Planken
AUSTRIA
47°10'
Schaan
VADUZ
SWITZERLAND
Triesenberg
Triesen
47°05'
Balzers
9°30'
9°35'

0   4 km
0   4 miles

FACTFILE

OFFICIAL NAME: Principality of Liechtenstein
DATE OF FORMATION: 1719
CAPITAL: Vaduz
POPULATION: 37,000
TOTAL AREA: 62 sq. miles (160 sq. km)
DENSITY: 597 people per sq. mile

LANGUAGES: German*, Alemannish dialect, Italian
RELIGIONS: Roman Catholic 79%, other 13%, Protestant 8%
ETHNIC MIX: Liechtensteiner 66%, other 12%, Swiss 10%, Austrian 6%, German 3%, Italian 3%
GOVERNMENT: Parliamentary system
CURRENCY: Swiss franc = 100 rappen/centimes

Lithuania

Lying on the eastern coast of the Baltic Sea, Lithuania is the largest of the Baltic states. The first Soviet republic to declare independence from Moscow in 1991, it joined the EU in 2004.

GEOGRAPHY

Mostly flat with moors, bogs, and an intensively farmed central lowland. Numerous lakes and forested sandy ridges in the east.

CLIMATE

Coastal location moderates continental extremes. Cold winters, cool summers, and steady rainfall.

PEOPLE & SOCIETY

Homogeneous population, with Lithuanians forming a large majority. Only 1200 Jews, known as Litvaks, remain in Lithuania. Strong Roman Catholic tradition and historic links with Poland. There are better relations among ethnic groups than in other Baltic states and interethnic marriages are fairly common. However, ethnic Russians and Poles see a threat from "Lithuanianization." A large income gap has grown since independence.

THE ECONOMY

High-tech and heavy industries: engineering, shipbuilding, food processing. Bounced back from deep recession in 2009. Litas pegged to euro; adoption of euro set for 2015.

◆ **INSIGHT:** *The "amber coast" of Lithuania produces most of the world's amber – fossilized resin*

■ 200m/656ft
Sea Level

0   50 km
0   50 miles

FACTFILE

OFFICIAL NAME: Republic of Lithuania
DATE OF FORMATION: 1991
CAPITAL: Vilnius
POPULATION: 3 million
TOTAL AREA: 25,174 sq. miles (65,200 sq. km)
DENSITY: 119 people per sq. mile

LANGUAGES: Lithuanian*, Russian
RELIGIONS: Roman Catholic 77%, other and nonreligious 17%, Russian Orthodox 4%, Protestant 1%, Old Believers 1%
ETHNIC MIX: Lithuanian 85%, Polish 7%, Russian 6%, Belarussian 1%, other 1%
GOVERNMENT: Parliamentary system
CURRENCY: Litas = 100 centu

Luxembourg

Part of the plateau of the Ardennes in western Europe, Luxembourg is one of Europe's richest states. A tax haven and banking center, it is also home to key EU institutions.

GEOGRAPHY
Dense Ardennes forests in the north, with a low, open plateau to the south. Undulating terrain throughout.

CLIMATE
The climate is moist, with warm summers and mild winters. Snow is common only in the Ardennes.

PEOPLE & SOCIETY
Ethnic tensions are rare, despite a large proportion of foreigners (over a third of residents). Integration has been straightforward; most are fellow western Europeans and Catholics, mainly from Italy and Portugal. Low unemployment and high salaries promote stability. Divorce rates are rising and marriage is becoming less common.

◆ **INSIGHT:** *Luxembourg's capital is home to around 2000 investment funds and 150 banks*

THE ECONOMY
Traditional industries such as steelmaking have given way to the banking and service sectors. Low taxes and banking secrecy laws attract foreign investors.

500m/1640ft
200m/656ft
Sea Level

Clervaux

GERMANY

Ettelbrück

Echternach

Mersch

BELGIUM

LUXEMBOURG

Pétange

Differdange

Esch-sur-Alzette

Dudelange

FRANCE

0   10 km
0   10 miles

FACTFILE

OFFICIAL NAME: Grand Duchy of Luxembourg
DATE OF FORMATION: 1867
CAPITAL: Luxembourg-Ville
POPULATION: 500,000
TOTAL AREA: 998 sq. miles (2586 sq. km)
DENSITY: 501 people per sq. mile

LANGUAGES: Luxembourgish*, German*, French*
RELIGIONS: Roman Catholic 97%, Protestant, Orthodox Christian, and Jewish 3%
ETHNIC MIX: Luxembourger 62%, foreign residents 38%
GOVERNMENT: Parliamentary system
CURRENCY: Euro = 100 cents

Macedonia

Landlocked Macedonia, formerly part of Yugoslavia, was hit hard in the 1990s by sanctions on its northern trading partners, and in 2001 by conflict with its Albanian minority.

GEOGRAPHY

Mainly mountainous or hilly, with deep river basins in the center. Plains in the northeast and southwest.

CLIMATE

Continental climate with wet springs and dry autumns. Heavy snowfalls in northern mountains.

PEOPLE & SOCIETY

Slav Macedonians are mostly Orthodox Christians, with some Muslims. Officially, Muslim Albanians account for 25% of the population, but they claim to number a third. Albanian militants fought a bitter war against the state in 2001. A peace deal promised greater equality, but is yet to be fully implemented. A major stumbling block to EU and NATO accession is Greece's objection to the name Macedonia, in order to prevent any possibility of claims to historic "Macedonian" lands in north Greece.

THE ECONOMY

Steel, minerals, clothing, shoes, and tobacco exported. High unemployment. Organized crime and large gray economy. Progress with reforms. Investment boosted by EU candidate status.

 INSIGHT: *Ohrid is the deepest lake in Europe at 964 ft (294 m)*

FACTFILE

OFFICIAL NAME: Republic of Macedonia

DATE OF FORMATION: 1991

CAPITAL: Skopje

POPULATION: 2.1 million

TOTAL AREA: 9781 sq. miles (25,333 sq. km)

DENSITY: 212 people per sq. mile

LANGUAGES: Macedonian*, Albanian*, Turkish, Romani, Serbian

RELIGIONS: Orthodox Christian 65%, Muslim 29%, Roman Catholic 4%, other 2%

ETHNIC MIX: Macedonian 64%, Albanian 25%, Turkish 4%, Roma 3%, Serb 2%, other 2%

GOVERNMENT: Mixed presidential–parliamentary system

CURRENCY: Macedonian denar = 100 deni

Madagascar

Lying off east Africa in the Indian Ocean, the former French colony of Madagascar is the world's fourth-largest island. Power struggles erupted onto the streets in 2002 and 2009.

GEOGRAPHY

More than two-thirds is a savanna-covered plateau, which drops in the east through rainforests to the coast.

CLIMATE

Tropical and often hit by cyclones. Monsoons affect the east coast. The southwest is much drier.

PEOPLE & SOCIETY

People are Malay-Indonesian in origin, intermixed with later migrants from Africa. The main ethnic division is between the Merina of the central plateau and the poorer côtier (coastal) peoples. The Merina were the country's historic rulers, and remain the social elite. The 2009 unrest led to four years of political transition.

◆ **INSIGHT:** *80% of Madagascar's plants and many of its animal species are found nowhere else*

THE ECONOMY

Most people are farmers. Cash crops are vanilla, coffee, and cloves. Garments and shrimp also exported. Political crises deter investors.

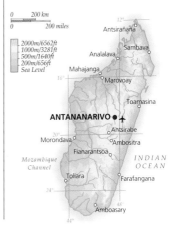

2000m/6562ft	
1000m/3281ft	
500m/1640ft	
200m/656ft	
Sea Level	

0   200 km
0   200 miles

Antsiranana
Analalava
Sambava
Mahajanga
Marovoay
Toamasina
ANTANANARIVO
Antsirabe
Morondava
Ambositra
Fianarantsoa
Mozambique Channel
Toliara
Farafangana
Amboasary

INDIAN OCEAN

FACTFILE

OFFICIAL NAME: Republic of Madagascar
DATE OF FORMATION: 1960
CAPITAL: Antananarivo
POPULATION: 22.9 million
TOTAL AREA: 226,656 sq. miles (587,040 sq. km)
DENSITY: 102 people per sq. mile

LANGUAGES: Malagasy*, French*, English*
RELIGIONS: Traditional beliefs 52%, Christian (mainly Roman Catholic) 41%, Muslim 7%
ETHNIC MIX: Other Malay 46%, Merina 26%, Betsimisaraka 15%, Betsileo 12%, other 1%
GOVERNMENT: Mixed presidential–parliamentary system
CURRENCY: Ariary = 5 iraimbilanja

Malawi

A former colony of the UK, Malawi lies landlocked in southeast Africa, following the Great Rift Valley. Its name means "the land where the sun is reflected in the water like fire."

 ## GEOGRAPHY
Lake Nyasa takes up one-fifth of the landscape. Highlands lie west of the lake. Much of the land is covered by forests and savanna.

 ## CLIMATE
Mainly subtropical. The south is hot and humid. Highlands are cooler.

 ## PEOPLE & SOCIETY
Most Malawians share a common Bantu origin. Protestant Chewa live in central regions, while Muslim Yao live along the lake and in the south. Unlike neighboring states, ethnicity has not been exploited for political ends. Multiparty elections in 1994 ended the 30-year dictatorship of Dr. Banda. Half of the population lives in poverty.

 INSIGHT: *Lake Nyasa is 353 miles (568 km) in length and contains at least 500 species of fish*

 ## THE ECONOMY
Mainly subsistence farming. Tobacco accounts for over half of export earnings. Tea and sugar are grown. Drought and corruption are problems.

2000m/6562ft
1000m/3281ft
500m/1640ft
200m/656ft
Sea Level

FACTFILE

OFFICIAL NAME: Republic of Malawi
DATE OF FORMATION: 1964
CAPITAL: Lilongwe
POPULATION: 16.4 million
TOTAL AREA: 45,745 sq. miles (118,480 sq. km)
DENSITY: 451 people per sq. mile

LANGUAGES: Chewa, Lomwe, Yao, Ngoni, English*
RELIGIONS: Protestant 55%, Muslim 20%, Roman Catholic 20%, traditional beliefs 5%
ETHNIC MIX: Bantu 99%, other 1%
GOVERNMENT: Presidential system
CURRENCY: Malawi kwacha = 100 tambala

Malaysia

Malaysia stretches 1240 miles (2000 km) across southeast Asia from the Malay peninsula to Sabah in eastern Borneo. Federated in 1963, it included Singapore for two years.

GEOGRAPHY
The Malay Peninsula has central mountains, an eastern coastal belt, and fertile western plains. Swampy coastal plains rise to mountains on Borneo.

CLIMATE
Warm equatorial. Rainfall always heavy, but with distinct rainy seasons.

INSIGHT: *Malaysia is southeast Asia's major tourist destination, with over 25 million visitors a year*

PEOPLE & SOCIETY
The key distinction is between Malays (Bumiputras, literally "sons of the soil") and the Chinese, who traditionally controlled most economic activity. Since the 1970s, Malays have been favored for education and jobs, in order to address this imbalance.

THE ECONOMY
Successful industrial base includes electronics, manufacturing, and heavy industry. Tourism is a major earner. Leading producer of palm oil, tin, and tropical hardwoods.

2000m/6562ft
1000m/3281ft
500m/1640ft
200m/656ft
Sea Level

0 100 km
0 100 miles

FACTFILE

OFFICIAL NAME: Malaysia
DATE OF FORMATION: 1963
CAPITALS: Kuala Lumpur; Putrajaya (administrative)
POPULATION: 29.7 million
TOTAL AREA: 127,316 sq. miles (329,750 sq. km)
DENSITY: 234 people per sq. mile

LANGUAGES: Bahasa Malaysia*, Malay, Chinese, Tamil, English
RELIGIONS: Muslim 61%, Buddhist 19%, Christian 9%, Hindu 6%, other 5%
ETHNIC MIX: Malay 53%, Chinese 26%, indigenous tribes 12%, Indian 8%, other 1%
GOVERNMENT: Parliamentary system
CURRENCY: Ringgit = 100 sen

Maldives

Set in the Indian Ocean, southwest of Sri Lanka, the Maldives is an archipelago of 1191 small coral islands, or atolls. 200 are inhabited. The word atoll comes from the Dhivehi word "atolu."

 GEOGRAPHY
Consists of low-lying islands and coral atolls. The larger ones are covered in lush, tropical vegetation.

 CLIMATE
Tropical. Rain falls throughout the year, but is heaviest June–November, during the monsoon. Violent storms occasionally hit the northern islands.

 PEOPLE & SOCIETY
Maldivians, who are all Sunni Muslim, are descended from Sinhalese, Dravidian, Arab, and black ancestors. A third of the population live on Male'. Tourism has grown on separate resort islands away from residents. Politics was controlled by a group of influential families until young reformers pushed for parties to be legalized in 2005. However, legislative stalemate followed, and a controversial presidential election in 2013 returned the former elite to power.

$ THE ECONOMY
Fluctuating tourist industry is the economic mainstay. Fish, especially tuna, are the main export. Upgraded to a "middle income" country, despite 2004 tsunami damage.

◆ INSIGHT:
The islands, which all lie below 4 ft (1.2 m), are threatened by rising sea levels, brought about by global warming and climatic changes

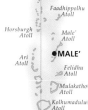

Ihavandippolhu Atoll

Faadhippolhu Atoll

Horsburgh Atoll

Male' Atoll

Ari Atoll

●**MALE'**

Felidhu Atoll

Mulakatho... Atoll

Kolhumadulu Atoll

Hadhdhunmathi Atoll

One and Half Degree Channel

North Huvadhu Atoll

INDIAN
OCEAN

South Huvadhu Atoll

Equator

☐ Sea Level

0   100 km
0   100 miles

○ *Addu Atoll*
○*Gan*

6°

73°

FACTFILE

OFFICIAL NAME: Republic of Maldives
DATE OF FORMATION: 1965
CAPITAL: Male'
POPULATION: 300,000
TOTAL AREA: 116 sq. miles
(300 sq. km)
DENSITY: 2586 people per sq. mile

LANGUAGES: Dhivehi* (Maldivian), Sinhala, Tamil, Arabic
RELIGIONS: Sunni Muslim 100%
ETHNIC MIX: All Maldivians are of Arab–Sinhalese–Malay descent
GOVERNMENT: Presidential system
CURRENCY: Rufiyaa
= 100 laari

A former French colony, Mali is landlocked in the heart of west Africa. The 1991 coup ended the 23-year dictatorship of Moussa Traoré and ushered in multiparty elections from 1992.

GEOGRAPHY

The northern half lies in the Sahara. The inland delta of the Niger River flows through grassy savanna in the south.

CLIMATE

In the south, intensely hot, dry weather precedes the westerly rains. The north is almost rainless.

PEOPLE & SOCIETY

Most people live in the south and are farmers, herders, or river fishermen. Nomadic Fulani and Tuareg herders travel the northern plains. Rebellion broke out there in 2012, initially Tuareg-led, but Islamist insurgents soon seized key towns. They were pushed back with international help, but low-level conflict continues. Women have little status.

◆ **INSIGHT:** *Tombouctou (Timbuktu) was the center of the 14th-century Malinké trading empire*

THE ECONOMY

Widespread poverty. Less than 2% of land can be cultivated. Vulnerable to drought. Gold, high-quality cotton, and livestock account for 90% of exports. Tourism held back by instability and kidnappings by Al-Qaeda in the Maghreb.

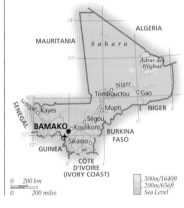

	500m/1640ft
	200m/656ft
	Sea Level

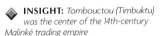

FACTFILE

OFFICIAL NAME: Republic of Mali

DATE OF FORMATION: 1960

CAPITAL: Bamako

POPULATION: 15.3 million

TOTAL AREA: 478,764 sq. miles (1,240,000 sq. km)

DENSITY: 32 people per sq. mile

LANGUAGES: Bambara, Fulani, Senufo, Soninke, French*

RELIGIONS: Muslim (mainly Sunni) 90%, traditional beliefs 6%, Christian 4%

ETHNIC MIX: Bambara 52%, other 18%, Fulani 11%, Saracolé 7%, Soninka 7%, Tuareg 5%

GOVERNMENT: Presidential system

CURRENCY: CFA franc = 100 centimes

Malta

The densely populated Maltese archipelago lies between Africa and Europe. Controlled throughout its history by successive colonial powers, it gained independence from the UK in 1964.

GEOGRAPHY
The main island of Malta has low hills and a ragged coastline with numerous harbors, bays, sandy beaches, and rocky coves. The island of Gozo is more densely vegetated.

CLIMATE
Mediterranean climate. There are many hours of sunshine all year round, with very little rainfall.

PEOPLE & SOCIETY
Over the centuries, the Maltese have been subject to Arab, Sicilian, Spanish, French, and British influences. Today, the population is socially conservative and devoutly Roman Catholic – Malta only legalized divorce in 2011, the last European country except the Vatican to do so. Population density is among the highest in the world. Illegal migration from Africa has increased since Malta joined the EU in 2004.

THE ECONOMY
Tourism provides 30% of GDP. Joined eurozone in 2008. Developing offshore banking, high-tech industry. Semiconductors exported. Most goods have to be imported.

◆ **INSIGHT:** *Malta is the only country to receive the George Cross for gallantry, in 1942 for national resilience to relentless German bombardment*

FACTFILE

OFFICIAL NAME: Republic of Malta
DATE OF FORMATION: 1964
CAPITAL: Valletta
POPULATION: 400,000
TOTAL AREA: 122 sq. miles (316 sq. km)
DENSITY: 3226 people per sq. mile

LANGUAGES: Maltese*, English*
RELIGIONS: Roman Catholic 98%, other and nonreligious 2%
ETHNIC MIX: Maltese 96%, other 4%
GOVERNMENT: Parliamentary system
CURRENCY: Euro = 100 cents

Marshall Islands

Under US rule as part of the UN Trust Territory of the Pacific Islands until independence in 1986, the Marshall Islands comprises a group of 34 widely scattered atolls.

GEOGRAPHY

Narrow coral rings with sandy beaches enclosing lagoons. Those in the south have thicker vegetation. Kwajalein is the world's largest atoll.

CLIMATE
Tropical oceanic, cooled year round by northeast trade winds.

PEOPLE & SOCIETY
Over half the poplulation live in Majuro, the capital and commercial center. Life on the outlying islands is still traditional, based around subsistence agriculture and fishing. Tensions are high due to poor living conditions, especially in periods of drought or flooding. Society is matrilineal, with land and titles handed down through the mother's clan.

◆ **INSIGHT:** *In 1954, Bikini Atoll was the site for the testing of the largest US H-bomb – the 18–22 megaton Bravo*

THE ECONOMY
Almost totally dependent on US aid and the rent paid by the US for its missile base on Kwajalein Atoll. High unemployment. Revenue from licenses to fish in Marshallese waters for tuna. Copra and coconut oil are the only significant agricultural exports.

All land under 100m/328ft

PACIFIC OCEAN

Bokaak

Enewetak Rongelap Ratak Chain
Bikini
Ujelang Likiep Wotje
Kwajalein Maloelap
Ralik Chain Jabat ● MAJURO
Majuro
Jaluit
Narikrik
Ebon

0   200 km
0   200 miles

FACTFILE

OFFICIAL NAME: Republic of the Marshall Islands

DATE OF FORMATION: 1986

CAPITAL: Majuro

POPULATION: 69,747

TOTAL AREA: 70 sq. miles (181 sq. km)

DENSITY: 996 people per sq. mile

LANGUAGES: Marshallese*, English*, Japanese, German

RELIGIONS: Protestant 90%, Roman Catholic 8%, other 2%

ETHNIC MIX: Micronesian 90%, other 10%

GOVERNMENT: Presidential system

CURRENCY: US dollar = 100 cents

Mauritania

Two-thirds of Mauritania's territory is desert – the only productive land is that drained by the Senegal River. The country has taken a strongly Arab direction since 1964.

 GEOGRAPHY
The Sahara, barren except for some scattered oases, covers the north. Savanna lands lie to the south.

 CLIMATE
The climate is generally hot and dry, aggravated by the dusty *harmattan* wind. Summer rain in the south, virtually none in the north.

 PEOPLE & SOCIETY
The Maures control political and economic life. Family solidarity among nomadic peoples is particularly strong. Ethnic tension centers on the oppression of the black minority. Tens of thousands of blacks are estimated to be in illegal slavery. Coups have interrupted civilian rule in recent years.

 INSIGHT: *Slavery officially became illegal in Mauritania in 1980, but de facto slavery still persists*

THE ECONOMY
Agriculture and herding. Iron, copper, and gold mining. World's largest gypsum deposits. Offshore oil from 2006. Rich fishing grounds.

500m/1640ft
200m/656ft
Sea Level

ALGERIA
WESTERN SAHARA
Zouérat
Nouâdhibou
Sahara
MALI
Atâr
ATLANTIC OCEAN
Tidjikja
NOUAKCHOTT
Rosso
Kaédi
Kiffa
Néma
Senegal
SENEGAL
MALI

0   200 km
0   200 miles

FACTFILE

OFFICIAL NAME: Islamic Republic of Mauritania
DATE OF FORMATION: 1960
CAPITAL: Nouakchott
POPULATION: 3.9 million
TOTAL AREA: 397,953 sq. miles (1,030,700 sq. km)

DENSITY: 10 people per sq. mile
LANGUAGES: Arabic*, Hassaniyah Arabic, Wolof, French
RELIGIONS: Sunni Muslim 100%
ETHNIC MIX: Maure 81%, Wolof 7%, Tukolor 5%, other 4%, Soninka 3%
GOVERNMENT: Presidential system
CURRENCY: Ouguiya = 5 khoums

Mauritius

The islands that make up Mauritius lie in the Indian Ocean east of Madagascar. They have enjoyed considerable economic success following recent industrial diversification and expansion.

GEOGRAPHY
The volcanic main island of Mauritius is ringed by coral reefs, and rises from the coast to a fertile central plateau. The outer islands – Rodrigues, the Agalega Islands, and the Cargados Carajos Shoals – lie some 300 miles (500 km) to the north.

CLIMATE
Warm and humid. Tropical storms are frequent December–March, the hottest and wettest months.

PEOPLE & SOCIETY
Most people are descendants of laborers brought over from India in the 19th century. A small minority of French descent form the wealthiest group. Creoles (descendants of African slaves) complain of discrimination. Literacy is high. Health care is free. Crime rates are low. Less-developed Rodrigues has been self-governing since 2001.

THE ECONOMY
Clothing manufacture, tourism, and sugar. Loss of preferential trade terms for sugar and textiles. Offshore financial center. Growing outsourcing and ICT industries. Most food is imported.

INSIGHT: *The islands form part of the Mascarene Archipelago – once a land bridge between Asia and Africa*

FACTFILE

OFFICIAL NAME: Republic of Mauritius
DATE OF FORMATION: 1968
CAPITAL: Port Louis
POPULATION: 1.2 million
TOTAL AREA: 718 sq. miles (1860 sq. km)
DENSITY: 1671 people per sq. mile
LANGUAGES: French Creole, Hindi, Urdu, Tamil, Chinese, English*, French
RELIGIONS: Hindu 48%, Roman Catholic 24%, Muslim 17%, Protestant 9%, other 2%
ETHNIC MIX: Indo-Mauritian 68%, Creole 27%, Sino-Mauritian 3%, Franco-Mauritian 2%
GOVERNMENT: Parliamentary system
CURRENCY: Mauritian rupee = 100 cents

Mexico

Mexico stretches from the US border southward into the ancient Aztec and Mayan heartlands. Independence from Spain came in 1836. One in five Mexicans lives in the sprawling capital.

GEOGRAPHY

Coastal plains along the Pacific and Atlantic seaboards rise to a high arid central plateau. To the east and west are the Sierra Madre mountain ranges. Limestone lowlands form the projecting Yucatan peninsula.

CLIMATE

The plateau and high mountains are warm for much of the year. Pacific coast is tropical: storms occur mostly March–December. Northwest is dry.

PEOPLE & SOCIETY

Most Mexicans are *mestizos* of Spanish–Amerindian descent. Rural Amerindians are largely segregated from Hispanic society and most live in poverty, though the state promotes their culture. The Zapatista movement backs indigenous rights. Few women in male-dominated politics and business. Narcotics-related violent crime is rising.

THE ECONOMY

One of world's largest oil producers. Corn, fruit, vegetables, sugar are cash crops. NAFTA has boosted exports, but exposes farmers to subsidized US competition. Wealth disparity. Hit hard by 2008–2009 global downturn and swine flu crisis.

INSIGHT: *More people cross the US–Mexican border each year – illegally or legally – than any other border in the world*

FACTFILE

OFFICIAL NAME: United Mexican States
DATE OF FORMATION: 1836
CAPITAL: Mexico City
POPULATION: 122 million
TOTAL AREA: 761,602 sq. miles (1,972,550 sq. km
DENSITY: 166 people per sq. mile

LANGUAGES: Spanish*, Nahuatl, Mayan, Zapotec, Mixtec, Otomi, Totonac, Tzotzil
RELIGIONS: Roman Catholic 77%, other 14%, Protestant 6%, nonreligious 3%
ETHNIC MIX: *Mestizo* 60%, Amerindian 30%, European 9%, other 1%
GOVERNMENT: Presidential system
CURRENCY: Mexican peso = 100 centavos

Micronesia

The Federated States of Micronesia (FSM), situated in the western Pacific, comprise 607 islands and atolls grouped into four main island states: Pohnpei, Kosrae, Chuuk, and Yap.

GEOGRAPHY
Mixture of high volcanic islands with forested interiors, and low-lying coral atolls. Some of the islands have coastal mangrove swamps.

CLIMATE
Tropical, with high humidity. There is very heavy rainfall outside the January–March dry season.

INSIGHT: *Chuuk's lagoon contains the sunken wrecks of over 100 Japanese ships and 270 planes from World War II*

PEOPLE & SOCIETY
Micronesians are physically, culturally, and linguistically diverse. Melanesians live on Yap, Polynesians in Pohnpei. The supply of electricity and running water is limited. Society is based on matrilineal clans.

THE ECONOMY
Dependent on US aid. Fishing licenses are a key source of foreign revenue. Tourism, fishing, betel nuts, copra are economic mainstays. Trust fund created to reduce aid reliance.

FACTFILE

OFFICIAL NAME: Federated States of Micronesia

DATE OF FORMATION: 1986

CAPITAL: Palikir (Pohnpei Island)

POPULATION: 106,104

TOTAL AREA: 271 sq. miles (702 sq. km)

DENSITY: 392 people per sq. mile

LANGUAGES: Trukese, Pohnpeian, Kosraean, Yapese, English*

RELIGIONS: Roman Catholic 50%, Protestant 47%, other 3%

ETHNIC MIX: Chuukese 49%, Pohnpeian 24%, other 14%, Kosraean 6%, Yapese 5%, Asian 2%

GOVERNMENT: Nonparty system

CURRENCY: US dollar = 100 cents

Moldova

The most densely populated of the former Soviet republics, Moldova has strong ethnic, linguistic, and cultural links with Romania, but relations with Russia remain paramount.

GEOGRAPHY

Steppes and hilly plains are drained by the Dniester and Prut rivers.

CLIMATE

Warm summers and relatively mild winters. Moderate rainfall is evenly spread throughout the year.

PEOPLE & SOCIETY

A shared heritage with Romania defines national identity, though in 1994 Moldovans voted against possible reunification with Romania. Most of the population is engaged in intensive agriculture. Transnistria is a breakaway state along the east bank of the Dniester, home to a largely ethnic Slav population. The Gagauz, in the south, have accepted autonomy.

INSIGHT: *Vast underground wine vaults contain entire "streets" of bottles built into rock quarries*

THE ECONOMY

Poorest country in Europe. Mainly agricultural: produces wine, tobacco, fruit. Food processing and textiles. Depends on Russia for raw materials, fuel, exports. Political instability.

FACTFILE

OFFICIAL NAME: Republic of Moldova

DATE OF FORMATION: 1991

CAPITAL: Chisinau

POPULATION: 3.5 million

TOTAL AREA: 13,067 sq. miles (33,843 sq. km)

DENSITY: 269 people per sq. mile

LANGUAGES: Moldovan*, Ukrainian, Russian

RELIGIONS: Orthodox Christian 93%, other 6%, Baptist 1%

ETHNIC MIX: Moldovan 84%, Ukrainian 7%, Gagauz 5%, Russian 2%, Bulgarian 1%, other 1%

GOVERNMENT: Parliamentary system

CURRENCY: Moldovan leu = 100 bani

Monaco

Monaco is a tiny principality on the Côte d'Azur. Its destiny changed radically when the casino was opened in 1863. Today, it promotes its image as an upmarket, glamorous destination.

 GEOGRAPHY
A rocky promontory overlooking a narrow coastal strip that has been enlarged through land reclamation.

 CLIMATE
Mediterranean. Summers are hot and dry; days with 12 hours of sunshine are not uncommon. Winters are mild and sunny.

PEOPLE & SOCIETY
Less than 20% of residents are Monégasques. Almost half are French, the rest Italian, American, British, Belgian, and others. Nationals enjoy considerable privileges, including housing subsidies to protect them from Monaco's high property prices, and the right of first refusal before a job can be offered to a foreigner. Women have equal status, but only acquired the vote in 1962. Prince Albert married South African swimmer Charlene Wittstock in 2011.

THE ECONOMY
Tourism, gambling, financial services. Banking secrecy laws and tax-haven conditions attract foreign investment. Close links and customs union with France (but not in EU). No resources: depends on imports.

INSIGHT: *High-profile social and sporting events attract large crowds each spring, including the Rose Ball, Tennis Open, and Grand Prix*

FACTFILE

OFFICIAL NAME: Principality of Monaco
DATE OF FORMATION: 1861
CAPITAL: Monaco-Ville
POPULATION: 36,136
TOTAL AREA: 0.75 sq. miles (1.95 sq. km)
DENSITY: 48,181 people per sq. mile

LANGUAGES: French*, Italian, Monégasque, English
RELIGIONS: Roman Catholic 89%, Protestant 6%, other 5%
ETHNIC MIX: French 47%, other 21%, Italian 16%, Monégasque 16%
GOVERNMENT: Mixed monarchical–parliamentary system
CURRENCY: Euro = 100 cents

Mongolia

Landlocked between Russia and China, Mongolia is a huge, isolated, and sparsely populated nation. Over two-thirds of the country is part of the Gobi Desert.

GEOGRAPHY

A mountainous steppe plateau in the north, with lakes in the north and west. The desert region of the Gobi dominates the south.

CLIMATE

Continental. Mild summers and long, dry, very cold winters, with heavy snowfall. Temperatures can drop as low as −22°F (−30°C).

PEOPLE & SOCIETY

Mongolia was unified by Genghis Khan in 1206 and was later absorbed into Manchu China. A majority of ethnic Mongolians live within China in Inner Mongolia. Tibetan Buddhism dominates. The traditional, nomadic way of life has been eroded as urban migration continues, spurred by ferocious winters, known as *zud*, which can devastate the rural economy.

THE ECONOMY

Rich deposits of oil, coal, copper, uranium, and other minerals remain largely untapped. Cashmere exports. Democracy, from 1990, brought a shift toward a market economy, but also rising poverty. State involvement in mining is an issue. Agriculture uses a third of the workforce, mainly as herders.

◆ **INSIGHT:** *Horseracing, wrestling, and archery are the national sports*

■	3000m/9843ft
■	2000m/6562ft
■	1000m/3281ft
■	500m/1640ft

0   400 km

0   400 miles

FACTFILE

OFFICIAL NAME: Mongolia

DATE OF FORMATION: 1924

CAPITAL: Ulan Bator

POPULATION: 2.8 million

TOTAL AREA: 604,247 sq. miles (1,565,000 sq. km)

DENSITY: 5 people per sq. mile

LANGUAGES: Khalkha Mongolian*, Kazakh

RELIGIONS: Tibetan Buddhist 50%, nonreligious 40%, Shamanist and Christian 6%, Muslim 4%

ETHNIC MIX: Khalkh 95%, Kazakh 4%, other 1%

GOVERNMENT: Mixed presidential–parliamentary system

CURRENCY: Tugrik (tögrög) = 100 möngo

Montenegro

Perched on the Adriatic coast, this tiny republic became a separate state in 2006, after 88 years of federation with its neighbors in various forms of the state of Yugoslavia.

GEOGRAPHY

A narrow coastal strip on the Adriatic. Fertile lowland plains around Lake Scutari. Mountainous interior with deep canyons.

CLIMATE

The lowlands have hot, dry summers and mild winters. Heavy snow in winter in the mountains.

PEOPLE & SOCIETY

Most Montenegrins are Orthodox Christians. They speak a language closely related to Serbian, that also uses Cyrillic script. Muslim Albanians, who make up 70% of the population of the southern Ulcinj region, supported independence. Foreigners, particularly Russians, British, and Serbs, are buying Adriatic real estate.

 INSIGHT: *Dark forests once cloaked Montenegro's mountains; its name means "Black Mountain"*

THE ECONOMY
Tourism (along Adriatic) drives growth. Bauxite reserves, aluminum industry. Return of investment, foreign aid. Crackdown on cigarette smuggling, black market, corruption led to approval in 2010 as candidate for EU membership. Uses euro, though not part of eurozone.

FACTFILE

OFFICIAL NAME: Montenegro

DATE OF FORMATION: 2006

CAPITAL: Podgorica

POPULATION: 600,000

TOTAL AREA: 5332 sq. miles (13,812 sq. km)

DENSITY: 113 people per sq. mile

LANGUAGES: Montenegrin*, Serbian, Albanian, Bosniak, Croatian

RELIGIONS: Orthodox Christian 74%, Muslim 18%, Roman Catholic 4%, other 4%

ETHNIC MIX: Montenegrin 43%, Serb 32%, other 12%, Bosniak 8%, Albanian 5%

GOVERNMENT: Parliamentary system

CURRENCY: Euro = 100 cents

Morocco

Morocco is a former French colony in northwest Africa. Since 1975, it has occupied the territory of Western Sahara, the future of which is yet to be determined by UN-supervised referendum.

GEOGRAPHY
Fertile coastal plain is interrupted in the east by the Rif Mountains. Atlas Mountain ranges to the south. Beyond lies the outer fringe of the Sahara.

CLIMATE
Ranges from temperate and warm in the north, to semiarid in the south. Cooler in the mountains.

PEOPLE & SOCIETY
The Berber minority descend from north Africa's original inhabitants, and live mainly in mountain villages. The Arab majority inhabits the lowlands. Morocco is unusual among Arab states in granting Jews religious freedom and civil rights. The king is spiritual leader and head of state. During the 2011 "Arab Spring" protesters called for more democracy. Islamists have gained influence in politics. Islamist militancy and the emergence of terrorist cells are of concern.

THE ECONOMY
Major exporter of phosphates. Investment in tourism and agriculture. Fishing. Relations with EU strained over illegal immigrants and cannabis trade.

INSIGHT: *Karueein University in Fès, founded in 859 CE, is the world's oldest existing educational institution*

FACTFILE

OFFICIAL NAME: Kingdom of Morocco
DATE OF FORMATION: 1956
CAPITAL: Rabat
POPULATION: 33 million
TOTAL AREA: 172,316 sq. miles (446,300 sq. km)
DENSITY: 192 people per sq. mile
LANGUAGES: Arabic*, Tamazight (Berber), French, Spanish
RELIGIONS: Muslim (mainly Sunni) 99%, other (mostly Christian) 1%
ETHNIC MIX: Arab 70%, Berber 29%, European 1%
GOVERNMENT: Mixed monarchical–parliamentary system
CURRENCY: Mor. dirham = 100 centimes

Mozambique

Mozambique lies on the southeast African coast. It was torn apart by a savage and devastating civil war between the Marxist government and a rebel faction between 1977 and 1992.

GEOGRAPHY
Largely a savanna-covered plateau. The coast is fringed by coral reefs and lagoons. The Zambezi River bisects the country.

CLIMATE
Tropical. Temperatures are hottest on the coast. Extremes of rainfall: drought and flood.

PEOPLE & SOCIETY
Tensions exist between north and south, rather than between ethnic groups. Life is centered on the extended family. Polygamy is fairly common. The country is struggling with the legacy of a war that killed around a million people, and the effects of frequent floods and droughts. Half the population lives in abject poverty.

INSIGHT: *Maputo's busy port serves Zimbabwe and South Africa*

THE ECONOMY
Extremely dependent on aid. Mineral potential. Cashew nuts, shrimp, cotton exported. Debt relief.

FACTFILE

OFFICIAL NAME: Republic of Mozambique

DATE OF FORMATION: 1975

CAPITAL: Maputo

POPULATION: 25.8 million

TOTAL AREA: 309,494 sq. miles
(801,590 sq. km)

DENSITY: 85 people per sq. mile

LANGUAGES: Makua, Xitsonga, Sena, Lomwe, Portuguese*

RELIGIONS: Traditional beliefs 56%, Christian 30%, Muslim 14%

ETHNIC MIX: Makua Lomwe 47%, Tsonga 23%, Malawi 12%, Shona 11%, Yao 4%, other 3%

GOVERNMENT: Presidential system

CURRENCY: New metical = 100 centavos

Myanmar (Burma)

Forming the eastern shores of the Bay of Bengal and the Andaman Sea in southeast Asia, Myanmar has suffered from isolation, political repression, and ethnic conflict.

GEOGRAPHY

The fertile Irrawaddy basin lies at the center. Mountains to the west, Shan plateau to the east. Tropical rainforest covers much of the land.

CLIMATE

Tropical. Hot summers, with high humidity, and warm winters.

PEOPLE & SOCIETY

The military, in power from 1962, paid little regard to human rights, and didn't tolerate opposition. The National League for Democracy won elections in 1990, but was kept from power. Elections in 2010, nominally restoring civilian rule, were dominated by the new military-backed party. Ethnic minorities are fighting for independence.

◆ **INSIGHT:** *Myanmar is one of the world's biggest teak exporters, though reserves are diminishing rapidly*

THE ECONOMY

Corrupt, mismanaged, subject to sanctions – but gas, teak, and gems are exported. One of world's largest illegal opium producers. Goods sold on black market carry high prices.

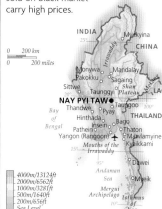

FACTFILE

OFFICIAL NAME: Republic of the Union of Myanmar

DATE OF FORMATION: 1948

CAPITAL: Nay Pyi Taw

POPULATION: 53.3 million

TOTAL AREA: 261,969 sq. miles (678,500 sq. km)

DENSITY: 210 people per sq. mile

LANGUAGES: Burmese (Myanmar)*, Shan, Karen, Rakhine, Chin, Yangbye, Kachin, Mon

RELIGIONS: Buddhist 89%, Christian 4%, Muslim 4%, other 2%, Animist 1%

ETHNIC MIX: Burman (Bamah) 68%, other 12%, Shan 9%, Karen 7%, Rakhine 4%

GOVERNMENT: Presidential system

CURRENCY: Kyat = 100 pyas

Namibia

Located in southwestern Africa, Namibia gained
independence from South Africa in 1990, after 24 years of
armed struggle. It regained the territory of Walvis Bay in 1994.

GEOGRAPHY
The Namib Desert stretches along
the coastal strip. Inland, a ridge of
mountains rises to 8000 ft (2500 m).
The Kalahari Desert lies in the east.

CLIMATE
Almost rainless. The coast is usually
shrouded in thick fog, unless the hot, dry
berg wind is blowing.

PEOPLE & SOCIETY
The Ovambo, the main ethnic
group, live mainly in the more populous
north. Some 100,000 whites, many of
German descent, are centered around
Windhoek and still control the economy.
The minority San and Khoi bushmen are
among the oldest human communities
in the world. Homosexual rights
are restricted.

INSIGHT: *The Namib is the Earth's
oldest, and one of its driest, deserts*

THE ECONOMY
Varied mineral resources, notably
uranium and diamonds. Rich offshore
fishing grounds. High unemployment.
HIV/AIDS epidemic. One of Africa's most
skewed distributions of wealth.

2000m/6562ft	
1000m/3281ft	
500m/1640ft	
200m/656ft	
Sea Level	

0   200 km
0   200 miles

FACTFILE

OFFICIAL NAME: Republic of Namibia

DATE OF FORMATION: 1990

CAPITAL: Windhoek

POPULATION: 2.3 million

TOTAL AREA: 318,694 sq. miles
(825,418 sq. km)

DENSITY: 7 people per sq. mile

LANGUAGES: Ovambo, Kavango, English*,
Bergdama, German, Afrikaans

RELIGIONS: Christian 90%, traditional
beliefs 10%

ETHNIC MIX: Ovambo 50%, other tribes 22%,
Kavango 9%, Damara 7%, Herero 7%, other 5%

GOVERNMENT: Presidential system

CURRENCY: Namibian dollar = 100 cents

Nauru

Nauru lies in the Pacific, northeast of Australia. Phosphate deposits gave its inhabitants huge temporary wealth, but economic mismanagement has left them facing ruin.

GEOGRAPHY

A single low-lying coral atoll, with a fertile coastal belt. Coral cliffs encircle an elevated interior plateau.

CLIMATE

Equatorial, moderated by sea breezes. Occasional long droughts.

PEOPLE & SOCIETY

Native Nauruans are of mixed Micronesian and Polynesian origin. Most live in simple, traditional houses and spend their money on luxury cars and consumer goods. Welfare and education are free. A diet of imported processed foods has caused widespread obesity and diabetes. Mining was left to imported laborers, mainly from Kiribati, who lived in enclaves of male-only barracks and had few rights. Many young Nauruans leave to seek a better life in Australia or New Zealand.

THE ECONOMY

Phosphate revenues diminished. Sale of fishing rights only other resource. State trust fund invested badly overseas. Offshore banking facilities closed after international pressure.

INSIGHT: *Phosphate mining has left 80% of the island uninhabitable*

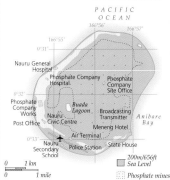

FACTFILE

OFFICIAL NAME: Republic of Nauru
DATE OF FORMATION: 1968
CAPITAL: None
POPULATION: 9434
TOTAL AREA: 8.1 sq. miles (21 sq. km)
DENSITY: 1165 people per sq. mile

LANGUAGES: Nauruan*, Kiribati, Chinese, Tuvaluan, English
RELIGIONS: Nauruan Congregational Church 60%, Roman Catholic 35%, other 5%
ETHNIC MIX: Nauruan 93%, Chinese 5%, other Pacific islanders 1%, European 1%
GOVERNMENT: Nonparty system
CURRENCY: Australian dollar = 100 cents

Nepal

Nepal, lying between India and China on the southern shoulder of the Himalayas, is one of the world's poorest countries. Its agricultural economy is heavily dependent on the monsoon.

GEOGRAPHY

Mainly mountainous. The area includes some of the highest mountains in the world, including Mount Everest. Flat, fertile river plains form the south.

CLIMATE

Warm monsoon season from July to October. The rest of the year is dry, sunny, and mild. Winter temperatures in the Himalayas average 14°F (–10°C).

PEOPLE & SOCIETY

Tensions are few between the diverse ethnic groups. Buddhist women, including Sherpas, face fewer social restrictions than Hindus. Trafficking of women and child labor are problems. Human rights violations rose during the 1999–2006 Maoist insurgency. The peace deal led to the abolition of the monarchy and the Maoists joining the political mainstream, but fractious coalitions mean instability continues.

THE ECONOMY

Agriculture employs two-thirds of workforce. Crops include rice and wheat. Tourism and investment affected by instability. Reliant on aid and overseas remittances. Hydropower potential.

INSIGHT: *Southern Nepal was the birthplace of Buddha (Prince Siddhartha Gautama) in 563 BCE*

Dadeldhurā
Jumlā
CHINA
Salyān
Baglung
Pokhara
Mt. Everest
29,029ft (8848m)
(Sagarmatha)
KATHMANDU
Lalitpur
Bhaktapur
INDIA
Janakpur
Ilām
Birātnagar

4000m/13124ft
3000m/9843ft
2000m/6562ft
1000m/3281ft
500m/1640ft
200m/656ft
Sea Level

0   100 km
0   100 miles

FACTFILE

OFFICIAL NAME: Federal Democratic Republic of Nepal
DATE OF FORMATION: 1769
CAPITAL: Kathmandu
POPULATION: 27.8 million
TOTAL AREA: 54,363 sq. miles (140,800 sq. km)
DENSITY: 526 people per sq. mile

LANGUAGES: Nepali*, Maithili, Bhojpuri
RELIGIONS: Hindu 81%, Buddhist 11%, Muslim 4%, other (including Christian) 4%
ETHNIC MIX: Other 52%, Chhetri 16%, Hill Brahman 13%, Tharu 7%, Magar 7%, Tamang 5%
GOVERNMENT: Transitional regime
CURRENCY: Nepalese rupee = 100 paisa

Netherlands

Astride the delta of five major rivers in northwest Europe, the Netherlands built its historic wealth on maritime trade. Rotterdam is Europe's largest port.

GEOGRAPHY

Mainly flat, with 27% of the land below sea level and protected by dunes, dikes, and canals. There are a few low hills in the south and east.

CLIMATE

Mild, rainy winters and cool summers. Gales from the North Sea are common in fall and winter.

PEOPLE & SOCIETY
The Dutch have a long history of welcoming immigrants from former colonies and refugees seeking asylum. However, lack of integration is now raising fears about the failing asylum system, immigrant crime, and militant Islam. Population is mostly urban and the density is high. The state does not try to impose a particular morality on its citizens. Laws concerning sexuality, narcotics-taking, and euthanasia are among the world's most liberal.

THE ECONOMY
Major trading hub. High-profile multinationals. Diverse industrial base: chemicals, machinery, electronics, and metals. Costly social welfare system.

INSIGHT: *In 2002, the Netherlands became the first country in the world to legalize euthanasia*

FACTFILE

OFFICIAL NAME: Kingdom of the Netherlands
DATE OF FORMATION: 1648
CAPITALS: Amsterdam and The Hague
POPULATION: 16.8 million
TOTAL AREA: 16,033 sq. miles (41,526 sq. km)

DENSITY: 1283 people per sq. mile
LANGUAGES: Dutch*, Frisian
RELIGIONS: Roman Catholic 36%, other 34%, Protestant 27%, Muslim 3%
ETHNIC MIX: Dutch 82%, other 12%, Turkish 2%, Surinamese 2%, Moroccan 2%
GOVERNMENT: Parliamentary system
CURRENCY: Euro = 100 cents

New Zealand

Lying in the South Pacific, 990 miles (1600 km) southeast of Australia, New Zealand comprises North and South Islands, separated by the Cook Strait, and many smaller islands.

GEOGRAPHY

North Island, noted for hot springs and geysers, has the bulk of the population. South Island is mostly mountainous, with eastern lowlands.

CLIMATE

Generally temperate and damp. The far north is almost subtropical, whereas southern winters are cold.

PEOPLE & SOCIETY

Maoris were the first settlers, 1200 years ago. Today's majority European population is descended mainly from British migrants who settled after 1840. Maoris' living and education standards are generally lower than average. The government is continuing to negotiate the settlement of Maori land claims.

INSIGHT: *New Zealand was the first country to give women the vote (1893)*

THE ECONOMY

Modern agricultural sector; world's top exporter of dairy products. Dairy vies with tourism as the biggest foreign-exchange earner. Hi-tech manufacturing. Open economy. Strong trade links.

2000m/6562ft
1000m/3281ft
500m/1640ft
200m/656ft
Sea Level

North Island

Auckland
Hamilton
Tauranga
Rotorua
New Plymouth
Hastings
Palmerston North
Tasman Sea
Blenheim
WELLINGTON
Greymouth
Cook Strait
South Island
Southern Alps
Christchurch
Timaru
Queenstown
Dunedin
PACIFIC OCEAN
Invercargill
Stewart Island

0   200 km
0   200 miles

FACTFILE

OFFICIAL NAME: New Zealand

DATE OF FORMATION: 1947

CAPITAL: Wellington

POPULATION: 4.5 million

TOTAL AREA: 103,737 sq. miles (268,680 sq. km)

DENSITY: 43 people per sq. mile

LANGUAGES: English*, Maori*

RELIGIONS: Anglican 24%, other 22%, Presbyterian 18%, nonreligious 16%, Roman Catholic 15%, Methodist 5%

ETHNIC MIX: European 75%, Maori 15%, other 7%, Samoan 3%

GOVERNMENT: Parliamentary system

CURRENCY: New Zealand dollar = 100 cents

Nicaragua

Nicaragua lies at the heart of Central America. The Sandinista revolution of 1978 led to 11 years of civil war between the left-wing Sandinistas and the right-wing US-backed Contras.

GEOGRAPHY

Extensive forested plains in the east. Central mountain region with many active volcanoes. The Pacific coastlands are dominated by lakes.

CLIMATE

Tropical. The lowlands are hot all year round. The mountains are cooler. Prone to occasional hurricanes.

PEOPLE & SOCIETY

Most people are *mestizo* (mixed Spanish–Amerindian), and there is a large white elite. Caribbean regions are home to communities of Miskito Amerindians and blacks, who gained autonomy in 1987. The revolution improved the status of women, but these gains have been undone by rampant poverty.

INSIGHT: *Lake Nicaragua is the only freshwater lake in the world to contain marine animals*

THE ECONOMY

Textiles, coffee, meat, tobacco are main exports: affected by world price fluctuations. Remittances from abroad. Substantial debt relief has cut debt to around 50% of GDP. Corruption.

1000m/3281ft
500m/1640ft
200m/656ft
Sea Level

FACTFILE

OFFICIAL NAME: Republic of Nicaragua

DATE OF FORMATION: 1838

CAPITAL: Managua

POPULATION: 6.1 million

TOTAL AREA: 49,998 sq. miles (129,494 sq. km)

DENSITY: 133 people per sq. mile)

LANGUAGES: Spanish*, English Creole, Miskito

RELIGIONS: Roman Catholic 80%, Protestant Evangelical 17%, other 3%

ETHNIC MIX: *Mestizo* 69%, White 17%, Black 9%, Amerindian 5%

GOVERNMENT: Presidential system

CURRENCY: Córdoba oro = 100 centavos

Niger

Niger lies in west Africa, upstream from Nigeria on the Niger River. One of the world's poorest states, it was ruled by one-party or military regimes until multipartyism was allowed in 1992.

GEOGRAPHY
The north and northeast regions are part of the Sahara. The Air Mountains in the center rise high above the desert. Savanna lies to the south.

CLIMATE
High temperatures persist for most of the year at around 95°F (35°C). The north is virtually rainless.

PEOPLE & SOCIETY
Tuareg nomads in the north feel excluded from politics and the benefits of their area's uranium resources. An early 1990s rebellion reignited briefly in 2007–2009. In the south, egalitarianism and a sense of community help to combat economic difficulties. Almost the entire urban population lives in slum conditions. Two-thirds of the population is under 25. Women have limited rights and restricted access to education. The army seized power briefly in 2010.

THE ECONOMY
Vast uranium deposits. Frequent droughts and food shortages. Banditry. Expansion of Sahara. Oil potential.

INSIGHT: *The name Niger comes from the Tuareg word* n'eghirren, *which means "flowing water"*

FACTFILE

OFFICIAL NAME: Republic of Niger
DATE OF FORMATION: 1960
CAPITAL: Niamey
POPULATION: 17.8 million
TOTAL AREA: 489,188 sq. miles (1,267,000 sq. km)
DENSITY: 36 people per sq. mile

LANGUAGES: Hausa, Djerma, Fulani, Tuareg, Teda, French*
RELIGIONS: Muslim 99%, other (including Christian) 1%
ETHNIC MIX: Hausa 53%, Djerma and Songhai 21%, Tuareg 11%, Fulani 7%, Kanuri 6%, other 2%
GOVERNMENT: Presidential system
CURRENCY: CFA franc = 100 centimes

Nigeria

West Africa's biggest nation, Nigeria is a federation of 36 states and the capital, Abuja. Dominated by military governments since 1966, democracy returned in 1999.

GEOGRAPHY

Coastal area of beaches, swamps, and lagoons gives way to rainforest, and then to savanna on the high plateaus. Semidesert to the north.

CLIMATE

The south is hot, rainy and humid for most of the year. The arid north has one very humid wet season. The Jos Plateau and highlands are cooler.

PEOPLE & SOCIETY

Some 250 ethnic groups: tensions threaten national unity, with sporadic intercommunal violence. The mainly Muslim north has introduced *sharia* (Islamic law); Boko Haram militants use bombings, assassinations, and abductions to fight for an Islamic state. Women have more economic independence in the south. Militants in the oil-rich Niger Delta demand a share in the oil wealth for the region's impoverished population.

THE ECONOMY

Overdependent on oil, principal export since 1970s. Mismanagement and corruption. Debt reduced. Instability.

INSIGHT: *Nigeria is Africa's most populous state – one in every six Africans is Nigerian*

FACTFILE

OFFICIAL NAME: Federal Republic of Nigeria
DATE OF FORMATION: 1960
CAPITAL: Abuja
POPULATION: 174 million
TOTAL AREA: 356,667 sq. miles (923,768 sq. km)
DENSITY: 494 people per sq. mile

LANGUAGES: Hausa, English*, Yoruba, Ibo
RELIGIONS: Muslim 50%, Christian 40%, traditional beliefs 10%
ETHNIC MIX: Other 29%, Hausa 21%, Yoruba 21%, Ibo 18%, Fulani 11%
GOVERNMENT: Presidential system
CURRENCY: Naira = 100 kobo

Norway

The Kingdom of Norway traces the rugged western coast of Scandinavia. Settlements are largely restricted to southern and coastal areas. Vast oil and natural gas revenues bring prosperity.

GEOGRAPHY

The western coast is indented with numerous fjords and features tens of thousands of islands. Mountains and plateaus cover most of the country.

CLIMATE

Mild coastal climate. Inland, the weather is more extreme, with warmer summers and cold, snowy winters.

PEOPLE & SOCIETY

Fairly homogeneous, but has welcomed refugees from Iraq, Somalia, Bosnia, Sri Lanka, and elsewhere. Strong family tradition, but divorce is common. Fair-minded consensus promotes female equality, boosted by the generous childcare provision. Wealth is more evenly distributed than in most countries. Voted against joining the EU in 1994.

 INSIGHT: Near Narvik, mainland Norway is only 4 miles (7 km) wide

THE ECONOMY

Western Europe's top oil and natural gas producer: trust fund saves for post-oil future. Metal, chemical, and engineering industries. Generous aid donor. High cost of living.

2000m/6562ft
1000m/3281ft
500m/1640ft
200m/656ft
Sea Level

Hammerfest
RUSS. FED.
Tromsø
FINLAND
Narvik
Bodø
Arctic Circle
Norwegian Sea
SWEDEN
Trondheim
Ålesund
Lillehammer
Bergen Hønefoss
North Sea
OSLO
Stavanger Moss
Kristiansand
Skagerrak

0   200 km
0   200 miles

FACTFILE

OFFICIAL NAME: Kingdom of Norway
DATE OF FORMATION: 1905
CAPITAL: Oslo
POPULATION: 5 million
TOTAL AREA: 125,181 sq. miles (324,220 sq. km)
DENSITY: 42 people per sq. mile
LANGUAGES: Norwegian* (*Bokmål* "book

language" and *Nynorsk* "new Norsk"), Sámi
RELIGIONS: Evangelical Lutheran 88%, other and nonreligious 8%, Muslim 2%, Pentecostal 1%, Roman Catholic 1%
ETHNIC MIX: Norwegian 93%, other 6%, Sámi 1%
GOVERNMENT: Parliamentary system
CURRENCY: Norwegian krone = 100 øre

Oman

Oman occupies a strategic position on the Arabian Peninsula, at the entrance to the Persian Gulf. It is the least developed Gulf state, despite modest oil exports.

GEOGRAPHY

Mostly gravelly desert, with mountains in the north and south. Some narrow fertile coastal strips.

CLIMATE

Blistering heat in the west. Summer temperatures often climb above 113°F (45°C). Southern uplands receive rains June–September.

PEOPLE & SOCIETY

Urban drift has seen most Omanis move to northern towns. The majority are Ibadi Muslims who follow an appointed leader, the imam. Ibadism is not opposed to freedom for women, and a few women hold positions of authority. Baluchi from Pakistan are the largest group of foreign workers.

INSIGHT: *Until the late 1980s, Oman was closed to all but business or official visitors*

THE ECONOMY

Oil and natural gas account for almost all export revenue. Commercially extractable reserves are limited. Other exports include fish, animals, and dates. Foreigners work in all sectors.

FACTFILE

OFFICIAL NAME: Sultanate of Oman

DATE OF FORMATION: 1951

CAPITAL: Muscat

POPULATION: 3.6 million

TOTAL AREA: 82,031 sq. miles (212,460 sq. km)

DENSITY: 44 people per sq. mile

LANGUAGES: Arabic*, Baluchi, Farsi, Hindi, Punjabi

RELIGIONS: Ibadi Muslim 75%, other Muslim and Hindu 25%

ETHNIC MIX: Arab 88%, Baluchi 4%, Persian 3%, Indian and Pakistani 3%, African 2%

GOVERNMENT: Monarchy

CURRENCY: Omani rial = 1000 baisa

Pakistan

Once a part of British India, Pakistan was created in 1947 in response to demands for an independent Muslim state. In 1971, Bangladesh (former East Pakistan) became a separate state.

GEOGRAPHY
Indus floodplain across east and south. Hindu Kush mountains in north. Semidesert plateau, mountains in west.

CLIMATE
Temperatures can soar to 122°F (50°C) in south and west, and fall to –4°F (–20°C) in the Hindu Kush.

PEOPLE & SOCIETY
Punjabis dominate government and the army. Tensions with minority groups, exacerbated by the vast gap between rich and poor. Strong family ties permeate politics and business. Relations with India are tense over Kashmir and terrorism. Islamist *taliban* insurgency in tribal areas on Afghan border: fighting has displaced millions.

◆ **INSIGHT:** *In 1988, Pakistan elected Benazir Bhutto as the first female prime minister in the Muslim world*

THE ECONOMY
Major cotton and rice producer, but unpredictable weather conditions often affect crop. Textiles. Instability. Corruption. Aid to fight terrorism and for earthquake reconstruction.

5000m/16405ft	
4000m/13124ft	
3000m/9843ft	
2000m/6562ft	
1000m/3281ft	
500m/1640ft	
200m/656ft	
Sea Level	

0   200 km
0   200 miles

FACTFILE

OFFICIAL NAME: Islamic Republic of Pakistan

DATE OF FORMATION: 1947

CAPITAL: Islamabad

POPULATION: 182 million

TOTAL AREA: 310,401 sq. miles (803,940 sq. km)

DENSITY: 612 people per sq. mile

LANGUAGES: Punjabi, Sindhi, Pashtu, Urdu*, Baluchi, Brahui

RELIGIONS: Sunni Muslim 77%, Shi'a Muslim 20%, Hindu 2%, Christian 1%

ETHNIC MIX: Punjabi 56%, Pathan (Pashtun) 15%, Sindhi 14%, Mohajir 7%, Baluchi 4%, other 4%

GOVERNMENT: Parliamentary system

CURRENCY: Pakistani rupee = 100 paisa

Palau

The 300-island Palau archipelago (known locally as Belau) lies in the western Pacific Ocean. It achieved independence in 1994, and is gradually reducing its aid dependence.

GEOGRAPHY

Terrain varies from thickly forested mountains to limestone and coral reefs. Babeldaob, the largest island, is volcanic, with many rivers and waterfalls.

CLIMATE

Hot and wet. Little variation in daily and seasonal temperatures. February–April is the dry season.

PEOPLE & SOCIETY

Native Palauans are a mix of the original Southeast Asian migrants and Pacific settlers. A modern influx from Asia, particularly the Philippines, China, and Bangladesh, has led to tension. As 70% of the population live on the island-city of Koror, a new capital was constructed recently on Babeldaob. Native culture is preserved on outer islands despite strong influence from the US and Japan. Modekngei is a blend of Christianity and local beliefs.

THE ECONOMY

Tourism and fishing licenses are main earners. Coconuts, bananas, and taro. New 15-year US aid plan to 2024.

INSIGHT: *Palau's reefs contain 1500 species of fish and 700 types of coral*

FACTFILE

OFFICIAL NAME: Republic of Palau
DATE OF FORMATION: 1994
CAPITAL: Ngerulmud
POPULATION: 21,108
TOTAL AREA: 177 sq. miles (458 sq. km)
DENSITY: 108 people per sq. mile

LANGUAGES: Palauan*, English*, Japanese, Angaur, Tobi, Sonsorolese
RELIGIONS: Christian 66%, Modekngei 34%
ETHNIC MIX: Palauan 74%, Filipino 16%, other 6%, Chinese and other Asian 4%
GOVERNMENT: Nonparty system
CURRENCY: US dollar = 100 cents

Panama

A Spanish colony until 1821, Panama is the southernmost country in Central America. The colossal Panama Canal (which was under US control until 2000) links the Pacific and Atlantic oceans.

GEOGRAPHY
Lowlands along both coasts, with savanna-covered plains and rolling hills. Mountainous interior. Swamps and rainforests in the east.

CLIMATE
Hot and humid, with heavy rainfall in the May–December wet season. Cooler at high altitudes.

PEOPLE & SOCIETY
A multiethnic society, dominated by people of mixed Spanish–Amerindian origin (*mestizo*). Amerindians live in remote areas. The Panama Canal and former US military bases (the last of which closed in 1999) have given society a cosmopolitan outlook, but Catholicism and the extended family remain strong. Wealth is unevenly divided. Money-laundering, narcotics trafficking, and corruption are rife.

THE ECONOMY
Colón Free Trade Zone: world's second-largest. Income from the canal (expansion project underway) and merchant ships sailing under flag of Panama. Banana and shrimp exports.

INSIGHT: *The Panama Canal shortens the sea route between the east coast of the US and Japan by 3000 miles (4800 km)*

FACTFILE

OFFICIAL NAME: Republic of Panama

DATE OF FORMATION: 1903

CAPITAL: Panama City

POPULATION: 3.9 million

TOTAL AREA: 30,193 sq. miles (78,200 sq. km)

DENSITY: 133 people per sq. mile

LANGUAGES: English Creole, Spanish*, Amerindian languages, Chibchan languages

RELIGIONS: Roman Catholic 84%, Protestant 15%, other 1%

ETHNIC MIX: *Mestizo* 70%, Black 14%, White 10%, Amerindian 6%

GOVERNMENT: Presidential system

CURRENCY: Balboa = 100 centésimos; US dollar is also legal tender

Papua New Guinea

A former Australian colony, Papua New Guinea (PNG) occupies the eastern section of the island of New Guinea and several other island groups. Much of the country is isolated.

GEOGRAPHY
Mountainous and forested mainland, with broad, swampy river valleys. 40 active volcanoes in the north. Around 600 outer islands.

CLIMATE
Hot and humid in lowlands, cooling toward highlands, where snow can fall on highest peaks.

PEOPLE & SOCIETY
Around 800 language groups and even more tribes. The main social distinction is between lowlanders, who have frequent contact with the outside world, and the very isolated, but increasingly threatened, highlanders. Great tensions exist between highland tribes, and vendettas can often last several generations. The island of Bougainville has been granted autonomy and promised a referendum on independence by 2020.

THE ECONOMY
Minerals: gold, copper, oil, and natural gas. High government spending almost led to national bankruptcy in 2002. Strong GDP growth since 2007.

INSIGHT: *PNG is home to the only known poisonous birds; contact with the feathers of some species of pitohui produces skin blisters*

FACTFILE
OFFICIAL NAME: Independent State of Papua New Guinea
DATE OF FORMATION: 1975
CAPITAL: Port Moresby
POPULATION: 7.3 million
TOTAL AREA: 178,703 sq. miles (462,840 sq. km)

DENSITY: 42 people per sq. mile
LANGUAGES: Pidgin English, Papuan, English*, Motu, c.800 native languages
RELIGIONS: Protestant 60%, Roman Catholic 37%, other 3%
ETHNIC MIX: Melanesian or mixed race 100%
GOVERNMENT: Parliamentary system
CURRENCY: Kina = 100 toea

Paraguay

Landlocked in central South America, and once a Spanish colony, Paraguay's post independence history has included periods of military rule. Free elections held since 1993.

GEOGRAPHY

The Paraguay River divides the hilly and forested east from a flat alluvial plain, with marsh and semidesert scrub land in the west.

CLIMATE

Subtropical. The Gran Chaco is generally hotter and drier. All areas experience floods and droughts.

PEOPLE & SOCIETY

The population is mainly *mestizo* (mixed Spanish and native Guaraní origin). Most people are bilingual, though in rural areas Guaraní is more widely used. Cattle ranchers populate the Chaco, along with communities of the German-origin Mennonite Church. Right-wing Colorados in power for decades, except 2008–2012.

◆ **INSIGHT:** *The War of the Triple Alliance (1864–1870) killed almost 90% of Paraguay's male population*

THE ECONOMY

Agriculture: soybeans are the main export. Electricity exported from massive hydroelectric dams, including Itaipú (world's second-largest, jointly run with Brazil). Large informal economy. Corruption and smuggling.

FACTFILE

OFFICIAL NAME: Republic of Paraguay

DATE OF FORMATION: 1811

CAPITAL: Asunción

POPULATION: 6.8 million

TOTAL AREA: 157,046 sq. miles (406,750 sq. km)

DENSITY: 44 people per sq. mile

LANGUAGES: Guaraní*, Spanish*, German

RELIGIONS: Roman Catholic 90%, Protestant (including Mennonite) 10%

ETHNIC MIX: Mestizo 91%, other 7%, Amerindian 2%

GOVERNMENT: Presidential system

CURRENCY: Guaraní = 100 céntimos

Peru

Once the heart of the Inca Empire, before the Spanish conquest in the 16th century, Peru lies on the Pacific coast of South America, just south of the equator.

 GEOGRAPHY
Coastal plain rises to Andes Mountains. Uplands, dissected by fertile valleys, lie east of the Andes. Tropical forest in extreme east.

 CLIMATE
Coast is mainly arid. Middle slopes of the Andes are temperate; higher peaks are snow-covered. East is hot, humid, and very wet.

 PEOPLE & SOCIETY
Though most people are Amerindians or mixed-race *mestizos*, society is dominated by a small group of Spanish descendants. Amerindians, and the small black community, suffer discrimination in towns, but access to information and political power are growing; the first Amerindian president was elected in 2001–2006. Clashes with left-wing militants killed almost 70,000 people between 1980 and 2000.

THE ECONOMY
Abundant mineral resources: notably copper and gold. Rich Pacific fish stocks. World's largest cocaine producer.

◆ **INSIGHT:** *Lake Titicaca is the world's highest navigable lake*

FACTFILE

OFFICIAL NAME: Republic of Peru
DATE OF FORMATION: 1824
CAPITAL: Lima
POPULATION: 30.4 million
TOTAL AREA: 496,223 sq. miles (1,285,200 sq. km)
DENSITY: 62 people per sq. mile

LANGUAGES: Spanish*, Quechua*, Aymara
RELIGIONS: Roman Catholic 81%, other 19%
ETHNIC MIX: Amerindian 45%, *Mestizo* (European–Amerindian) 37%, White 15%, other 3%
GOVERNMENT: Presidential system
CURRENCY: New sol = 100 céntimos

Philippines

Lying in the western Pacific Ocean, the Philippines is the world's second-largest archipelago, with 7107 islands, of which 4600 are named but only around 1000 inhabited.

GEOGRAPHY
Larger islands are forested and mountainous. Over 20 active volcanoes. Frequent earthquakes.

CLIMATE
Tropical. Warm and humid all year round. Typhoons occur in the rainy season: June–October.

PEOPLE & SOCIETY
Over 100 ethnic groups, most of which are of Malay origin. The Catholic Church is a dominant cultural force; it opposes family-planning, despite high population growth. The Chinese minority has been established for 400 years. Women play a prominent part in society. High literacy levels. Islamist separatists and communist insurgents undermine stability.

◆ **INSIGHT:** *Mass "People Power" demonstrations have brought down two presidents, in 1986 and 2001*

THE ECONOMY
Coconuts, bananas, pineapples exported. Growing outsourcing center. Remittances from abroad. Corruption and poor infrastructure limit growth.

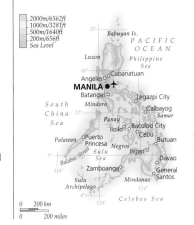

- 2000m/6562ft
- 1000m/3281ft
- 500m/1640ft
- 200m/656ft
- Sea Level

Babuyan Is.

PACIFIC OCEAN

Luzon

Philippine Sea

Angeles○ ○Cabanatuan

MANILA ●✈

Batangas○

South China Sea

Mindoro

Legazpi City

○Calbayog

Panay *Samar*

Iloilo○ ○Bacolod City

Palawan ○Puerto Princesa *Negros* ○Cebu

○Butuan

Sulu Sea ○Iligan

○Davao

Zamboanga○

Balabac Strait

Mindanao ○General Santos

Sulu Archipelago

Celebes Sea

0 ___ 200 km

0 ___ 200 miles

FACTFILE

OFFICIAL NAME: Republic of the Philippines
DATE OF FORMATION: 1946
CAPITAL: Manila
POPULATION: 98.4 million
TOTAL AREA: 115,830 sq. miles (300,000 sq. km)
DENSITY: 855 people per sq. mile

LANGUAGES: Filipino*, English*, Tagalog, Cebuano, Ilocano, Hiligaynon, many others
RELIGIONS: Roman Catholic 81%, Protestant 9%, Muslim 5%, other (including Buddhist) 5%
ETHNIC MIX: Other 34%, Tagalog 28%, Cebuano 13%, Ilocano 9%, Hiligaynon 8%, Bisaya 8%
GOVERNMENT: Presidential system
CURRENCY: Philippine peso = 100 centavos

Poland

Located in the heart of Europe, Poland has undergone massive social, economic, and political change since the collapse of communism in 1989. It joined the EU in 2004.

GEOGRAPHY

Lowlands, part of the North European Plain, cover most of the country. The Tatra Mountains run along the southern border.

CLIMATE

Rainfall peaks during the hot summers. Cold winters with snow, especially in mountains.

PEOPLE & SOCIETY

Ethnic homogeneity masks social tensions. Secular liberals criticize the semiofficial status of the Roman Catholic Church, though its influence is now waning. Abortion is banned, except for special cases. Growing wealth disparities are resented. The German minority in the west is becoming more assertive.

INSIGHT: *Wild wisent (European bison) live in the Bialowieza Forest straddling the Poland–Belarus border*

THE ECONOMY

Heavy industries dominate; services growing. Foreign investment reflects large potential market. Rapid privatization. Only EU state to avoid recession in 2007–2009 global downturn. Not adopting euro yet.

FACTFILE

OFFICIAL NAME: Republic of Poland

DATE OF FORMATION: 1918

CAPITAL: Warsaw

POPULATION: 38.2 million

TOTAL AREA: 120,728 sq. miles (312,685 sq. km)

DENSITY: 325 people per sq. mile

LANGUAGES: Polish*

RELIGIONS: Roman Catholic 93%, other and nonreligious 5%, Orthodox Christian 2%

ETHNIC MIX: Polish 98%, other 2%

GOVERNMENT: Parliamentary system

CURRENCY: Zloty = 100 groszy

Portugal

Portugal, with its long Atlantic coast, lies on the western side of the Iberian Peninsula, which it shares with Spain. It is the most westerly country on the European mainland.

GEOGRAPHY
The Tagus River bisects the country roughly east to west, dividing mountainous north from lower and more undulating south.

CLIMATE
North is cool and moist. South is warmer, with dry, mild winters.

PEOPLE & SOCIETY
A homogeneous and stable society, which is losing some of its conservative traditions. History of immigration from former colonies, and recently from eastern Europe. Urban areas and the south are more socially liberal. The north is more responsive to traditional Roman Catholic values. Family ties remain important.

INSIGHT: *Portugal is the world's leading producer of cork, which comes from the bark of the cork oak*

THE ECONOMY
Tourism. Exports of vegetables, fruit, wine, cars, and clothing. Mounting debt forced EU bailout in 2011 and tough cuts to reduce the budget deficit.

1000m/3281ft
500m/1640ft
200m/656ft
Sea Level

FACTFILE

OFFICIAL NAME: Portuguese Republic
DATE OF FORMATION: 1139
CAPITAL: Lisbon
POPULATION: 10.6 million
TOTAL AREA: 35,672 sq. miles (92,391 sq. km)
DENSITY: 299 people per sq. mile

LANGUAGES: Portuguese*
RELIGIONS: Roman Catholic 92%, Protestant 4%, nonreligious 3%, other 1%
ETHNIC MIX: Portuguese 98%, African and other 2%
GOVERNMENT: Parliamentary system
CURRENCY: Euro = 100 cents

Qatar

Qatar projects from the Arabian Peninsula into the Persian Gulf. A founding member of OPEC, it is one of the region's wealthiest states due to oil and natural gas exports.

GEOGRAPHY

Flat, semiarid desert with dunes and salt pans. Vegetation is limited to small patches of scrub.

CLIMATE

Hot and humid. Temperatures in summer can soar to over 104°F (40°C). Rainfall is rare.

PEOPLE & SOCIETY

Only one in five residents is native-born; the rest are guest workers from across the Middle East, the Indian subcontinent, Southeast Asia, and north Africa. Qataris were once nomadic Bedouins, but since the advent of oil wealth, most now live in Doha and its suburbs, leaving the north dotted with abandoned villages. Women enjoy relative freedom; most wear the veil.

◆ **INSIGHT:** *There are three times as many men as women in Qatar*

THE ECONOMY
Steady supply of crude oil and huge natural gas reserves, plus related industries. All other raw materials and most foods are imported. Strong GDP growth in 2004–2011. Economy is heavily dependent on foreign workforce.

FACTFILE

OFFICIAL NAME: State of Qatar
DATE OF FORMATION: 1971
CAPITAL: Doha
POPULATION: 2.2 million
TOTAL AREA: 4416 sq. miles (11,437 sq. km)
DENSITY: 518 people per sq. mile

LANGUAGES: Arabic*
RELIGIONS: Muslim (mainly Sunni) 95%, other 5%
ETHNIC MIX: Qatari 20%, other Arab 20%, Indian 20%, Nepalese 13%, Filipino 10%, other 10%, Pakistani 7%
GOVERNMENT: Monarchy
CURRENCY: Qatar riyal = 100 dirhams

Romania

Once dominated by Poles, Hungarians, and Ottomans, Romania has been slowly converting to a market economy since the 1989 overthrow of its communist regime. It joined the EU in 2007.

GEOGRAPHY

Carpathian Mountains encircle the Transylvanian plateau. Wide plains to the south and east. Danube River forms southern border.

CLIMATE

Continental. Summers are hot and humid, winters are cold and snowy. Very heavy spring rains.

PEOPLE & SOCIETY

Romanians are ethnically distinct from their Slav and Hungarian (Magyar) neighbors. Hungarians are the largest minority, living mainly in Transylvania. They are protected by the influence of Hungary, unlike the Roma, who suffer discrimination. Net emigration (since EU membership) is slowing. Low birth rate.

 INSIGHT: *In 2001, Romania became the last country in Europe to lift its ban on homosexuality*

THE ECONOMY

Polluting, outdated heavy industry. Unmechanized agricultural sector. Textile and metal exports led growth in 2000s. High budget deficits exposed economy in 2007–2009 global downturn: IMF bailout, austerity measures. Plans to join euro in 2019. Privatization continues.

■	2000m/6562ft
■	1000m/3281ft
■	500m/1640ft
■	200m/656ft
□	Sea Level

FACTFILE

OFFICIAL NAME: Romania

DATE OF FORMATION: 1878

CAPITAL: Bucharest

POPULATION: 21.7 million

TOTAL AREA: 91,699 sq. miles (237,500 sq. km)

DENSITY: 244 people per sq. mile

LANGUAGES: Romanian*, Hungarian (Magyar), Romani, German

RELIGIONS: Romanian Orthodox 87%, Roman Catholic 5%, Protestant 5%, Greek Orthodox 1%, Uniate 1%, other 1%

ETHNIC MIX: Romanian 89%, Magyar 7%, Roma 3%, other 1%

GOVERNMENT: Presidential system

CURRENCY: New Romanian leu = 100 bani

Russian Federation

The Russian Federation was the core of the old Soviet Union, which broke up in 1991. Russia is still the world's largest state. Its diversity is a source of both strength and problems.

 GEOGRAPHY
The Ural Mountains divide the European steppes and forests from the tundra and forests of Siberia. South-central deserts and mountains.

CLIMATE
Continental in European Russia, with warm summers and freezing winters. Elsewhere climate ranges from sub-arctic to Mediterranean and hot desert.

PEOPLE & SOCIETY
57 "nationalities" and 95 minorities in addition to ethnic Russians. Separatism suppressed. Population predicted to fall to 107 million by 2050. HIV/AIDS rising.

THE ECONOMY
Vast resources (oil, gas, metals, timber). Inefficient industry, agriculture. Tax evasion. Black market and organized crime. Wealth disparities. 2009 recession

INSIGHT: *The Trans-Siberian Railroad, running 5578 miles (9297 km) from Moscow to Vladivostok, is the longest in the world, traversing eight time zones*

3000m/9843ft
2000m/6562ft
1000m/3281ft
500m/1640ft
200m/656ft
Sea Level
Below Sea Level

0   1000 km
0   1000 miles

FACTFILE

OFFICIAL NAME: Russian Federation
DATE OF FORMATION: 1480
CAPITAL: Moscow
POPULATION: 143 million
TOTAL AREA: 6,592,735 sq. miles (17,075,200 sq. km)
DENSITY: 22 people per sq. mile

LANGUAGES: Russian*, Tatar, Ukrainian, other
RELIGIONS: Orthodox Christian 75%, Muslim 14%, other 11%
ETHNIC MIX: Russian 80%, other 12%, Tatar 4%, Ukrainian 2%, Chavash 1%, Bashkir 1%
GOVERNMENT: Mixed presidential–parliamentary system
CURRENCY: Russian rouble = 100 kopeks

Rwanda

Rwanda lies just south of the equator in east central Africa, far from the nearest sea port. Since independence from France in 1962, ethnic tensions have dominated politics.

GEOGRAPHY

A series of plateaus descend from the ridge of volcanic peaks in the west to the Akagera River on the eastern border. The Great Rift Valley also passes through this region.

CLIMATE

Tropical, though tempered by the altitude. Two wet seasons are separated by a dry season, from June to August. Heaviest rain in the west.

PEOPLE & SOCIETY

For over 500 years the cattle-owning Tutsi minority were politically dominant over the land-owning Hutu. In 1959, violent revolt led to a reversal of the roles. Ethnic tensions are fierce; in the most recent violence, in 1994, over 800,000 people, mostly Tutsi, were massacred in an act of state-backed genocide; trials are ongoing. Most people live at subsistence level.

THE ECONOMY

Reliant on aid. Production of tea and speciality coffee is booming. Decade of strong GDP growth. Exports tin, coltan, and iron ore. Ecotourism is growing. Possible oil and gas reserves. Landlocked: high transportation costs.

INSIGHT: *Rwanda's parliament in 2008 was the first in the world to have more women members than men*

FACTFILE

OFFICIAL NAME: Republic of Rwanda

DATE OF FORMATION: 1962

CAPITAL: Kigali

POPULATION: 11.8 million

TOTAL AREA: 10,169 sq. miles (26,338 sq. km)

DENSITY: 1225 people per sq. mile

LANGUAGES: Kinyarwanda*, French*, Kiswahili, English*

RELIGIONS: Christian 94%, Muslim 5% traditional beliefs 1%

ETHNIC MIX: Hutu 85%, Tutsi 14%, other (including Twa) 1%

GOVERNMENT: Presidential system

CURRENCY: Rwanda franc = 100 centimes

St. Kitts & Nevis

A popular Caribbean tourist destination, St. Kitts and Nevis lies in the northern part of the Leeward Island chain. Nevis is the smaller and less developed of the two islands.

GEOGRAPHY

Volcanic in origin, with forested, mountainous interiors. Nevis has hot and cold springs.

CLIMATE

Tropical, tempered by trade winds. Little seasonal variation in temperature. Moderate rainfall.

PEOPLE & SOCIETY

The majority of the population are descended from former African slaves. There are small numbers of Europeans, and South Asians, and a community of Lebanese. Levels of emigration are high, and overseas remittances are an important source of national income. The government has pledged to retrain sugar workers. Native professionals and civil servants have largely replaced the former expatriate elite. The secessionist movement on Nevis remains an issue.

THE ECONOMY

Successful tourist industry is vulnerable to downturns in US market. Financial services. Once-key sugar industry closed down in 2005.

INSIGHT: *Nevis has been renowned as a spa since the 18th century, and is known as the "Queen of the Caribbean"*

FACTFILE

OFFICIAL NAME: Federation of Saint Christopher and Nevis

DATE OF FORMATION: 1983

CAPITAL: Basseterre

POPULATION: 51,134

TOTAL AREA: 101 sq. miles (261 sq. km)

DENSITY: 368 people per sq. mile

LANGUAGES: English*, English Creole

RELIGIONS: Anglican 33%, Methodist 29%, other 22%, Moravian 9%, Roman Catholic 7%

ETHNIC MIX: Black 95%, mixed race 3%, White 1%, other and Amerindian 1%

GOVERNMENT: Parliamentary system

CURRENCY: East Caribbean dollar = 100 cents

St. Lucia

St. Lucia is one of the most beautiful of the Caribbean Windward Islands. Ruled by France and the UK at different times in its past, the island retains the influences of both.

GEOGRAPHY

Volcanic and mountainous, with some broad fertile valleys. The Pitons, ancient lava cones, rise from the sea on the forested west coast.

CLIMATE

Tropical, moderated by trade winds. May–October wet season brings daily warm showers. Rainfall is highest in the mountains.

PEOPLE & SOCIETY

The population is a tension-free mixture of descendants of Africans, Caribs, and Europeans. Family life and the Roman Catholic Church are important to most St. Lucians. In rural areas, women often head the households and run much of the farming. Plantation and hotel owners are the richest group. There is growing local resistance to overdevelopment of the island for tourism.

THE ECONOMY

Bananas are still biggest export, but struggling to compete since loss of preferential access to EU market. Successful tourism. Offshore banking.

INSIGHT: *St. Lucia has two Nobel laureates, the most per capita in the world*

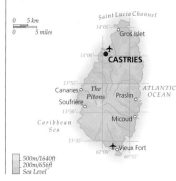

FACTFILE

OFFICIAL NAME: Saint Lucia
DATE OF FORMATION: 1979
CAPITAL: Castries
POPULATION: 162,781
TOTAL AREA: 239 sq. miles
(620 sq. km)
DENSITY: 690 people per sq. mile

LANGUAGES: English*, French Creole
RELIGIONS: Roman Catholic 90%, other 10%
ETHNIC MIX: Black 83%, *Mulatto* (mixed race) 13%, Asian 3%, other 1%
GOVERNMENT: Parliamentary system
CURRENCY: East Caribbean dollar = 100 cents

St. Vincent & the Grenadines

The islands of St. Vincent and the Grenadines form part of the Windward group in the Caribbean. St. Vincent is mostly volcanic, while the Grenadines are flat, mainly bare, coral reefs.

GEOGRAPHY
St. Vincent is mountainous and forested, with one of two active volcanoes in the Caribbean, La Soufrière. The Grenadines are 32 islands and cays, fringed by beaches.

CLIMATE
Tropical, with constant trade winds. Hurricanes are likely during the wet season in July–November.

PEOPLE & SOCIETY
Population is racially diverse; intermarriage has reduced tensions. Society is informal and relaxed, but family life is strongly influenced by the Christian Church. Locals fear that their traditional lifestyle is being threatened by the expanding tourist industry.

◈ **INSIGHT:** *The islands' precolonial inhabitants, the Carib, named them "Hairoun" – home of the blessed*

THE ECONOMY
Dependent on agriculture and tourism. Bananas are the main cash crop. Tourism, targeted at the jet-set and cruise-ship markets, is concentrated on the Grenadines.

FACTFILE

OFFICIAL NAME: Saint Vincent and the Grenadines
DATE OF FORMATION: 1979
CAPITAL: Kingstown
POPULATION: 103,220
TOTAL AREA: 150 sq. miles (389 sq. km)
DENSITY: 788 people per sq. mile

LANGUAGES: English*, English Creole
RELIGIONS: Anglican 47%, Methodist 28%, Roman Catholic 13%, other 12%
ETHNIC MIX: Black 66%, Mulatto (mixed race) 19%, other 12%, Carib 2%, Asian 1%
GOVERNMENT: Parliamentary system
CURRENCY: East Caribbean dollar = 100 cents

Samoa

The Pacific islands of Samoa gained independence from New Zealand in 1962. Four of the nine volcanic islands are inhabited – Apolima, Manono, Savai'i, and Upolu.

GEOGRAPHY

Comprises two large islands and seven smaller ones. The two largest islands have rainforested, mountainous interiors surrounded by coastal lowlands and coral reefs.

CLIMATE

Tropical, with high humidity. Cooler in May–November. Cyclone season is December–March.

PEOPLE & SOCIETY

Ethnic Samoans are the world's second-largest Polynesian group, after the Maoris. Their way of life is communal and formalized. Extended family groups own 80% of the land. Each family has an elected chief, who looks after its political and social interests. Large-scale migration to the US and New Zealand reflects the country's lack of jobs and the attractions of a Western lifestyle.

THE ECONOMY

Exports fish, coconut products (oil, cream, copra), and nonu fruit. Growth of tourism, offshore banking, and light manufacturing (Japanese car parts). Dependent on aid and expatriate remittances. Rainforests are increasingly exploited for timber.

INSIGHT: *Samoa was named for the sacred (sa) chickens (moa) of Lu, son of Tagaloa, the god of creation*

FACTFILE

OFFICIAL NAME: Independent State of Samoa

DATE OF FORMATION: 1962

CAPITAL: Apia

POPULATION: 200,000

TOTAL AREA: 1104 sq. miles (2860 sq. km)

DENSITY: 183 people per sq. mile

LANGUAGES: Samoan*, English*

RELIGIONS: Christian 99%, other 1%

ETHNIC MIX: Polynesian 91%, Euronesian (mixed European and Polynesian) 7%, other 2%

GOVERNMENT: Parliamentary system

CURRENCY: Tala = 100 sene

San Marino

Perched on the slopes of Monte Titano in the Italian Appennines, San Marino has maintained its independence since the 4th century CE, but Italy effectively controls most of its affairs.

GEOGRAPHY
Distinctive limestone outcrop of Monte Titano dominates wooded hills and pastures near Italy's Adriatic coast.

CLIMATE
High altitude and sea breezes moderate a Mediterranean climate. Hot summers and cool, wet winters.

PEOPLE & SOCIETY
Territory is divided into nine "castles," or districts. Tightly knit society, with 16 centuries of tradition. Strict immigration rules require 30-year residence before applying for citizenship. Living standards are similar to those in northern Italy. About 12,000 Sammarinesi live abroad, most in Italy.

INSIGHT: *Sales of postage stamps and coins contribute around 10% of the national income*

THE ECONOMY
Tourism, banking, manufacturing, and investment all hit by 2008–2009 global downturn. Banking transparency has improved. Lower tax rates than Italy. Wine, cheese, olive oil, textiles, and ceramics are exported. Also relies on Italian subsidy and infrastructure.

44°

Dogana
Serravalle
Fiorina
Cailungo
Gualdicciolo
Borgo Maggiore
ITALY   Monte Titano ▲   ● **SAN MARINO**
2424ft (739m)
Faetano
Murata   *A p p e n n i n o*   **ITALY**
Chiesanuova   Montegiardino

12°30′

500m/1640ft
200m/656ft
Sea Level

0   4 km
0   4 miles

FACTFILE

OFFICIAL NAME: Republic of San Marino
DATE OF FORMATION: 1631
CAPITAL: San Marino
POPULATION: 32,448
TOTAL AREA: 23.6 sq. miles (61 sq. km)

DENSITY: 1352 people per sq. mile
LANGUAGES: Italian*
RELIGIONS: Roman Catholic 93%, other and nonreligious 7%
ETHNIC MIX: Sammarinese 88%, Italian 10%, other 2%
GOVERNMENT: Parliamentary system
CURRENCY: Euro = 100 cents

Sao Tome & Principe

A former Portuguese colony, Sao Tome and Principe comprises two main islands and surrounding islets, off the west coast of Africa. Elections in 1991 ended 15 years of Marxism.

GEOGRAPHY
Islands scattered across the equator. Sao Tome and Principe are heavily forested and mountainous.

CLIMATE
Hot and humid, but cooled by the Benguela Current. Plentiful rainfall.

PEOPLE & SOCIETY
Population is mostly black, though Portuguese culture pre-dominates. Blacks run the political parties. Society is well integrated and free from racial prejudice. Príncipe assumed autonomous status in 1995. There is a growing business class. The extended family offers the main form of social security. One of Africa's highest aid-to-population ratios.

INSIGHT: *The population is entirely of immigrant descent: the islands were uninhabited when colonized in 1470*

THE ECONOMY
Cocoa provides 90% of export earnings. Coconuts, pepper, coffee also farmed. Tourism. Reliant on aid. Offshore oil expected to come onstream shortly.

FACTFILE

OFFICIAL NAME: Democratic Republic of Sao Tome and Principe

DATE OF FORMATION: 1975

CAPITAL: São Tomé

POPULATION: 200,000

TOTAL AREA: 386 sq. miles (1001 sq. km)

DENSITY: 539 people per sq. mile

LANGUAGES: Portuguese Creole, Portuguese*

RELIGIONS: Roman Catholic 84%, other 16%

ETHNIC MIX: Black 90%, Portuguese and Creole 10%

GOVERNMENT: Presidential system

CURRENCY: Dobra = 100 céntimos

Saudi Arabia

Occupying most of the Arabian Peninsula, Saudi Arabia covers an area the size of western Europe. It is the world's largest oil producer and has a major petrochemicals industry.

GEOGRAPHY

Mostly desert or semidesert plateau. Mountain ranges in the west run parallel to the Red Sea and drop steeply to a coastal plain.

CLIMATE

In summer, temperatures often soar above 118°F (48°C), but in winter they may fall below freezing. Rainfall is rare.

PEOPLE & SOCIETY

Most Saudis are Sunni Muslims who embrace *sharia* (Islamic law) and follow the strictly orthodox Wahhabi interpretation of Islam in their daily lives. Women are obliged to wear the veil, cannot hold a driver's license, and have no role in public life. The al-Sa'ud family rules with absolute power. Supported by the religious establishment, it controls all political life and makes few concessions to any calls for wider public participation.

THE ECONOMY

Vast oil and natural gas reserves. A third of workers are foreign. Attractive jobs for young Saudis are scarce, however.

INSIGHT: *Three million Muslims a year make the hajj (pilgrimage) to the holy city of Mecca. Only practicing Muslims are allowed inside the city*

FACTFILE

OFFICIAL NAME: Kingdom of Saudi Arabia
DATE OF FORMATION: 1932
CAPITAL: Riyadh
POPULATION: 28.8 million
TOTAL AREA: 756,981 sq. miles
(1,960,582 sq. km)
DENSITY: 35 people per sq. mile

LANGUAGES: Arabic*
RELIGIONS: (Native population) Sunni Muslim 85%, Shi'a Muslim 15%
ETHNIC MIX: Arab 72%, foreign residents (mostly south or southeast Asian) 20%, Afro-Asian 8%
GOVERNMENT: Monarchy
CURRENCY: Saudi riyal = 100 halalat

Senegal

Senegal's capital, Dakar, stands on the westernmost cape of Africa. After independence from France, Senegal became a single-party state, but it has had multiparty elections since 1981.

GEOGRAPHY

Arid semidesert in the north. The south is mainly savanna bushland. Plains in the southeast.

CLIMATE

Tropical, with humid rainy conditions June–October, and a drier season December–May. The coast is cooled by northern trade winds.

PEOPLE & SOCIETY

Interethnic marriage has reduced ethnic tensions. Groups can be identified regionally. Dakar is a Wolof area, with the Serer concentrated to the east and southeast of Dakar. The Senegal River is dominated by the Peul and Toucouleur. The Diola (Jola) in Casamance have felt politically excluded, prompting a long-running secessionist struggle; a cease-fire in 2014 may signal the conflict's end. A large diaspora raises global awareness of Senegalese culture and music.

THE ECONOMY

Good infrastructure, particularly port at Dakar. Fishing (though stocks diminishing). Remittances. Phosphate mining. Groundnuts. Development of tourism. Oil potential off Casamance.

INSIGHT: *Senegal's name derives from the Muslim Zenega Berbers who invaded in the 1300s*

FACTFILE

OFFICIAL NAME: Republic of Senegal

DATE OF FORMATION: 1960

CAPITAL: Dakar

POPULATION: 14.1 million

TOTAL AREA: 75,749 sq. miles (196,190 sq. km)

DENSITY: 190 people per sq. mile

LANGUAGES: Wolof, Serer, Pulaar, Diola, Mandinka, Malinké, Soninké, French*

RELIGIONS: Sunni Muslim 95%, Christian (mainly Catholic) 4%, traditional beliefs 1%

ETHNIC MIX: Wolof 43%, Serer 15%, other 14%, Peul 14%, Toucouleur 9%, Diola 5%

GOVERNMENT: Presidential system

CURRENCY: CFA franc = 100 centimes

Serbia

The central and eastern region of what was once Yugoslavia, Serbia was a pariah state until Slobodan Milosevic was ousted in 2000. Montenegro broke away in 2006, and Kosovo in 2008.

GEOGRAPHY

Landlocked since secession of Montenegro. Fertile Danube plain in the north, rolling uplands in the center and southeast. Mountains in southwest.

CLIMATE

Continental in north, with wet springs and warm summers. Colder winters with heavy snow in south.

PEOPLE & SOCIETY

Serbs are Orthodox Christian and use the Cyrillic script. Catholic Magyars (Hungarians) in Vojvodina have some autonomy. Society was severely shaken in the 1990s by interethnic conflict. After Serbia's cooperation in apprehending suspected war criminals, it was granted EU candidate status, but the issue of Kosovo is a major obstacle to accession.

INSIGHT: *The medieval Serbian Empire reached into northern Greece*

THE ECONOMY

Recovery from sanctions and 1999 NATO bombing: GDP only returned to pre-1990 level by 2006. Reserves of coal, oil. Strong industrial base. Privatization ongoing. Foreign investment growing. Danube is a key transportation link.

FACTFILE

OFFICIAL NAME: Republic of Serbia

DATE OF FORMATION: 2006

CAPITAL: Belgrade

POPULATION: 9.5 million

TOTAL AREA: 29,905 sq. miles (77,453 sq. km)

DENSITY: 318 people per sq. mile

LANGUAGES: Serbian*, Hungarian (Magyar)

RELIGIONS: Orthodox Christian 85%, other 6%, Roman Catholic 6%, Muslim 3%

ETHNIC MIX: Serb 83%, other 10%, Magyar 4%, Bosniak 2%, Roma 1%

GOVERNMENT: Parliamentary system

CURRENCY: Serbian dinar = 100 para

Seychelles

Formerly a UK colony, the Seychelles comprises 115 islands in the Indian Ocean. After 14 years as a one-party state, multiparty elections were introduced in 1993.

GEOGRAPHY

Mostly low-lying coral atolls, but 40, including the largest, Mahé, are mountainous and are the only granitic midocean islands in the world.

CLIMATE

Tropical oceanic climate. Hot and humid. Rainy season December–May.

PEOPLE & SOCIETY

The islands were uninhabited when French settlers arrived in the 18th century. Today, the population is homogeneous – a result of inter-marriage between ethnic groups. Almost 90% of people live on Mahé. Living standards are among Africa's highest. Poverty is rare and the welfare system caters to all.

INSIGHT: *The Seychelles' unique species include the coco-de-mer palm, which produces the world's largest seeds*

THE ECONOMY

Tourism is main sector, based on appeal of beaches and exotic wildlife. Tuna is fished and canned for export. Re-export trade. All domestic requirements are imported. Virtually no mineral resources. High debt-servicing burden. Lack of foreign exchange.

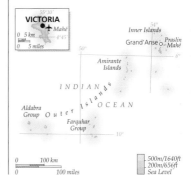

■	500m/1640ft
■	200m/656ft
	Sea Level

FACTFILE

OFFICIAL NAME: Republic of Seychelles

DATE OF FORMATION: 1976

CAPITAL: Victoria

POPULATION: 90,846

TOTAL AREA: 176 sq. miles (455 sq. km)

DENSITY: 874 people per sq. mile

LANGUAGES: French Creole*, English*, French*

RELIGIONS: Roman Catholic 82%, Anglican 6%, other (including Muslim) 6%, other Christian 4%, Hindu 2%

ETHNIC MIX: Creole 89%, Indian 5%, other 4%, Chinese 2%

GOVERNMENT: Presidential system

CURRENCY: Seychelles rupee = 100 cents

Sierra Leone

The west African state of Sierra Leone achieved independence from the UK in 1961. Today, trying to recover from ten years of devastating civil war, it is one of the world's poorest nations.

GEOGRAPHY

Flat plain, running the length of the coast, stretches inland for 83 miles (133 km). Beyond, forests rise to highlands near neighboring Guinea in the northeast.

CLIMATE

Hot tropical weather, with very high rainfall and humidity. The dusty, northeastern *harmattan* wind blows November–April.

PEOPLE & SOCIETY

Mende and Temne are the major ethnic groups. Freetown's citizens are largely descended from slaves freed from Britain and the US, resulting in a strongly Anglicized Creole culture in the capital. The countryside is less developed. A brutal civil war broke out in 1991 and was not properly resolved until a 2001 peace agreement. Two million people were displaced during the conflict. A deadly Ebola outbreak hit the country in 2014.

THE ECONOMY

Aid is vital: reconstruction will take years. Diamond exports, though smuggling is rife. Rutile and bauxite also mined. Coffee and cocoa are cash crops, but most farming is subsistence.

INSIGHT: *The British philanthropist Granville Sharp set up a settlement for freed slaves in Freetown in 1787*

FACTFILE

OFFICIAL NAME: Republic of Sierra Leone

DATE OF FORMATION: 1961

CAPITAL: Freetown

POPULATION: 6.1 million

TOTAL AREA: 27,698 sq. miles (71,740 sq. km)

DENSITY: 221 people per sq. mile

LANGUAGES: Mende, Temne, Krio, English*

RELIGIONS: Muslim 60%, Christian 30%, traditional beliefs 10%

ETHNIC MIX: Mende 35%, Temne 32%, other 21%, Limba 8%, Kuranko 4%

GOVERNMENT: Presidential system

CURRENCY: Leone = 100 cents

Singapore

Linked to the southernmost tip of the Malay peninsula by a causeway, Singapore was established as a trading settlement in 1819. It is now one of Asia's most important commercial centers.

GEOGRAPHY
Little remains of the original vegetation on Singapore Island. The other 54 much smaller islands are little more than swampy jungle.

CLIMATE
Equatorial. Hot and humid, with heavy rainfall all year round.

PEOPLE & SOCIETY
Chinese majority includes old-established English-speaking Straits Chinese and more recent immigrants. Median income is highest in Indian households and lowest in Malay households. Significant expatriate workforce. Aging population: cash incentives, longer maternity leave aim to boost birth rate. Society is highly regulated; official campaigns aim to improve public behavior. Crime is low; punishment can be severe. Living standards are among world's highest.

THE ECONOMY
Wealth from success as entrepôt and center of high-tech industries, such as electronics and pharmaceuticals. Leads research in new biotechnologies. All food, energy, and water imported. Worst-ever recession in 2008–2009.

INSIGHT: *Chewing gum was banned outright from 1992 to 2004*

Urban areas
Open areas
Nature reserves

FACTFILE
OFFICIAL NAME: Republic of Singapore
DATE OF FORMATION: 1965
CAPITAL: Singapore
POPULATION: 5.4 million
TOTAL AREA: 250 sq. miles (648 sq. km)
DENSITY: 22,881 people per sq. mile

LANGUAGES: Mandarin*, Malay*, Tamil*, English*
RELIGIONS: Buddhist 55%, Taoist 22%, Muslim 16%, Hindu, Christian, and Sikh 7%
ETHNIC MIX: Chinese 74%, Malay 14%, Indian 9%, other 3%
GOVERNMENT: Parliamentary system
CURRENCY: Singapore dollar = 100 cents

Slovakia

Landlocked in central Europe, Slovakia became a separate state in 1993, splitting ex-communist Czechoslovakia in two. It joined the EU in 2004 and the eurozone five years later.

GEOGRAPHY

The Tatra Mountains stretch along the northern border with Poland. Southern lowlands include the fertile Danube plain.

CLIMATE

Continental. Moderately warm summers and steady rainfall. Cold winters with heavy snowfalls.

PEOPLE & SOCIETY

The majority Slovaks are the dominant group. The Magyars (Hungarians) seek protection of their language and culture, backed by Hungary. Magyar parties exist in the political mainstream, and on occasion form part of the ruling coalition. Ethnic Czechs have dual citizenship. Roma are unrepresented and face significant discrimination. Rural eastern regions are least developed.

THE ECONOMY

Heavy industry, especially cars. Exports hit by 2007–2009 global downturn. Inexpensive workforce. Rising foreign investment. High unemployment, budget deficits. Successful privatizations.

INSIGHT: *From 1526 to 1784 Bratislava, then known as Pozsony, served as the capital of Hungary*

FACTFILE

OFFICIAL NAME: Slovak Republic
DATE OF FORMATION: 1993
CAPITAL: Bratislava
POPULATION: 5.5 million
TOTAL AREA: 18,859 sq. miles (48,845 sq. km)
DENSITY: 290 people per sq. mile

LANGUAGES: Slovak*, Hungarian (Magyar), Czech
RELIGIONS: Roman Catholic 69%, other 13%, nonreligious 13%, Greek Catholic (Uniate) 4%, Orthodox Christian 1%
ETHNIC MIX: Slovak 86%, Magyar 10%, Roma 2%, Czech 1%, other 1%
GOVERNMENT: Parliamentary system
CURRENCY: Euro = 100 cents

Slovenia

Lying at the junction of central Europe and the Balkans, Slovenia seceded from socialist Yugoslavia in 1991. In 2004, it became the first former Yugoslav state to join the EU.

GEOGRAPHY
Alpine terrain with hills and mountains. Forests cover almost half the country's area. There is a short coastline on the Adriatic Sea.

CLIMATE
Mediterranean climate on the small coastal strip. The alpine interior has continental extremes.

PEOPLE & SOCIETY
Long historical association with western Europe, accounts for the "Alpine" rather than "Balkan" outlook of Slovenia's people, despite close similarities to other former Yugoslavs. The absence of sizable Serb or Croat minorities made for a relatively peaceful secession from Yugoslavia. There are small communities of Italians and Magyars (Hungarians) in the southwest and east respectively.

THE ECONOMY
First new EU member to join eurozone (in 2007). Export-oriented, so vulnerable to global economic trends. Competitive manufacturing industry. Sizable state-owned sector remains.

INSIGHT: *A wheel found in a marsh in 2003 is claimed to be the world's oldest, pre-dating 3000 BCE*

1000m/3281ft
500m/1640ft
200m/656ft
Sea Level

HUNGARY
AUSTRIA
Murska Sobota
Maribor
Drava
Mura
Jesenice
Ptuj
Kranj
Celje
LJUBLJANA
Sava
Nova Gorica
Krško
Brežice
Postojna
ITALY
Adriatic Sea
Kolpa
CROATIA

0   25 km
0   25 miles

FACTFILE

OFFICIAL NAME: Republic of Slovenia
DATE OF FORMATION: 1991
CAPITAL: Ljubljana
POPULATION: 2.1 million
TOTAL AREA: 7820 sq. miles (20,253 sq. km)
DENSITY: 269 people per sq. mile

LANGUAGES: Slovenian*
RELIGIONS: Roman Catholic 58%, other 28%, Atheist 10%, Orthodox Christian 2%, Muslim 2%
ETHNIC MIX: Slovene 83%, other 12%, Serb 2%, Croat 2%, Bosniak 1%
GOVERNMENT: Parliamentary system
CURRENCY: Euro = 100 cents

Solomon Islands

The Solomons archipelago comprises several hundred coral reef islands scattered in the southwestern Pacific. Most of the population live on the six largest islands.

GEOGRAPHY
The six largest islands are volcanic, mountainous, and thickly forested. Flat coastal plains provide the only cultivable land.

CLIMATE
Northern islands are hot and humid all year round; farther south a cool season develops. November–April wet season brings cyclones.

PEOPLE & SOCIETY
Almost all Solomon Islanders are Melanesian. Animist beliefs exist alongside Christianity. Tensions are regional; Guadalcanal natives (Isatabu) fought against immigrant Malaitan workers in the 1998–2000 conflict, displacing thousands and ruining the economy. In 2003, Australian-led peacekeepers arrived. A new devolved "state system" has granted outlying islands more autonomy and brought a semblance of stability.

THE ECONOMY
Subsistence farming and fishing sustain 75% of people. Cash crops are copra and cocoa. Gold deposits. Civil conflict bankrupted the government, closed the main gold mine, and cut trade links. Forests have been depleted.

◆ INSIGHT: *The battle for Japanese-held Guadalcanal was the first major US offensive in the Pacific War during World War II*

FACTFILE

OFFICIAL NAME: Solomon Islands
DATE OF FORMATION: 1978
CAPITAL: Honiara
POPULATION: 600,000
TOTAL AREA: 10,985 sq. miles (28,450 sq. km)
DENSITY: 56 people per sq. mile
LANGUAGES: English*, Pidgin English, Melanesian Pidgin, c. 120 others
RELIGIONS: Church of Melanesia (Anglican) 34%, Roman Catholic 19%, other 19%, South Seas Evangelical Church 17%, Methodist 11%
ETHNIC MIX: Melanesian 93%, Polynesian 4%, Micronesian 2%, other 1%
GOVERNMENT: Parliamentary system
CURRENCY: Solomon Is. dollar = 100 cents

Somalia

A semiarid state occupying the Horn of Africa, Somalia was formed from the Italian and British colonies of Somaliland. Conflict has left it without effective government since 1991.

GEOGRAPHY

Highlands in the north, flatter scrub-covered land to the south. Coastal areas are more fertile.

CLIMATE

Very dry, except for the north coast, which is hot and humid. The interior has among the world's highest average annual temperatures.

PEOPLE & SOCIETY

The clan system forms the basis of commercial, political, and social life. The minority Bantu are traditionally seen as socially inferior to Somalis. Since the 1991 coup, Somalia has lacked strong central authority. Somaliland claims independence, and Puntland autonomy. Islamist militias controlled rump Somalia by 2009, and al-Shabab still holds sway in the south. A new federal structure was formulated in 2012 ahead of the latest attempt to form a national government.

THE ECONOMY

Ongoing war. All goods, except arms, are in short supply. Piracy, banditry. Few natural resources. Prone to drought; latest famine in 2011–2012 killed 260,000. Somaliland is more stable, but its trade is hampered by lack of global recognition.

◆ **INSIGHT:** *Until 1973, Somali was an unwritten language*

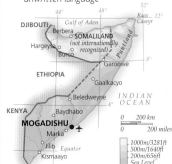

FACTFILE

OFFICIAL NAME: Federal Republic of Somalia
DATE OF FORMATION: 1960
CAPITAL: Mogadishu
POPULATION: 10.5 million
TOTAL AREA: 246,199 sq. miles (637,657 sq. km)

DENSITY: 43 people per sq. mile
LANGUAGES: Somali*, Arabic*, English, Italian
RELIGIONS: Sunni Muslim 99%, Christian 1%
ETHNIC MIX: Somali 85%, other 15%
GOVERNMENT: Nonparty system
CURRENCY: Somali shilin = 100 senti

South Africa

After 80 years of white minority rule, South Africa held its first multiracial, multiparty elections in 1994. Victory for the blacks marked the symbolic overturning of long years of apartheid.

GEOGRAPHY

Much of the interior is grassy *veld*. Desert in the west and far north. Mountains east, south, and west.

CLIMATE

Warm, temperate, and dry. Cape Town has a Mediterranean climate. Semiarid in the west.

PEOPLE & SOCIETY
The majority black population now dominates politically, but the minority white community still controls the economy. A small black middle class is growing, but unemployment among blacks remains high. Over five million people are HIV-positive, but the fight against AIDS is hampered by social attitudes. Violent crime is a problem.

INSIGHT: *Over the last century, South Africa has produced over half of the world's gold*

THE ECONOMY
Africa's largest, most developed economy. Leading mineral producer, notably metals, diamonds, coal. Tourism is also key. Wealth gap has widened: jobs, housing, and better access to basic services are needed to fight poverty.

FACTFILE

OFFICIAL NAME: Republic of South Africa
DATE OF FORMATION: 1934
CAPITALS: Pretoria; Cape Town; Bloemfontein
POPULATION: 52.8 million
TOTAL AREA: 471,008 sq. miles (1,219,912 sq. km)
DENSITY: 112 people per sq. mile

LANGUAGES: English*, isiZulu*, isiXhosa*, Afrikaans*, 7 other official languages*
RELIGIONS: Christian 68%, animist and traditional beliefs 29%, Muslim 2%, Hindu 1%
ETHNIC MIX: Black 80%, White 9%, Colored 9%, Asian 2%
GOVERNMENT: Presidential system
CURRENCY: Rand = 100 cents

South Sudan

A long civil war in Sudan led to independence in 2011 for the mainly Christian southern part. The landlocked new state is poor and lacks vital infrastructure, despite its oil reserves.

GEOGRAPHY
The White Nile flows through South Sudan, from remote forest areas into the world's largest swamp, the Sudd.

CLIMATE
Tropical South Sudan's long, heavy rains result in some areas getting cut off. January to March is drier.

PEOPLE & SOCIETY
Most people are subsistence farmers. Village life is based on extended families; arranged marriages involve the payment of bride-price. There are over 60 language groups. The common cause of independence engendered ethnic unity. The Sudanese People's Liberation Movement, whose armed wing led the fighting, holds power in the new country, but in 2013 a fallout between president and vice president spiraled into civil war, exposing ethnic divisions between Dinka and Nuer.

THE ECONOMY
Needs foreign aid for humanitarian crisis and development. Issues over oil revenue and borders remain unresolved with Sudan, which controls sole oil export pipeline. Inherited foreign debt.

INSIGHT: *Decades of fighting from 1983 left over four million internally displaced*

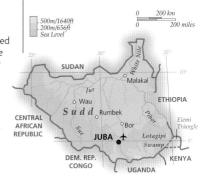

500m/1640ft
200m/656ft
Sea Level

0   200 km
0   200 miles

SUDAN

White Nile

Malakal

Jur

Wau

Sudd   Rumbek

Sue   Bor

Pibor

JUBA

CENTRAL
AFRICAN
REPUBLIC

ETHIOPIA

*Elemi
Triangle*

*Lotagipi
Swamp*

DEM. REP.
CONGO   UGANDA   KENYA

FACTFILE

OFFICIAL NAME: Republic of South Sudan

DATE OF FORMATION: 2011

CAPITAL: Juba

POPULATION: 11.3 million

TOTAL AREA: 248,777 sq. miles
(644,329 sq. km)

DENSITY: 45 people per sq. mile

LANGUAGES: English*, Arabic, Dinka, Nuer, Zande, Bari, Shilluk, Lotuko

RELIGIONS: Over half of the population follow Christian or traditional beliefs

ETHNIC MIX: Dinka 40%, Nuer 15%, Bari 10%, Azande 10%, Shilluk 10%, Arab 10%, other 5%

GOVERNMENT: Transitional regime

CURRENCY: South Sudan pound = 100 piastres

Spain

At its unification under Ferdinand and Isabella in 1492, Spain occupied a pivotal position between Europe, Africa, the North Atlantic, and the Mediterranean.

GEOGRAPHY

Mountain ranges in the north, center, and south, with a huge central plateau. Mediterranean lowlands. Verdant valleys in the northwest.

CLIMATE

Maritime in north. Hotter and drier in south. The central plateau has an extreme climate.

PEOPLE & SOCIETY

A vigorous ethnic regionalism, suppressed under Franco's fascist regime, now flourishes. There are 17 autonomous regions. People remain churchgoing, though Roman Catholic teachings on social issues are often flouted. Spanish women are increasingly emancipated, with strong political representation.

 INSIGHT: *Over 3000 festivals and feasts take place each year in Spain*

THE ECONOMY

Exports food, wine. Few natural resources. Large fishing fleet. Tourism, motor industry hit by global downturn; soaring unemployment since abrupt end of construction boom. Austerity program aim to cut debt and deficits. A target for economic migrants from Africa.

■	2000m/6562ft
■	1000m/3281ft
■	500m/1640ft
■	200m/656ft
■	Sea Level

FACTFILE

OFFICIAL NAME: Kingdom of Spain
DATE OF FORMATION: 1492
CAPITAL: Madrid
POPULATION: 46.9 million
TOTAL AREA: 194,896 sq. miles (504,782 sq. km)
DENSITY: 243 people per sq. mile

LANGUAGES: Spanish*, Catalan*, Galician*, Basque*
RELIGIONS: Roman Catholic 96%, other 4%
ETHNIC MIX: Castilian Spanish 72%, Catalan 17%, Galician 6%, Basque 2%, Roma 1%, other 2%
GOVERNMENT: Parliamentary system
CURRENCY: Euro = 100 cents

Sri Lanka

The teardrop-shaped island of Sri Lanka is separated from India by the Palk Strait. Ethnic Tamil rebels – the Tamil Tigers – were defeated in 2009, after a brutal 26-year civil war.

GEOGRAPHY

The main island is dominated by rugged central highlands. Fertile northern plains are dissected by rivers. Much of the land is tropical jungle.

CLIMATE

Tropical, with breezes on the coast and cooler air in highlands. Northeast is driest and hottest.

PEOPLE & SOCIETY

The Sinhalese are mostly Buddhist, while Tamils are mostly Hindu. Moors are the Muslim descendants of Arab traders. Tamils were the minority group favored by the British colonists. Majority-Sinhalese power since independence in 1948 fueled tensions, erupting into civil war in 1983. The eventual government victory in 2009 made this the only rebel insurgency ever defeated in modern times.

THE ECONOMY

Garment industry. Remittances. Major tea exporter. End of costly civil war; return of foreign investment and tourists. Decade of strong GDP growth.

INSIGHT: *Sri Lanka elected the world's first woman prime minister, Sirimavo Bandaranaike, in 1960*

Palk Strait

0   50 km
0   50 miles

Bay of Bengal

Jaffna

Gulf of Mannar

Mannar

Trincomalee

Puttalam

Anuradhapura

Batticaloa

INDIAN OCEAN

Negombo

Kandy

COLOMBO

SRI JAYEWARDENAPURA KOTTE

Kalutara

Ratnapura

2000m/6562ft
1000m/3281ft
500m/1640ft
200m/656ft
Sea Level

Galle

Matara

FACTFILE

OFFICIAL NAME: Democratic Socialist Republic of Sri Lanka

DATE OF FORMATION: 1948

CAPITAL: Colombo / Sri Jayewardenapura Kotte

POPULATION: 21.3 million

TOTAL AREA: 25,332 sq. miles (65,610 sq. km)

DENSITY: 852 people per sq. mile

LANGUAGES: Sinhala*, Tamil*, English

RELIGIONS: Buddhist 69%, Hindu 15%, Muslim 8%, Christian 8%

ETHNIC MIX: Sinhalese 74%, Tamil 18%, Moor 7%, other 1%

GOVERNMENT: Mixed presidential–parliamentary system

CURRENCY: Sri Lanka rupee = 100 cents

Sudan

The secession of the black African south in 2011 left Sudan as Africa's third-largest country. Darfur in the west is suffering a terrible humanitarian crisis.

GEOGRAPHY
Lies within the upper Nile basin. Mostly arid plains. Highlands border the Red Sea in the northeast.

CLIMATE
North is hot, arid desert with constant dry winds. Rainy season lasting a few months in the south.

PEOPLE & SOCIETY
About two million people are nomads. There are many ethnic groups. Islamic law, imposed by the Arab majority, restricts women's freedoms and alienated the non-Muslim south, which finally seceded in 2011 after prolonged conflict. Ethnic violence by Arab militias in Darfur since 2003 has killed 300,000 people and created a huge refugee crisis within Sudan and in neighboring Chad and CAR. President Bashir faces an international arrest warrant for crimes against humanity.

THE ECONOMY
Oil reserves reduced by secession of South. Cotton, sesame, gum arabic. Violence and drought hamper farming. Millions of people displaced. Large debt.

INSIGHT: *Sudan has more pyramids than Egypt: over 200 structures remain from ancient Nubian kingdoms on the Nile*

2000m/6562ft
1000m/3281ft
500m/1640ft
200m/656ft
Sea Level

0   400 km
0   400 miles

FACTFILE

OFFICIAL NAME: Republic of the Sudan
DATE OF FORMATION: 1956
CAPITAL: Khartoum
POPULATION: 38 million
TOTAL AREA: 718,722 sq. miles (1,861,481 sq. km)
DENSITY: 53 people per sq. mile

LANGUAGES: Arabic*, Nubian, Beja, Fur
RELIGIONS: Nearly the whole population is Muslim (mainly Sunni)
ETHNIC MIX: Arab 60%, other 18%, Nubian 10%, Beja 8%, Fur 3%, Zaghawa 1%
GOVERNMENT: Presidential system
CURRENCY: New Sudanese pound = 100 piastres

Suriname

Suriname is a former Dutch colony on the north coast of South America. Democracy was restored in 1991, after almost 11 years of military rule. The Netherlands is still a major donor of aid.

GEOGRAPHY

Mostly covered by tropical rainforest. Coastal plain rises to central plateaus and the Guiana Highlands.

CLIMATE

Tropical. Hot and humid, but cooled by trade winds. High rainfall, especially in the interior.

PEOPLE & SOCIETY

The Dutch brought laborers from South Asia and Java. Independence saw mass emigration: around 350,000 Surinamese live in the Netherlands. Of those left, over 85% live near the coast, the rest in scattered rainforest communities. Indigenous Amerindians only number a few thousand. *Bosnegers* – descended from runaway African slaves – fought the military government in the late 1980s. Under civilian rule, each group has had a political party representing its interests.

THE ECONOMY

Alumina and gold are the key exports. Rice and bananas are main cash crops. Oil production and tourism are growing. Excessive bureaucracy.

INSIGHT: *In a 1667 Anglo-Dutch deal, Holland gained Suriname but lost New Amsterdam (now New York)*

FACTFILE

OFFICIAL NAME: Republic of Suriname

DATE OF FORMATION: 1975

CAPITAL: Paramaribo

POPULATION: 500,000

TOTAL AREA: 63,039 sq. miles (163,270 sq. km)

DENSITY: 8 people per sq. mile

LANGUAGES: Sranan (Creole), Dutch*, Hindi, Javanese, Sarnami, Saramaccan, Chinese, Carib

RELIGIONS: Christian 48%, Hindu 27%, Muslim 20%, traditional beliefs 5%

ETHNIC MIX: E Indian 27%, Creole 18%, Black 15%, Javanese 15%, mixed race 13%, other 12%

GOVERNMENT: Mixed presidential–parliamentary system

CURRENCY: Surinamese dollar = 100 cents

Swaziland

The tiny southern African kingdom of Swaziland is crippled with HIV/AIDS and economically dependent on South Africa. Vocal demands for multiparty democracy have been ignored.

GEOGRAPHY

Mainly high plateaus and mountains. Rolling grasslands and low scrub plains to the east. Pine forests on western border.

CLIMATE

Temperatures rise and rainfall declines as the land descends eastward, from high to low grassy *veld*.

PEOPLE & SOCIETY

One of Africa's most conservative states, though there is pressure from urban-based modernizers. Political system promotes Swazi tradition and is dominated by powerful monarchy. Women face discrimination. Swaziland has the world's highest prevalence of HIV/AIDS: chastity is urged to combat its spread.

◆ **INSIGHT:** *Polygamy is practiced in Swaziland – when King Sobhuza died in 1982, he left 100 widows*

THE ECONOMY

Sugarcane is the main cash crop. Wood pulp and soft drink concentrates are also exported. Loss of workforce to HIV/AIDS, and high cost of health care.

FACTFILE

OFFICIAL NAME: Kingdom of Swaziland
DATE OF FORMATION: 1968
CAPITAL: Mbabane
POPULATION: 1.2 million
TOTAL AREA: 6704 sq. miles (17,363 sq. km)
DENSITY: 181 people per sq. mile

LANGUAGES: English*, siSwati*, isiZulu, Xitsonga
RELIGIONS: Traditional beliefs 40%, other 30%, Roman Catholic 20%, Muslim 10%
ETHNIC MIX: Swazi 97%, other 3%
GOVERNMENT: Monarchy
CURRENCY: Lilangeni = 100 cents

Sweden

The largest Scandinavian country by both population and area, Sweden has one of the world's most extensive welfare systems and is among the leading proponents of equal rights for women.

 GEOGRAPHY
Heavily forested, with many lakes. Northern plateau extends beyond the Arctic Circle. Southern lowlands are widely cultivated.

CLIMATE
Southern coasts warmed by Gulf Stream. Northern areas have more extreme continental climate.

PEOPLE & SOCIETY
The nuclear family forms the basis of society, but the marriage rate is one of the lowest in the world, and cohabitation is now common. The model welfare system is paid for by a high tax burden. Women are well represented at all levels. A minority of 30,000 Sámi lives in the far north. Most industries and the bulk of population are based in and around the southern cities. An EU member since 1995, Sweden has voted not to join the euro.

THE ECONOMY
Companies of global importance, including Volvo, Saab, SFK, Ericsson. Highly developed infrastructure. Up-to-date technology. Skilled workforce.

INSIGHT: *Sweden has maintained a position of armed neutrality since 1815*

1000m/3281ft
500m/1640ft
200m/656ft
Sea Level

0   100 km
0   100 miles

FINLAND
Lapland
Arctic Circle
NORWAY
Luleå
Umeå
Östersund
Gulf of Bothnia
Sundsvall
Uppsala
Västerås
Örebro
Vänern
STOCKHOLM
Norrköping
Vättern
Göteborg
Jönköping
Gotland
Öland
Kattegat
Helsingborg
Malmö
Baltic Sea

FACTFILE

OFFICIAL NAME: Kingdom of Sweden
DATE OF FORMATION: 1523
CAPITAL: Stockholm
POPULATION: 9.6 million
TOTAL AREA: 173,731 sq. miles (449,964 sq. km)
DENSITY: 60 people per sq. mile
LANGUAGES: Swedish*, Finnish, Sámi

RELIGIONS: Evangelical Lutheran 75%, other 13%, other Protestant 5%, Muslim 5%, Roman Catholic 2%
ETHNIC MIX: Swedish 86%, foreign-born or first-generation immigrant 12%, Finnish and Sámi 2%
GOVERNMENT: Parliamentary system
CURRENCY: Swedish krona = 100 öre

Switzerland

One of the world's most prosperous countries, Switzerland sits at the center of Europe. It has retained its neutral status through every major European conflict since 1815.

GEOGRAPHY

Mostly mountainous, with river valleys. The Alps cover 60% of its area; the Jura in the west cover 10%. Lowlands lie along the east–west axis.

CLIMATE

Most rain falls in the warm summer months. Winters are snowy, but milder and foggy away from the mountains. Avalanches are a problem.

PEOPLE & SOCIETY

Switzerland is composed of distinct German-Swiss, French-Swiss, and Italian-Swiss linguistic groups. In the east, a 60,000-strong minority speaks Romansch. The country is divided into 26 autonomous cantons (states), each with control over health care, education, housing, and taxation. Public referenda are widely used to decide policy. Society is conservative; marriage is common but divorce is above the EU average rate.

THE ECONOMY

Diversified economy relies on services – the banking sector manages over a quarter of the world's offshore private wealth – and specialized industries (engineering, watches, etc).

INSIGHT: *Famed for its neutrality, Switzerland only joined the UN in 2002, and remains outside the EU*

FACTFILE

OFFICIAL NAME: Swiss Confederation

DATE OF FORMATION: 1291

CAPITAL: Bern

POPULATION: 8.1 million

TOTAL AREA: 15,942 sq. miles (41,290 sq. km)

DENSITY: 528 people per sq. mile

LANGUAGES: German*, Swiss-German, French*, Italian*, Romansch*

RELIGIONS: Roman Catholic 42%, Protestant 35%, other and nonreligious 19%, Muslim 4%

ETHNIC MIX: German 64%, French 20%, other 9%, Italian 6%, Romansch 1%

GOVERNMENT: Parliamentary system

CURRENCY: Swiss franc = 100 rappen/centimes

Syria

Stretching from the eastern Mediterranean to the Tigris River, Syria's borders are regarded as an artificial creation of French colonial rule by many Syrians. Civil war erupted in 2011.

GEOGRAPHY
A short stretch of coastal plain is backed by a low range of hills. The Euphrates River cuts through a vast interior desert plateau.

CLIMATE
Mediterranean coastal climate. Inland areas are arid. In winter, snow is common on the mountains.

PEOPLE & SOCIETY
Towns tend to lie within 60 miles (100 km) of the coast. Most Syrians are Sunni Muslim, but the Shi'a Alawis control politics. The authoritarian Assad regime, in power since 1970, fiercely repressed pro-democracy "Arab Spring" protests in 2011, and brutal conflict soon broke out. 200,000 have been killed and ten million are displaced, including many Palestinians and Iraqis formerly sheltering in Syrian refugee camps. Islamic State (IS) jihadists control the Euphrates valley.

THE ECONOMY
Conflict has destroyed economy. Oil fields held by rebels, production down. Sanctions limit exports. Lack of food, medicines. Infrastructure bombed.

INSIGHT: *Syria is an ancient land; there are at least 3500 as yet unexcavated archaeological sites*

2000m/6562ft
1000m/3281ft
500m/1640ft
200m/656ft
Sea Level

FACTFILE

OFFICIAL NAME: Syrian Arab Republic
DATE OF FORMATION: 1941
CAPITAL: Damascus
POPULATION: 21.9 million
TOTAL AREA: 71,498 sq. miles (184,180 sq. km)
DENSITY: 308 people per sq. mile

LANGUAGES: Arabic*, French, Kurdish, Armenian, Circassian, Assyrian, Aramaic
RELIGIONS: Sunni Muslim 74%, Alawi (Shi'a sect) 12%, Christian 10%, Druze 3%, other 1%
ETHNIC MIX: Arab 90%, Kurdish 9%, Armenian, Turkmen, and Circassian 1%
GOVERNMENT: Presidential system
CURRENCY: Syrian pound = 100 piastres

Taiwan

The republic of Taiwan (formerly Formosa) is on an island 80 miles (130 km) off the southeast coast of mainland China, which still considers it to be a renegade province.

GEOGRAPHY
Mountain region covers two-thirds of the island. Highly fertile lowlands and coastal plains.

CLIMATE
Tropical monsoon. Hot and humid. Typhoons July–September. Snow falls in mountains in winter.

PEOPLE & SOCIETY
Most Taiwanese are Han Chinese, descendants of the 1644 migration of the Ming dynasty from the mainland. The modern republic was created in 1949, when the nationalist Kuomintang was expelled from the mainland following Communist victory in the civil war. 100,000 emigrés established themselves as a ruling class. Initial resentment has subsided as a new Taiwan-born generation has taken over the reins of power. The aboriginal minority suffers discrimination.

THE ECONOMY
Successful economy of small, adaptable companies. High-tech goods: TVs, computers, and semiconductors. Rising trade, investment with China.

INSIGHT: *Taiwan lost its seat at the UN to Beijing in 1971: both claim to represent "China"*

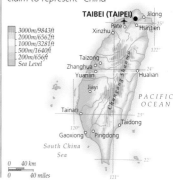

FACTFILE
OFFICIAL NAME: Republic of China (ROC)
DATE OF FORMATION: 1949
CAPITAL: Taibei (Taipei)
POPULATION: 23.3 million
TOTAL AREA: 13,892 sq. miles (35,980 sq. km)
DENSITY: 1871 people per sq. mile
LANGUAGES: Amoy Chinese, Mandarin Chinese*, Hakka Chinese
RELIGIONS: Buddhist, Confucianist, and Taoist 93%, Christian 5%, other 2%
ETHNIC MIX: Han Chinese (pre-20th-century migration) 84%, Han Chinese (20th-century migration) 14%, Aboriginal 2%
GOVERNMENT: Presidential system
CURRENCY: Taiwan dollar = 100 cents

Tajikistan

Tajikistan lies landlocked on the western slopes of the Pamirs in central Asia. Soon after the breakup of the USSR in 1991, civil war erupted between ruling communists and Islamists.

GEOGRAPHY

Mainly mountainous: bare slopes of the Pamir ranges, with fast-flowing rivers, cover most of the country. Small but fertile Fergana Valley in northwest.

CLIMATE

Continental extremes in the valleys. Bitterly cold winters in the mountains. Rainfall is low.

PEOPLE & SOCIETY

Unlike the other former Soviet republics of central Asia, Tajikistan is dominated by a people of Persian (Iranian) rather than Turkic origin. The main ethnic conflict is with the Turkic Uzbek minority. Russians are discriminated against; most fled in the 1992–1997 civil war, and standards of living fell dramatically. Islamist militants are active. Two million people work abroad, primarily in Russia.

THE ECONOMY

Mass poverty. Declining cotton revenue. Also exports aluminum. Uranium deposits. Transit route for illicit Afghan opium. Corruption. Needs reforms to attract foreign investment.

◆ **INSIGHT:** *Carpet-making, an ancient tradition learned from Persia, is still a major source of revenue*

FACTFILE

OFFICIAL NAME: Republic of Tajikistan

DATE OF FORMATION: 1991

CAPITAL: Dushanbe

POPULATION: 8.2 million

TOTAL AREA: 55,251 sq. miles (143,100 sq. km)

DENSITY: 148 people per sq. mile

LANGUAGES: Tajik*, Uzbek, Russian

RELIGIONS: Sunni Muslim 95%, Shi'a Muslim 3%, other 2%

ETHNIC MIX: Tajik 80%, Uzbek 15%, other 3%, Kyrgyz 1%, Russian 1%

GOVERNMENT: Presidential system

CURRENCY: Somoni = 100 diram

Tanzania

The east African state of Tanzania was formed in 1964 by the union of Tanganyika and the Zanzibar islands. A third of its area is game reserve or national park.

GEOGRAPHY

The mainland is mostly a high plateau lying to the east of the Great Rift Valley. Forested coastal plain. Highlands in the north and south.

CLIMATE

Tropical on the coast and Zanzibar. Semiarid on central plateau, semitemperate in the highlands. March–May rains.

PEOPLE & SOCIETY

99% of people belong to one of 120 small ethnic Bantu groups. Arabs, Asians, and Europeans make up the remaining population. Use of Kiswahili as the lingua franca has eliminated ethnic rivalries. The majority of Tanzanians are subsistence famers.

◆ INSIGHT: At 19,340 ft (5895 m), Kilimanjaro in northeast Tanzania is Africa's highest mountain

THE ECONOMY

Reliant on agriculture, including forestry and cattle. Coffee, cotton, tea, cashew nuts, sisal, and cloves are cash crops. Gold, diamonds, and gems mined. Safari and beach tourism. Debt relief.

FACTFILE

OFFICIAL NAME: United Republic of Tanzania

DATE OF FORMATION: 1964

CAPITAL: Dodoma

POPULATION: 49.3 million

TOTAL AREA: 364,898 sq. miles (945,087 sq. km)

DENSITY: 144 people per sq. mile

LANGUAGES: Kiswahili*, Sukuma, Chagga, Nyamwezi, Hehe, Makonde, Yao, English*

RELIGIONS: Christian 63%, Muslim 35%, other 2%

ETHNIC MIX: Native African (over 120 tribes) 99%, European, Asian, and Arab 1%

GOVERNMENT: Presidential system

CURRENCY: Tanzanian shilling = 100 cents

Thailand

Thailand lies at the heart of mainland southeast Asia. Continuing rapid industrialization has resulted in massive congestion in the capital and a serious depletion of natural resources.

GEOGRAPHY

One-third is low plateau, drained by tributaries of the Mekong River. Central plain is the most fertile area.

CLIMATE
Tropical. Hot, humid March–May; monsoon rains May–October; cooler season November–March.

PEOPLE & SOCIETY

Buddhism is a national binding force. 600,000 hill tribes-people live in the north and northeast. The Chinese minority is the most assimilated in the region. In the undeveloped far south, Malay Islamists are fighting for secession. Politics has been unstable since the 2006 fall of populist Prime Minister Thaksin; the military intervened again in 2014.

◆ INSIGHT: *Thailand, meaning "land of the free," is the only SE Asian nation never to have been colonized*

THE ECONOMY
Successful manufacturing. Natural gas reserves. Leading exporter of rice, rubber. Political turmoil. Tourism, though sex industry harms image. Damage from natural disasters.

FACTFILE

OFFICIAL NAME: Kingdom of Thailand
DATE OF FORMATION: 1238
CAPITAL: Bangkok
POPULATION: 67 million
TOTAL AREA: 198,455 sq. miles (514,000 sq. km)
DENSITY: 340 people per sq. mile

LANGUAGES: Thai*, Chinese, Malay, Khmer, Mon, Karen, Miao
RELIGIONS: Buddhist 95%, Muslim 4%, other (including Christian) 1%
ETHNIC MIX: Thai 83%, Chinese 12%, Malay 3%, Khmer and other 2%
GOVERNMENT: Transitional regime
CURRENCY: Baht = 100 satang

Togo

Togo lies sandwiched between Ghana and Benin in west Africa. General Eyadema ruled from 1967–2005; his son succeeded him. Lomé port is an important entrepôt for regional trade.

GEOGRAPHY
Central forested region bounded by savanna lands to the north and south. Mountain range stretches southwest to northeast.

CLIMATE
Coast hot and humid; drier inland. Rainy season March–July, with heaviest falls in the west.

PEOPLE & SOCIETY
Harsh resentment between Ewe in the south and Kabye in the north. Kabye control the military, but the north is less developed than the south. Extended family is important. Tribalism and nepotism are key factors in everyday life. Some ethnic groups, such as the Mina, have matriarchal societies.

◆ **INSIGHT:** *The "Nana Benz," the entrepreneurial market-women of Lomé, control Togo's retail trade*

THE ECONOMY
Most people are farmers. Self-sufficient in staple foods. Togo's main cash crops are coffee and cocoa: cotton has declined. Its phosphate deposits are the most mineral-rich in the world, but easily extractable reserves are depleted and the sector needs investment.

500m/1640ft
200m/656ft
Sea Level

0   50 km
0   50 miles

FACTFILE

OFFICIAL NAME: Togolese Republic
DATE OF FORMATION: 1960
CAPITAL: Lomé
POPULATION: 6.8 million
TOTAL AREA: 21,924 sq. miles (56,785 sq. km)
DENSITY: 324 people per sq. mile

LANGUAGES: Ewe, Kabye, Gurma, French*
RELIGIONS: Christian 47%, traditional beliefs 33%, Muslim 14%, other 6%
ETHNIC MIX: Ewe 46%, other African 41%, Kabye 12%, European 1%
GOVERNMENT: Presidential system
CURRENCY: CFA franc = 100 centimes

Tonga

Tonga is a South Pacific archipelago of 170 islands; only 45 of these islands are inhabited. The king retains significant powers though some democratic reforms were introduced in 2011.

GEOGRAPHY

Easterly islands are generally low and fertile. Those in the west are higher and volcanic in origin.

CLIMATE

Tropical oceanic. Temperatures range between 68°F (20°C) and 86°F (30°C) all year round. Heavy rainfall, especially February–March.

PEOPLE & SOCIETY

Tonga is the last remaining Polynesian monarchy. All land belongs to the crown, but is administered by nobles who allot it to the common people. Respect for traditional values is high, though younger, Westernized Tongans are starting to question some attitudes. The first elected commoner became prime minister in 2006.

◆ **INSIGHT:** *Unique in the Pacific, Tonga was never brought under foreign rule*

THE ECONOMY

Squashes and vanilla exported. Remittances. Potential for tourism and fisheries. Large debt owed to China for rebuilding of capital's business district, destroyed in 2006 prodemocracy riots.

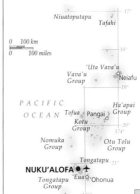

Niuatoputapu
Tafahi
0   100 km
0   100 miles
'Uta Vava'u
Vava'u Group
Neiafu
PACIFIC
OCEAN
Ha'apai Group
Tofua   Pangai
Kotu Group
Otu Tolu Group
Nomuka Group
Tongatapu
NUKU'ALOFA
Tongatapu Group
'Eua   Ohonua

200m/656ft
Sea Level

FACTFILE

OFFICIAL NAME: Kingdom of Tonga

DATE OF FORMATION: 1970

CAPITAL: Nuku'alofa

POPULATION: 106,322

TOTAL AREA: 289 sq. miles (748 sq. km)

DENSITY: 382 people per sq. mile

LANGUAGES: English*, Tongan*

RELIGIONS: Free Wesleyan 41%, other 17%, Roman Catholic 16%, Church of Jesus Christ of Latter-Day Saints 14%, Free Church of Tonga 12%

ETHNIC MIX: Tongan 98%, other 2%

GOVERNMENT: Monarchy

CURRENCY: Pa'anga (Tongan dollar) = 100 seniti

Trinidad & Tobago

The two islands of the former UK colony of Trinidad and Tobago are the most southerly of the Caribbean Windward Islands, lying just 9 miles (15 km) off the coast of Venezuela.

GEOGRAPHY
Both islands are hilly and wooded. Trinidad has a rugged mountain range in the north, and swamps on its east and west coasts.

CLIMATE
Tropical, with July–December wet season. Escapes the region's hurricanes, which pass to the north.

PEOPLE & SOCIETY
Trinidad's East Indian community is the Caribbean's largest and holds onto its Muslim and Hindu heritage. There are tensions with the mainly Christian blacks; political parties are divided along race lines. Blacks form the majority on Tobago. High rates of kidnapping and murder are an issue.

◆ **INSIGHT:** *Trinidad and Tobago is the birthplace of steel bands and Calypso music*

THE ECONOMY
Oil and natural gas: major provider of liquefied natural gas to US, but reserves are declining fast. Associated industries: second-largest producer of methanol. Tourism on wildlife-rich Tobago.

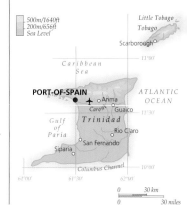

FACTFILE

OFFICIAL NAME: Republic of Trinidad and Tobago

DATE OF FORMATION: 1962

CAPITAL: Port-of-Spain

POPULATION: 1.3 million

TOTAL AREA: 1980 sq. miles (5128 sq. km)

DENSITY: 656 people per sq. mile

LANGUAGES: English Creole, English*, Hindi, French, Spanish

RELIGIONS: Roman Catholic 26%, Hindu 23%, other 23%, Protestant 22%, Muslim 6%

ETHNIC MIX: East Indian 40%, Black 38%, mixed race 20%, White, Chinese, other 2%

GOVERNMENT: Parliamentary system

CURRENCY: Trin. & Tob. dollar = 100 cents

Tunisia

A French north African colony until 1956, Tunisia was relatively liberal in social terms, but in 2011 protesters ousted the dictatorial president, triggering the "Arab Spring" across the region.

GEOGRAPHY
Mountains in the north are surrounded by plains. Vast, low-lying salt pans in the center. To the south lies the Sahara Desert.

CLIMATE
Summer temperatures are high. The north is often wet and windy in winter. Far south is arid.

PEOPLE & SOCIETY
The population is almost entirely of Arab-Berber descent, with Jewish and Christian minorities. Many still live in extended family groups of three or four generations. Women have better rights than in most other Arab countries and make up a quarter of the workforce. The low birth rate is a result of a long-standing family planning policy. The Islamist-led transitional government elected in 2011 was replaced by a consensus government in 2014.

THE ECONOMY
Competitive and diversified. Expanding manufacturing. Exports olives, dates, citrus fruit, phosphates. Instability, affecting tourism. Free trade with EU.

INSIGHT: *Tunisia was the center of trading empires from the 9th century BCE*

FACTFILE

OFFICIAL NAME: Tunisian Republic
DATE OF FORMATION: 1956
CAPITAL: Tunis
POPULATION: 11 million
TOTAL AREA: 63,169 sq. miles (163,610 sq. km)
DENSITY: 183 people per sq. mile

LANGUAGES: Arabic*, French
RELIGIONS: Muslim (mainly Sunni) 98%, Christian 1%, Jewish 1%
ETHNIC MIX: Arab and Berber 98%, Jewish 1%, European 1%
GOVERNMENT: Transitional regime
CURRENCY: Tunisian dinar = 1000 millimes

Turkey

Lying partly in the region of eastern Thrace in Europe, but mostly in Asia, Turkey's position gives it significant influence in the Mediterranean, the Black Sea, and the Middle East.

GEOGRAPHY

Asian Turkey (Anatolia) is dominated by two mountain ranges, separated by a high, semidesert plateau. Coastal regions are fertile.

CLIMATE

Coast has a Mediterranean climate. Interior has cold, snowy winters and hot, dry summers.

PEOPLE & SOCIETY

Despite racial diversity, Turkey has a strong sense of national identity, and close links with other Turkic states. Kurds, the largest minority, based in the southeast, have waged a violent campaign for greater autonomy intermittently since 1984. The current political dominance of Islamists challenges Turkey's cherished identity as a secular state. It has applied to join the EU, but progress will be slow.

THE ECONOMY

Liberalized economy, boosted by self-sufficient agriculture, and textiles, tourism, and manufacturing sectors. Route of Asian oil pipelines to Europe.

INSIGHT: *Turkey had two of the seven wonders of the ancient world: the tomb of King Mausolus at Halicarnassus (now Bodrum), and the temple of Artemis at Ephesus*

FACTFILE

OFFICIAL NAME: Republic of Turkey

DATE OF FORMATION: 1923

CAPITAL: Ankara

POPULATION: 74.9 million

TOTAL AREA: 301,382 sq. miles (780,580 sq. km)

DENSITY: 252 people per sq. mile

LANGUAGES: Turkish*, Kurdish, Arabic, Circassian, Armenian, Greek, Georgian, Ladino

RELIGIONS: Muslim (mainly Sunni) 99%, other 1%

ETHNIC MIX: Turkish 70%, Kurdish 20%, other 8%, Arab 2%

GOVERNMENT: Parliamentary system

CURRENCY: Turkish lira = 100 kurus

Turkmenistan

Stretching from the Caspian Sea into the central Asian desert, Turkmenistan has had less upheaval than most ex-Soviet states, under President Niyazov's dictatorial rule (1991–2006).

GEOGRAPHY

Low Garagum Desert covers 80% of the country. Mountains on southern border with Iran. Fertile Amu Darya Valley in north.

CLIMATE

Arid desert climate with extreme summer heat, but sub-freezing winter temperatures.

PEOPLE & SOCIETY

The Turkmen were once largely nomadic, and the tribal unit remains strong, with population clustered around desert oases. "Turkmenization" of government, education, and religion has strained relations with Uzbek and Russian minorities. Political reform since Niyazov's sudden death in 2006 led to multiparty elections in 2013, though all seats were won by the former sole party and Niyazov's successor runs a similarly authoritarian regime.

THE ECONOMY
State-controlled, though there is some private investment. Natural gas and oil are main resources. Overintensive farming of cotton. Black market.

◆ **INSIGHT:** *President Niyazov created an elaborate personality cult, styling himself as Turkmenbashi – "head" of all Turkmen*

FACTFILE

OFFICIAL NAME: Turkmenistan

DATE OF FORMATION: 1991

CAPITAL: Asgabat

POPULATION: 5.2 million

TOTAL AREA: 188,455 sq. miles (488,100 sq. km)

DENSITY: 28 people per sq. mile

LANGUAGES: Turkmen*, Uzbek, Russian, Kazakh, Tatar

RELIGIONS: Sunni Muslim 89%, Orthodox Christian 9%, other 2%

ETHNIC MIX: Turkmen 85%, other 6%, Uzbek 5%, Russian 4%

GOVERNMENT: Presidential system

CURRENCY: New manat = 100 tenge

Tuvalu

One of the world's smallest, most isolated states, Tuvalu lies in the central Pacific. The nine islands were linked to the Gilbert Islands (Kiribati) as a UK colony until independence.

GEOGRAPHY
A series of coral atolls, none more than 15 ft (4.6 m) above sea level. Poor soils restrict vegetation to bush, coconut palms, and breadfruit trees.

CLIMATE
Hot all year round. Heavy annual rainfall. Hurricane season brings many violent storms.

PEOPLE & SOCIETY
People are mostly Polynesian. Around half the population lives on Funafuti, where government jobs are based. Life is communal and traditional. Most people live by subsistence farming, digging pits out of the coral to grow crops. Fresh water is precious, due to frequent droughts.

INSIGHT: *Low-lying Tuvalu, like the Maldives, is set to disappear with rising sea levels*

THE ECONOMY
World's smallest economy. Remittances from Tuvaluan seafarers. Sale of fishing licenses. Copra, stamps, and coins exported. Income from trust fund and the lease of .tv Internet suffix.

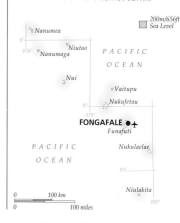

FACTFILE

OFFICIAL NAME: Tuvalu
DATE OF FORMATION: 1978
CAPITAL: Fongafale, on Funafuti Atoll
POPULATION: 10,698
TOTAL AREA: 10 sq. miles (26 sq. km)
DENSITY: 1070 people per sq. mile

LANGUAGES: Tuvaluan, Kiribati, English*
RELIGIONS: Church of Tuvalu 97%, Baha'i 1%, Seventh-day Adventist 1%, other 1%
ETHNIC MIX: Polynesian 96%, Micronesian 4%
GOVERNMENT: Nonparty system
CURRENCY: Australian dollar and Tuvaluan dollar = 100 cents each

Uganda

Landlocked in east Africa, Uganda has a history of ethnic strife. Under President Museveni, steps have been taken to restore peace and to rebuild the economy and democracy.

GEOGRAPHY
Predominantly a large plateau with the Ruwenzori mountain range and the Great Rift Valley in the west. Lake Victoria lies to the southeast. Vegetation is of savanna type.

CLIMATE
Altitude and the influence of the lakes modify the equatorial climate. Rain falls throughout the year; spring is the wettest period.

PEOPLE & SOCIETY
Mostly rural population comprising 13 main ethnic groups. President Museveni has worked hard to break down ethnic animosities, but a noticeable north–south divide persists, with most development in the south. After two decades of brutal clashes (1987–2008), the Ugandan army is still pursuing remnants of the Lord's Resistance Army across the DRC, South Sudan, and the CAR.

THE ECONOMY
Resource-rich, but undeveloped and poor. Exports coffee, fish, tea, and flowers. Oil exploration. Hydroelectric power is reducing oil imports. Great potential from mining. Debt relief.

INSIGHT: *Lake Victoria is the world's third-largest lake*

FACTFILE

OFFICIAL NAME: Republic of Uganda
DATE OF FORMATION: 1962
CAPITAL: Kampala
POPULATION: 37.6 million
TOTAL AREA: 91,135 sq. miles (236,040 sq. km)
DENSITY: 488 people per sq. mile

LANGUAGES: Luganda, Nkole, Chiga, Lango, Acholi, Teso, Lugbara, English*
RELIGIONS: Christian 85%, Muslim (mainly Sunni) 12%, other 3%
ETHNIC MIX: Other 50%, Baganda 17%, Banyakole 10%, Basoga 9%, Iteso 7%, Bakiga 7%
GOVERNMENT: Presidential system
CURRENCY: Uganda shilling = 100 cents

Ukraine

The former "breadbasket of the Soviet Union," Ukraine lies on the Black Sea. Divisions between pro-Russian sentiments and pro-European nationalism erupted into civil war in 2014.

GEOGRAPHY

Mainly fertile steppes and forests. Carpathian Mountains in west, Crimean chain in south. Pripet Marshes in northwest.

CLIMATE

Mainly continental climate, with distinct seasons. Southern Crimea has Mediterranean climate.

PEOPLE & SOCIETY
Over 90% of people in the west are Ukrainian, but in cities in the east and south, and in Crimea, Russians form a majority. Tatars returned to Crimea after the Soviet Union's collapse and comprise around 12% of the population there. Pro-Russian president Yanukovych's refusal to sign an EU deal provoked protests that ousted him from power in 2014. Russia responded by backing eastern rebels and annexing Crimea.

THE ECONOMY
Minerals: 5% of global reserves. Political instability, and conflict in the east. Slow reform of land laws, holding back agriculture. Oil/natural gas transit from Russia and the Caspian to Europe: natural gas price disputes with Russia.

◆ **INSIGHT:** *Ukraine means "on the border," referring to its position on the edge of the old Russian Empire*

2000m/6562ft
1000m/3281ft
500m/1640ft
200m/656ft
Sea Level

BELARUS
Pripet Marshes
Luts'k
POLAND
L'viv
Chernivtsi
HUNGARY
Zhytomyr
Vinnytsya
Chornobyl'
KIEV
Cherkasy
MOLDOVA
Mykolayiv
ROMANIA
Odesa
Danube
Chernihiv
RUSS. FED.
Kremenchuts'ke Vdskh.
Kharkiv
Dnipro-petrovs'k
Luhans'k
Zaporizhzhya
Donets'k
Dnieper
Mariupol'
Sea of Azov
Crimea
Sevastopol'
Black Sea

0   100 km
0   100 miles

(the Ukrainian territory of Crimea was annexed by Russia in 2014)

FACTFILE

OFFICIAL NAME: Ukraine
DATE OF FORMATION: 1991
CAPITAL: Kiev
POPULATION: 45.2 million
TOTAL AREA: 223,089 sq. miles (603,700 sq. km)
DENSITY: 194 people per sq. mile

LANGUAGES: Ukrainian*, Russian, Tatar
RELIGIONS: Christian (mainly Orthodox) 95%, other 5%
ETHNIC MIX: Ukrainian 78%, Russian 17%, other 5%
GOVERNMENT: Presidential system
CURRENCY: Hryvna = 100 kopiykas

United Arab Emirates

Bordering the Gulf on the northern coast of the Arabian Peninsula, the seven states of the UAE are Abu Dhabi, Dubai, Sharjah, Ajman, Umm al Qaywayn, Ras al Khaymah, and Fujayrah.

GEOGRAPHY

Mostly flat, semiarid desert with dunes, salt pans, and occasional oases. Cities are watered by extensive irrigation systems.

CLIMATE

Summers are humid, despite minimal rainfall. Sand-laden *shamal* winds blow in winter and spring.

PEOPLE & SOCIETY
Emirians, who make up just a quarter of the population, are mostly Sunni Muslims of Bedouin descent, and largely city dwellers. In theory, women enjoy equal rights with men. Poverty is rare and there is no income tax. The 1970s oil boom encouraged the immigration of workers, mostly from Asia. Western expatriates are permitted a virtually unrestricted lifestyle. Islamism, however, is a growing force among the young.

THE ECONOMY
Major oil and natural gas exporter; plentiful reserves. Dynamic Dubai: free trade zone, financial center (but 2008 global downturn caught overextended banks). Water is scarce. Imports most food. Some emirates are less developed.

 INSIGHT: *Mina Jabal Ali, in Dubai, is the largest man-made port in the world*

FACTFILE

OFFICIAL NAME: United Arab Emirates
DATE OF FORMATION: 1971
CAPITAL: Abu Dhabi
POPULATION: 9.3 million
TOTAL AREA: 32,000 sq. miles (82,880 sq. km)
DENSITY: 288 people per sq. mile

LANGUAGES: Arabic*, Farsi, Indian and Pakistani languages, English
RELIGIONS: Muslim (mainly Sunni) 96%, Christian, Hindu, and other 4%
ETHNIC MIX: Asian 60%, Emirian 25%, other Arab 12%, European 3%
GOVERNMENT: Monarchy
CURRENCY: UAE dirham = 100 fils

United Kingdom

Separated from continental Europe by the English Channel, the UK consists of Great Britain (England, Wales, and Scotland), several smaller islands, and Northern Ireland.

GEOGRAPHY

Rugged uplands dominate the landscape of Scotland, Wales, and northern England. All of the peaks in the United Kingdom over 4000 ft (1219 m) are in highland Scotland. The Pennine mountains, known as the "backbone of England," run the length of northern England. Lowland England rises into several ranges of rolling hills, and there is an interconnected system of rivers and canals. Over 600 islands, many uninhabited, lie west and north of the Scottish mainland.

CLIMATE

Generally mild, temperate, and highly changeable. Rain is fairly well distributed throughout the year. The west is generally wetter than the east, and the south warmer than the north. Winter snow is common in upland areas.

PEOPLE & SOCIETY

Scottish and Welsh people have a stronger sense of separate identity than the English; both Scotland and Wales have some self-government, as does Northern Ireland. In 2014 Scotland rejected independence in a referendum. Other ethnic minorities account for 5% of the population; more than half of them were born in the UK. Asian women in particular can be socially isolated. Asians and West Indians in most cities face deprivation and social stress, but white working-class youths were also evident when innercity rioting erupted in 2011. Income inequality is greater now than in 1884, when records began. In key areas such as policing, multiethnic recruitment has made little progress. Marriage is in decline. Over 40% of all births occur outside marriage, but most of them to cohabiting couples. Single-parent households account for just over a quarter of all families.

FACTFILE

OFFICIAL NAME: United Kingdom of Great Britain and Northern Ireland

DATE OF FORMATION: 1707

CAPITAL: London

POPULATION: 63.1 million

TOTAL AREA: 94,525 sq. miles (244,820 sq. km)

DENSITY: 676 people per sq. mile

LANGUAGES: English*, Welsh, Scottish Gaelic

RELIGIONS: Anglican 45%, other and nonreligious 37%, Roman Catholic 9%, Presbyterian 4%, Muslim 3%, Methodist 2%

ETHNIC MIX: English 80%, Scottish 9%, other 5%, Welsh 3%, Northern Irish 3%

GOVERNMENT: Parliamentary system

CURRENCY: Pound sterling = 100 pence

THE ECONOMY

World leader in financial services, pharmaceuticals, and defense industries. Strong multinationals. Precision engineering and high-tech industries, including biotechnology and telecommunications. Energy sector based on declining North Sea oil and natural gas reserves. Innovative in computer software development. Flexible working practices. Long-term decline of manufacturing sector, particularly heavy industries and car manufacture, partially offset by rise in financial and other services.

Nonparticipant in euro. High levels of government, corporate, and consumer debt: banks made major losses in 2007–2009 global downturn. Bailouts and stimulus packages pushed the government's finances further into the red. Tackling the deficit by cuts in spending puts pressure on growth strategy and social programs, with rising unemployment.

INSIGHT: *The UK has no formal written constitution, but a stable government system based on Parliament, which originated as a check on royal power in the 13th century*

1000m/3281ft
500m/1640ft
200m/656ft
Sea Level

0   100 km
0   100 miles

United States of America

Stretching across the most temperate part of North America, and with many natural resources, the US is the world's leading economic power and third-largest country.

GEOGRAPHY

The US has a varied topography. Forested mountains stretch from New England in the far northeast, giving way to lowlands and swamps in the extreme south. The central plains are dominated by the Mississippi–Missouri River system and the Great Lakes on the Canadian border. The Rocky Mountains in the west contain active volcanoes and drop to the coast across the earthquake-prone San Andreas Fault. The southwest is arid desert. Mountainous Alaska is mostly Arctic tundra.

CLIMATE

There are four main climatic zones. The north and east are continental and temperate, with heavy rainfall, warm summers, and cold winters. Florida and the Deep South are tropical and prone to hurricanes. The southwest is arid desert, with searing summer heat and low rainfall. Southern California is Mediterranean, with hot summers and mild winters.

INSIGHT: *The United States of America has the world's oldest constitution. Drafted in 1787, it has operated continuously ever since, albeit with numerous amendments*

CANADA

NORTH DAKOTA
Bismark

MINNESOTA
Lake Superior

MICHIGAN

Saint Lawrence

VERMONT
MAINE

NEW HAMPSHIRE

Pierre
SOUTH DAKOTA
Sioux Falls

Minneapolis
Saint Paul
WISCONSIN

Lake Huron
Lake Michigan
Lake Ontario

NEW YORK

Boston
MASSACHUSETTS
RHODE I.
CONNECTICUT

heyenne
NEBRASKA

Milwaukee
Chicago
IOWA
Des Moines

Detroit
Lake Erie
Buffalo

New York
NEW JERSEY

Omaha
Fort Wayne

Cleveland
Pittsburgh

PENNSYLVANIA
Philadelphia

Denver
ORADO
ueblo
Kansas City
KANSAS

Missouri
ILLINOIS
Springfield
INDIANA
Indianapolis

Cincinnati
OHIO
Columbus

Baltimore
WASHINGTON, D.C.
DELAWARE
MARYLAND

MISSOURI
Saint Louis
Louisville
WEST VIRGINIA

Richmond
VIRGINIA
Newport News

Tulsa
OKLAHOMA
Oklahoma City

KENTUCKY
Nashville
TENNESSEE

NORTH CAROLINA
Charlotte

ARKANSAS
Little Rock
Memphis

SOUTH CAROLINA

ATLANTIC

bock
Fort Worth
Dallas
Mississippi
MISSISSIPPI
Birmingham
Atlanta
GEORGIA
Savannah

OCEAN

Shreveport
Jackson
ALABAMA
Mobile
Jacksonville

TEXAS
LOUISIANA
Baton Rouge
New Orleans
Tallahassee
FLORIDA

Colorado
San Antonio
Houston

Tampa
Orlando

ards
ean

Corpus Christi

Gulf of Mexico

The Everglades
Miami

Rio Grande

3000m/9843ft
2000m/6562ft
1000m/3281ft
500m/1640ft
200m/656ft
Sea Level

0   400 km
0   400 miles

United States of America

◆ **INSIGHT:** By law, the actual records collected in a United States census must remain confidential for 72 years

 PEOPLE & SOCIETY

Although the demographic, economic, and cultural dominance of White Americans is firmly entrenched after over 400 years of settlement, the ethnic balance of the country is shifting. Barack Obama, whose father was African, became the first non-White US president in 2009. The African-American community, originally uprooted by the slave trade, has a strong consciousness. Less well organized socially but more numerous, and faster-growing, the Hispanic community is predicted to number over 30% of the population by 2050. Native Americans, dispossessed in the 19th century, are now among the poorest people. Constitutionally, state and religion are clearly separated. Conservative Christianity, however, is increasingly dominant politically. Living standards are high, but bad diet and insufficient exercise have left over a third of Americans obese.

$ THE ECONOMY

World's largest economy: huge resource base; well-established high-tech, engineering, and entertainment industries; global spread of US culture. Decline of manufacturing as jobs lost to low-wage economies. The combination of the "war on terrorism" launched after 9/11, military involvement in Afghanistan and Iraq, and a drive to cut taxes sent government debt spiraling. Hurricane Katrina hit oil production in 2005. Then a "bubble" of excessive risky mortgage lending burst and the financial and stock market crisis went global after the Lehman Brothers bank crashed in 2008. An economic incentive programme, combining tax cuts with more public spending, helped lift the economy out of recession, but widened the gaping budget deficit. Obama's government has struggled with this ever since. Pressures to cut spending hurt its social agenda, while conservative opponents denounce tax increases as a threat to growth.

FACTFILE

OFFICIAL NAME: United States of America
DATE OF FORMATION: 1776
CAPITAL: Washington, D.C.
POPULATION: 320 million
TOTAL AREA: 3,717,792 sq. miles (9,626,091 sq. km)
DENSITY: 90 people per sq. mile

LANGUAGES: English*, Spanish, other
RELIGIONS: Protestant 52%, Roman Catholic 25%, other 20%, Jewish 2%, Muslim 1%
ETHNIC MIX: White 60%, Hispanic 17%, African American 14%, Asian 6%, Native American 2%, Hawaiian or Pacific Islander 1%
GOVERNMENT: Presidential system
CURRENCY: US dollar = 100 cents

Uruguay

Situated in southeastern South America, Uruguay returned to civilian government in 1985, after 12 years of military rule. Most land is used for farming: Uruguay is a major wool exporter.

GEOGRAPHY

Low, rolling grasslands cover 80% of the country. Narrow coastal plain. Alluvial floodplain in southwest. Five rivers flow westward and drain into the Uruguay River.

CLIMATE

Temperate throughout the country. Warm summers, mild winters, and moderate rainfall.

PEOPLE & SOCIETY

Uruguayans are largely second-or third-generation Italians or Spaniards. Wealth derived from cattle ranching enabled the country to establish the first welfare state in South America. Waves of emigration occurred during the economic decline of the 1960s, the period of military rule, and the 1999–2002 economic crisis. Though a Roman Catholic country, Uruguay is liberal in its attitude to religion and all forms are tolerated.

THE ECONOMY

Exports wool, meat, hides, rice, wood, soy. Well-educated workforce. Banking services. Mineral potential.

INSIGHT: *Uruguay's rich pastures are ideal for raising livestock; animal products bring in over 40% of export earnings*

200m/656ft
Sea Level

0   100 km
0   100 miles

FACTFILE

OFFICIAL NAME: Eastern Republic of Uruguay

DATE OF FORMATION: 1828

CAPITAL: Montevideo

POPULATION: 3.4 million

TOTAL AREA: 68,039 sq. miles (176,220 sq. km)

DENSITY: 50 people per sq. mile

LANGUAGES: Spanish*

RELIGIONS: Roman Catholic 66%, other and nonreligious 30%, Jewish 2%, Protestant 2%

ETHNIC MIX: White 90%, *Mestizo* (European–Amerindian) 6%, Black 4%

GOVERNMENT: Presidential system

CURRENCY: Uruguayan peso = 100 centésimos

Uzbekistan

Sharing what is left of the Aral Sea with neighboring Kazakhstan, Uzbekistan lies on the ancient Silk Road between Asia and Europe. It is the most populous central Asian republic.

GEOGRAPHY

Arid and semiarid plains in much of the west. Fertile, irrigated farmland in the east lies below the peaks of the western Pamirs.

CLIMATE

Harsh continental climate. Summers can be extremely hot and dry; winters are cold.

PEOPLE & SOCIETY

Complex ethnic makeup. Ex-Communists are in firm control, but traditional social patterns based on clan, religion, and region have reemerged. Constitutional measures aim to control the influence of Islam: activities against Islamists have drawn international condemnation. Most people live in the fertile east. Birth rates are high, and the status of women continues to be low.

THE ECONOMY

Highly regulated. Reserves of natural gas, oil, coal, gold (has one of the world's largest gold mines), and other minerals. Cash crop is cotton: requires much irrigation. Grain imports necessary.

INSIGHT: *The Aral Sea holds just a tenth of its former volume of water, due to diversion of rivers for irrigation*

FACTFILE

OFFICIAL NAME: Republic of Uzbekistan

DATE OF FORMATION: 1991

CAPITAL: Tashkent

POPULATION: 28.9 million

TOTAL AREA: 172,741 sq. miles (447,400 sq. km)

DENSITY: 167 people per sq. mile

LANGUAGES: Uzbek*, Russian, Tajik, Kazakh

RELIGIONS: Sunni Muslim 88%, Orthodox Christian 9%, other 3%

ETHNIC MIX: Uzbek 80%, other 6%, Russian 6%, Tajik 5%, Kazakh 3%

GOVERNMENT: Presidential system

CURRENCY: Som = 100 tiyin

Vanuatu

An archipelago of 82 islands and islets in the South Pacific, Vanuatu was ruled jointly by the UK and France from 1906 until independence in 1980. Politics is democratic but volatile.

GEOGRAPHY
Mountainous and volcanic, with coral beaches and dense rainforest. Cultivated land along the coasts.

CLIMATE
Tropical. Temperatures and rainfall decline from north to south.

PEOPLE & SOCIETY
Indigenous Melanesians form a majority. Ni-Vanuatu culture is traditional; local social and religious customs are strong, despite centuries of missionary influence. Subsistence farming and fishing are the main activities. 80% of the population lives on the 12 main islands. Women have lower social status than men and payment of bride-price is common.

INSIGHT: *With 112 indigenous tongues, Vanuatu has the world's highest per capita density of languages*

THE ECONOMY
Reliant on aid. Main exports are copra (dried coconut), kava, cocoa, and beef. Tourism. Offshore banking: rules tightened after international pressure.

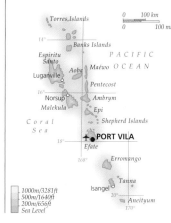

Torres Islands

Banks Islands

Espiritu Santo

Aoba   Maëwo

Luganville

Pentecost

Norsup   Ambrym

Malekula   Epi

Shepherd Islands

Coral Sea

PACIFIC OCEAN

✈ PORT VILA

Efate

Erromango

Tanna

Isangel

Aneityum

1000m/3281ft
500m/1640ft
200m/656ft
Sea Level

0   100 km
0   100 miles

FACTFILE

OFFICIAL NAME: Republic of Vanuatu
DATE OF FORMATION: 1980
CAPITAL: Port Vila
POPULATION: 300,000
TOTAL AREA: 4710 sq. miles (12,200 sq. km)
DENSITY: 64 people per sq. mile
LANGUAGES: Bislama (Melanesian pidgin)*,

English*, French*, other indigenous languages
RELIGIONS: Presbyterian 37%, other 19%, Anglican 15%, Roman Catholic 15%, traditional beliefs 8%, Seventh-day Adventist 6%
ETHNIC MIX: ni-Vanuatu 94%, European 4%, other 2%
GOVERNMENT: Parliamentary system
CURRENCY: Vatu = 100 centimes

Vatican City

The Vatican City, or Holy See, the seat of the Roman Catholic Church, is a walled enclave in the Italian city of Rome. It is the world's smallest fully independent state.

GEOGRAPHY

The Vatican's territory includes 10 other buildings in Rome, plus the papal residence. The Vatican Gardens cover half the City's area.

CLIMATE

Mild winters with regular rainfall. Hot, dry summers with occasional thunderstorms.

PEOPLE & SOCIETY

The Vatican has about 800 permanent inhabitants, including over 100 lay persons. Thousands of lay staff are also employed. Citizenship can be acquired through long-term residence and holding a position within the City. The reigning pope has supreme legislative and judicial powers, and holds office for life. Though the Vatican City is officially neutral, papal opinion has a great influence on the world's 1.2 billion Roman Catholics.

THE ECONOMY

Investments and voluntary contributions made by Catholics worldwide (known as Peter's Pence) are backed up by tourist revenue and the issue of Vatican stamps and coins.

INSIGHT: *The Vatican City is the spiritual center for one in six of the world's population*

FACTFILE

OFFICIAL NAME: State of the Vatican City
DATE OF FORMATION: 1929
CAPITAL: Vatican City
POPULATION: 839
TOTAL AREA: 0.17 sq. miles (0.44 sq. km)
DENSITY: 4935 people per sq. mile

LANGUAGES: Italian*, Latin*
RELIGIONS: Roman Catholic 100%
ETHNIC MIX: Cardinals are from many nationalities, but Italians form the largest group. Most resident lay persons are Italian. The current pope is from Argentina.
GOVERNMENT: Papal state
CURRENCY: Euro = 100 cents

Venezuela

Lying on the southern shores of the Caribbean, Venezuela was the first of Spain's colonies to seek independence. Despite large oil revenues, many Venezuelans still live in poverty.

GEOGRAPHY
Andes Mountains and the Maracaibo lowlands in the northwest. Central grassy plains are drained by the Orinoco River system. Forested Guiana Highlands in the southeast.

CLIMATE
Tropical. Hot and humid. Uplands are cooler. Orinoco plains are alternately parched or flooded.

PEOPLE & SOCIETY
Venezuela is historically a "melting pot," with immigrants from Europe and all over Latin America. The few indigenous Amerindians live in remote areas. Venezuela has one of the most urbanized societies in the region, with most of its population living in the northern cities. The left-wing rhetoric of President Hugo Chávez (1999–2013) raised opposition within Venezuela from urban society, and from the US.

THE ECONOMY
Oil accounts for 95% of exports; world's largest reserves. Coal, gold, other minerals. Nationalizations are enlarging the inefficient, corruption-prone state sector and deterring foreign investors.

INSIGHT: *Venezuela's Angel Falls is the world's tallest waterfall, with a total drop of 3210 ft (979 m)*

FACTFILE

OFFICIAL NAME: Bolivarian Republic of Venezuela

DATE OF FORMATION: 1830

CAPITAL: Caracas

POPULATION: 30.4 million

TOTAL AREA: 352,143 sq. miles (912,050 sq. km)

DENSITY: 89 people per sq. mile

LANGUAGES: Spanish*, Amerindian languages

RELIGIONS: Roman Catholic 96%, Protestant 2%, other 2%

ETHNIC MIX: *Mestizo* (European–Amerindian) 69%, White 20%, Black 9%, Amerindian 2%

GOVERNMENT: Presidential system

CURRENCY: Bolívar fuerte = 100 céntimos

Vietnam

French rule of Vietnam ended in 1954. Divided at 17°N, the US-backed South fought the Communist North. Reunified after the North's 1975 victory, it is run as a single-party state.

GEOGRAPHY

A heavily forested mountain range separates the northern Red River delta lowlands from the Mekong Delta in the south.

CLIMATE

Cool winters in north; south is tropical, with even temperatures.

PEOPLE & SOCIETY

Ethnic Vietnamese dominate; the Chinese minority was viewed as a corrupt bourgeoisie by the victorious Communists after the war. Mountain-based minorities *(montagnards)* were also sidelined; tensions persist over the settling of highlands by lowlanders. Women play an active role in society. There is no political or press freedom.

INSIGHT: *Intense US bombing and defoliant spraying in the 1962–1975 Vietnam War has scarred the landscape*

THE ECONOMY

Liberal economic policy *(doi moi)* from 1986: now one of fastest-growing economies. Major rice exporter. Cheap labor. Strong manufacturing: textiles, electrical goods. Diverse resource base.

FACTFILE

OFFICIAL NAME: Socialist Republic of Vietnam

DATE OF FORMATION: 1976

CAPITAL: Hanoi

POPULATION: 91.7 million

TOTAL AREA: 127,243 sq. miles (329,560 sq. km)

DENSITY: 730 people per sq. mile

LANGUAGES: Vietnamese*, Chinese, Thai, Khmer, Muong, Nung, Miao, Yao, Jarai

RELIGIONS: Other 74%, Buddhist 14%, Roman Catholic 7%, Cao Dai 3%, Protestant 2%

ETHNIC MIX: Vietnamese 86%, other 8%, Tay 2%, Thai 2%, Muong 2%

GOVERNMENT: One-party state

CURRENCY: Dông = 10 hao = 100 xu

Yemen

Located in southern Arabia, Yemen was formerly two countries: the People's Democratic Republic of Yemen (south and east) and the Yemen Arab Republic (northwest) were united in 1990.

GEOGRAPHY
Mountainous west with a fertile strip along the Red Sea. Arid desert and mountains elsewhere.

CLIMATE
Desert climate, modified by altitude, which affects temperatures by as much as 54°F (30°C).

PEOPLE & SOCIETY
Almost entirely of Arab and Bedouin descent, most Yemenis are Sunni Muslims, of the Shafi sect. In rural and northern areas, tribalism and Islamic orthodoxy are strong and most women wear the veil. Tension continues between cosmopolitan Aden and the more conservative north. Islamists have a growing political role. Popular protests that began in the 2011 "Arab Spring" ended President Saleh's 33-year rule. Foreigners are subject to attacks and kidnappings.

THE ECONOMY
Instability deters investment. Considerable oil and natural gas reserves. Agriculture is the largest employer: qat (mild narcotic), coffee, and cotton.

INSIGHT: *Mokha, on the Red Sea, gave its name to the first coffee beans exported to Europe in the 1600s*

3000m/9843ft	
2000m/6562ft	
1000m/3281ft	
500m/1640ft	
200m/656ft	
Sea Level	

0 100 km
0 100 miles

SAUDI ARABIA

OMAN

Ar Rub' al Khālī

Say'ūn

SANA

Sayhūt

Red Sea Al Hudaydah

Ash Shihr

Bayt al Faqīh

Ḥaḍramawt

Al Mukallā

Ta'izz

Al Mukhā (Mokha)

'Adan (Aden)

Gulf of Aden

Suquṭrā

'Abd al Kūrī

16°

48°

44°

52°

12°

FACTFILE
OFFICIAL NAME: Republic of Yemen
DATE OF FORMATION: 1990
CAPITAL: Sana
POPULATION: 24.4 million
TOTAL AREA: 203,849 sq. miles (527,970 sq. km)
DENSITY: 112 people per sq. mile

LANGUAGES: Arabic*
RELIGIONS: Sunni Muslim 55%, Shi'a Muslim 42%, Christian, Hindu, and Jewish 3%
ETHNIC MIX: Arab 99%, Afro-Arab, Indian, Somali, and European 1%
GOVERNMENT: Transitional regime
CURRENCY: Yemeni rial = 100 fils

Zambia

Bordered to the south by the Zambezi River, Zambia lies at the heart of southern Africa. In 1991, it made a peaceful transition from single-party rule to multiparty democracy.

GEOGRAPHY
A high savanna plateau, broken by mountains in northeast. Vegetation mainly trees and scrub.

CLIMATE
Tropical, with three seasons: cool and dry, hot and dry, and wet. Southwest is prone to drought.

PEOPLE & SOCIETY
There are more than 70 different ethnic groups, but there are fewer tensions than in many African states. Major groups are the Bemba (in the northeast), Tonga (south), Nyanja (east), and Lozi (west). There are also thousands of refugees, mostly from the DRC and Angola. A National Gender Policy was issued in 2000 to redress inequalities between the sexes. The standard of living has fallen in real terms since independence. One in seven adults is infected with HIV/AIDS.

THE ECONOMY
Copper: output has risen since 2000, when decades of falling global prices ended. New agricultural exports, notably flowers. Debt relief.

◆ **INSIGHT:** *Spray from Musi-o-Tunya (Victoria Falls) can be seen up to 20 miles (35 km) away*

FACTFILE

OFFICIAL NAME: Republic of Zambia

DATE OF FORMATION: 1964

CAPITAL: Lusaka

POPULATION: 14.5 million

TOTAL AREA: 290,584 sq. miles (752,614 sq. km)

DENSITY: 51 people per sq. mile

LANGUAGES: Bemba, Tonga, Nyanja, Lozi, Lala-Bisa, Nsenga, English*

RELIGIONS: Christian 63%, traditional beliefs 36%, Muslim and Hindu 1%

ETHNIC MIX: Bemba 34%, other African 26%, Tonga 16%, Nyanja 14%, Lozi 9%, European 1%

GOVERNMENT: Presidential system

CURRENCY: New Zamb. kwacha = 100 ngwee

Zimbabwe

Situated in southern Africa, Zimbabwe achieved independence from the UK in 1980. President Robert Mugabe, in power since then, has become increasingly authoritarian.

GEOGRAPHY

High plateaus in center bordered by Zambezi River in the north and Limpopo in the south. Rivers crisscross central area.

CLIMATE

Tropical, though moderated by the high altitude. Wet season November–March. Drought is common in the eastern highlands.

PEOPLE & SOCIETY

Two main ethnic groups: Shona in the north and east, and Ndebele in the south. Shona outnumber Ndebele by four to one. Whites are generally far more affluent than blacks. Official efforts to redress this imbalance (such as land redistribution) have become increasingly aggressive. The political opposition to Mugabe joined him in a fractious unity government from 2009 to 2013 in an attempt to rebuild the country.

THE ECONOMY

Undermined by mismanagement, corruption, and international isolation. High unemployment. Abandoned own currency in 2009 after hyperinflation.

INSIGHT: *The ruins of the 1000-year-old city of Great Zimbabwe, after which the country is named, are near modern-day Masvingo*

FACTFILE

OFFICIAL NAME: Republic of Zimbabwe
DATE OF FORMATION: 1980
CAPITAL: Harare
POPULATION: 14.1 million
TOTAL AREA: 150,803 sq. miles (390,580 sq. km)
DENSITY: 94 people per sq. mile
LANGUAGES: Shona, isiNdebele, English*

RELIGIONS: Syncretic 50%, Christian 25%, traditional beliefs 24%, other 1%
ETHNIC MIX: Shona 71%, Ndebele 16%, other African 11%, White 1%, Asian 1%
GOVERNMENT: Presidential system
CURRENCY: US $, S African rand, euro, UK £, Botswanan pula, Australian $, Chinese yuan, Indian rupee, Japanese yen all legal tender

Overseas territories

Despite the rapid process of global decolonization since World War II, around eight million people in more than 50 territories around the world continue to live under the protection of Australia, Denmark, France, the Netherlands, New Zealand, Norway, the UK, or the USA. These remnants of former colonial empires may have persisted for economic, strategic, or political reasons and are administered by the protecting country in a variety of ways.

AUSTRALIA

Australia's overseas territories have not been an issue since Papua New Guinea became independent in 1975. Consequently, there is no overriding policy toward them.

Ashmore & Cartier Is. *Ref: 124 A3*
STATUS: External territory
CLAIMED: 1931
POPULATION: None
AREA: 2 sq miles (5.2 sq km)

Christmas Island *Ref: 123 E5*

STATUS: External territory
CLAIMED: 1958
CAPITAL: The Settlement
POPULATION: 1530
AREA: 52 sq miles (135 sq km)

Cocos Islands *Ref: 123 D5*
STATUS: External territory
CLAIMED: 1955
CAPITAL: West Island
POPULATION: 596
AREA: 5.5 sq miles (14 sq km)

Coral Sea Islands *Ref: 126 B4*
STATUS: External territory
CLAIMED: 1969
POPULATION: 8 (Meteorologists)
AREA: 1.2 sq miles (3 sq km)

Heard & McDonald Is. *Ref: 123 C7*
STATUS: External territory
CLAIMED: 1947
POPULATION: None
AREA: 161 sq miles (417 sq km)

Norfolk Island *Ref: 124 D4*

STATUS: External territory
CLAIMED: 1774
CAPITAL: Kingston
POPULATION: 2210
AREA: 13 sq miles (34 sq km)

DENMARK

The Faroes and Greenland have had home rule since 1948 and 1979 respectively.

Faroe Islands *Ref: 65 F5*

STATUS: External territory
CLAIMED: 1380
CAPITAL: Tórshavn
POPULATION: 49,469
AREA: 540 sq miles (1399 sq km)

Greenland *Ref: 64 D3*

STATUS: External territory
CLAIMED: 1380
CAPITAL: Nuuk
POPULATION: 56,483
AREA: 836,109 sq miles (2,166,086 sq km)

Overseas territories

FRANCE

France's relations with *L'Outre-Mer* stress interdependence rather than independence. *Départements* have their own governments. *Collectivités* have some autonomy.

Clipperton Island *Ref: 135 F3*
STATUS: Dependency of French Polynesia
CLAIMED: 1935
POPULATION: None
AREA: 3.4 sq miles (9 sq km)

French Guiana *Ref: 41 H3*
STATUS: Overseas department
CLAIMED: 1817
CAPITAL: Cayenne
POPULATION: 250,109
AREA: 35,135 sq miles (91,000 sq km)

French Polynesia *Ref: 127 H4*
STATUS: Overseas collectivity
CLAIMED: 1843
CAPITAL: Papeete
POPULATION: 276,831
AREA: 1608 sq miles (4165 sq km)

French Southern & Antarctic Lands
Ref: 123 B6
STATUS: Overseas territory
CLAIMED: 1772, 1840, 1843, 1924
CAPITAL: Port-aux-Français
POPULATION: 140
AREA: 169,800 sq miles (439,781 sq km)

Guadeloupe *Ref: 37 G4*
STATUS: Overseas department
CLAIMED: 1635
CAPITAL: Basse-Terre
POPULATION: 405,739
AREA: 629 sq miles (1628 sq km)

Martinique *Ref: 37 G4*
STATUS: Overseas department
CLAIMED: 1635
CAPITAL: Fort-de-France
POPULATION: 386,486
AREA: 425 sq miles (1100 sq km)

Mayotte *Ref: 61 G2*
STATUS: Overseas department
CLAIMED: 1843
CAPITAL: Mamoudzou
POPULATION: 212,645
AREA: 144 sq miles (374 sq km)

New Caledonia *Ref: 126 D5*
STATUS: Special collectivity
CLAIMED: 1853
CAPITAL: Nouméa
POPULATION: 262,000
AREA: 7347 sq miles (19,100 sq km)

Réunion *Ref: 61 H4*
STATUS: Overseas department
CLAIMED: 1638
CAPITAL: Saint-Denis
POPULATION: 840,974
AREA: 970 sq miles (2500 sq km)

St Barthélemy *Ref: 37 G3*
STATUS: Overseas collectivity
CLAIMED: 1878
CAPITAL: Gustavia
POPULATION: 7267
AREA: 8 sq miles (21 sq km)

St Martin *Ref: 37 E5*
STATUS: Overseas collectivity
CLAIMED: 1648
CAPITAL: Marigot
POPULATION: 31,264
AREA: 20 sq miles (53 sq km)

Overseas territories

St Pierre & Miquelon *Ref: 21 G4*
STATUS: Overseas collectivity
CLAIMED: 1604
CAPITAL: Saint-Pierre
POPULATION: 5716
AREA: 93 sq miles (242 sq km)

Wallis & Futuna *Ref: 127 E4*
STATUS: Overseas collectivity
CLAIMED: 1842
CAPITAL: Mata'Utu
POPULATION: 15,561
AREA: 106 sq miles (274 sq km)

NETHERLANDS
These islands were once part of the Dutch West Indies. They are now self-governing,

Aruba *Ref: 37 E5*

STATUS: Constituent country
CLAIMED: 1636
CAPITAL: Oranjestad
POPULATION: 102,911
AREA: 75 sq miles (194 sq km)

Bonaire *Ref: 37 E5*

STATUS: Special municipality
CLAIMED: 1816
CAPITAL: Kralendijk
POPULATION: 18,413
AREA: 113 sq miles (294 sq km)

Curaçao *Ref: 37 E5*
STATUS: Constituent country
CLAIMED: 1815
CAPITAL: Willemstad
POPULATION: 153,500
AREA: 171 sq miles (444 sq km)

Saba *Ref: 37 G3*
STATUS: Special municipality
CLAIMED: 1816
CAPITAL: The Bottom
POPULATION: 1846
AREA: 5 sq miles (13 sq km)

Sint-Eustatius *Ref: 37 G3*
STATUS: Special municipality
CLAIMED: 1784
CAPITAL: Oranjestad
POPULATION: 4020
AREA: 8 sq miles (21 sq km)

Sint-Maarten *Ref: 37 G3*
STATUS: Constituent country
CLAIMED: 1648
CAPITAL: Phillipsburg
POPULATION: 39,689
AREA: 13 sq miles (34 sq km)

NEW ZEALAND

New Zealand remains responsible for its territories' foreign policy and defense.

Cook Islands *Ref: 127 G4*

STATUS: Associated territory
CLAIMED: 1901
CAPITAL: Avarua
POPULATION: 13,700
AREA: 91 sq miles (235 sq km)

Niue *Ref: 127 F5*

STATUS: Associated territory
CLAIMED: 1901
CAPITAL: Alofi
POPULATION: 1190
AREA: 102 sq miles (264 sq km)

Overseas territories

Tokelau *Ref: 127 F3*
- STATUS: Dependent territory
- CLAIMED: 1926
- CAPITAL: None
- POPULATION: 1337
- AREA: 4 sq miles (10 sq km)

NORWAY

There is a NATO base on Jan Mayen. Bouvet Island is a nature reserve.

Bouvet Island *Ref: 49 D7*
- STATUS: Dependency
- CLAIMED: 1928
- POPULATION: None
- AREA: 22 sq miles (58 sq km)

Jan Mayen *Ref: 65 F3*
- STATUS: Dependency
- CLAIMED: 1929
- POPULATION: 18 (Meteorologists)
- AREA: 147 sq miles (381 sq km)

Peter I. Island *Ref: 136 A3*
- STATUS: Dependency
- CLAIMED: 1931
- POPULATION: None
- AREA: 69 sq miles (180 sq km)

Svalbard *Ref: 65 F2*
- STATUS: Dependency
- CLAIMED: 1920
- CAPITAL: Longyearbyen
- POPULATION: 1872
- AREA: 24,289 sq miles (62,906 sq km)

UNITED KINGDOM

The UK's dependencies are locally governed by a mix of elected and appointed officials.

Anguilla *Ref: 37 G3*

- STATUS: Overseas territory
- CLAIMED: 1650
- CAPITAL: The Valley
- POPULATION: 16,086
- AREA: 37 sq miles (96 sq km)

Ascension Island *Ref: 49 C5*
- STATUS: Overseas territory
- CLAIMED: 1673
- CAPITAL: Georgetown
- POPULATION: 880
- AREA: 34 sq miles (88 sq km)

Bermuda *Ref: 17 E6*
- STATUS: Overseas territory
- CLAIMED: 1612
- CAPITAL: Hamilton
- POPULATION: 65,024
- AREA: 20 sq miles (53 sq km)

British Indian Ocean Territory
Ref: 122 C4
- STATUS: Overseas territory
- CLAIMED: 1814
- CAPITAL: Diego Garcia
- POPULATION: 4000
- AREA: 23 sq miles (60 sq km)

British Virgin Islands *Ref: 37 F3*
- STATUS: Overseas territory
- CLAIMED: 1672
- CAPITAL: Road Town
- POPULATION: 32,680
- AREA: 59 sq miles (153 sq km)

Cayman Islands *Ref: 36 B3*

- STATUS: Overseas territory
- CLAIMED: 1670
- CAPITAL: George Town
- POPULATION: 58,435
- AREA: 100 sq miles (259 sq km)

Overseas territories

Falkland Islands *Ref: 47 D7*

STATUS: Overseas territory
CLAIMED: 1832
CAPITAL: Stanley
POPULATION: 2840
AREA: 4699 sq miles (12,173 sq km)

Gibraltar *Ref: 74 D5*

STATUS: Overseas territory
CLAIMED: 1713
CAPITAL: Gibraltar
POPULATION: 29,185
AREA: 2.5 sq miles (6.5 sq km)

Guernsey *Ref: 71 D8*

STATUS: Crown Dependency
CLAIMED: 1066
CAPITAL: St. Peter Port
POPULATION: 65,849
AREA: 25 sq miles (65 sq km)

Isle of Man *Ref: 71 C5*

STATUS: Crown Dependency
CLAIMED: 1765
CAPITAL: Douglas
POPULATION: 85,888
AREA: 221 sq miles (572 sq km)

Jersey *Ref: 71 D8*

STATUS: Crown Dependency
CLAIMED: 1066
CAPITAL: St. Helier
POPULATION: 96,513
AREA: 45 sq miles (116 sq km)

Montserrat *Ref: 37 G4*

STATUS: Overseas territory
CLAIMED: 1632
CAPITAL: Brades *(de facto)*
POPULATION: 5215
AREA: 40 sq miles (102 sq km)

Pitcairn Group of Is. *Ref: 125 G4*

STATUS: Overseas territory
CLAIMED: 1887
CAPITAL: Adamstown
POPULATION: 48
AREA: 18 sq miles (47 sq km)

Saint Helena *Ref: 49 D5*

STATUS: Overseas territory
CLAIMED: 1673
CAPITAL: Jamestown
POPULATION: 7776
AREA: 47 sq miles (122 sq km)

South Georgia & the South Sandwich Islands *Ref: 49 C7*

STATUS: Overseas territory
CLAIMED: 1775
POPULATION: None
AREA: 1387 sq miles (3592 sq km)

Tristan da Cunha *Ref: 49 D6*

STATUS: Overseas territory
CLAIMED: 1612
CAPITAL: Edinburgh
POPULATION: 264
AREA: 38 sq miles (98 sq km)

Turks & Caicos Islands *Ref: 37 E2*

STATUS: Overseas territory
CLAIMED: 1766
CAPITAL: Cockburn Town
POPULATION: 33,098
AREA: 166 sq miles (430 sq km)

UNITED STATES

Commonwealth territories are self-governing and an integral part of the US. Unincorporated territories have varying degrees of autonomy.

Overseas territories

American Samoa *Ref: 127 F4*

STATUS: Unincorp. territory
CLAIMED: 1900
CAPITAL: Pago Pago
POPULATION: 55,165
AREA: 75 sq miles (195 sq km)

Baker & Howland Islands *Ref: 127 E2*

STATUS: Unincorporated territory
CLAIMED: 1856
POPULATION: None
AREA: 0.5 sq miles (1.4 sq km)

Guam *Ref: 126 B1*

STATUS: Unincorp. territory
CLAIMED: 1898
CAPITAL: Hagåtña
POPULATION: 165,124
AREA: 212 sq miles (549 sq km)

Jarvis Island *Ref: 127 G2*

STATUS: Unincorporated territory
CLAIMED: 1856
POPULATION: None
AREA: 1.7 sq miles (4.5 sq km)

Johnston Atoll *Ref: 125 E1*

STATUS: Unincorporated territory
CLAIMED: 1858
POPULATION: None
AREA: 1 sq mile (2.8 sq km)

Kingman Reef *Ref: 127 F2*

STATUS: Unincorporated territory
CLAIMED: 1856
POPULATION: None
AREA: 0.4 sq miles (1 sq km)

Midway Islands *Ref: 134 D2*

STATUS: Unincorporated territory
CLAIMED: 1867
CAPITAL: None
POPULATION: 40
AREA: 2 sq miles (5.2 sq km)

Navassa Island *Ref: 36 D3*

STATUS: Unincorporated territory
CLAIMED: 1856
POPULATION: None
AREA: 2 sq miles (5.2 sq km)

Northern Mariana Islands *Ref: 124 C1*

STATUS: Comm. territory
CLAIMED: 1947
CAPITAL: Saipan
POPULATION: 53,855
AREA: 177 sq miles (457 sq km)

Palmyra Atoll *Ref: 127 G2*

STATUS: Incorporated territory
CLAIMED: 1898
POPULATION: None
AREA: 5 sq miles (12 sq km)

Puerto Rico *Ref: 37 F3*

STATUS: Comm. territory
CLAIMED: 1898
CAPITAL: San Juan
POPULATION: 3.62 million
AREA: 3515 sq miles (9104 sq km)

Virgin Islands *Ref: 37 F3*

STATUS: Unincorp. territory
CLAIMED: 1917
CAPITAL: Charlotte Amalie
POPULATION: 104,737
AREA: 137 sq miles (355 sq km)

Wake Island *Ref: 124 D1*

STATUS: Unincorporated territory
CLAIMED: 1898
CAPITAL: None
POPULATION: 150 (US air base)
AREA: 2.5 sq miles (6.5 sq km)

International organizations

This listing provides acronym definitions for the main international organizations concerned with worldwide economics, trade, and defense, plus an indication of membership.

ASEAN
Association of Southeast Asian Nations
ESTABLISHED: 1967
MEMBERS: Brunei, Cambodia, Indonesia, Laos, Malaysia, Myanmar, Philippines, Singapore, Thailand, Vietnam

CIS
Commonwealth of Independent States
ESTABLISHED: 1991
MEMBERS: Arm., Az., Belarus, Kaz., Kyrgy., Mold., Russia, Tajik., Turkmen.*, Ukraine*, Uzbek. *Unofficial members*

COMM   *The Commonwealth of Nations*
ESTABLISHED: 1931; evolved out of the British Empire. Formerly known as the British Commonwealth of Nations.
MEMBERS: 53   *(Fiji currently suspended)*

EU   *European Union*
ESTABLISHED: 1965; formerly known as EEC (European Economic Community) and EC (Economic Community)
MEMBERS: Austria, Belg., Bulg., Croatia, Cyprus, Czech Rep., Denmark, Est., Fin., Fr., Ger., Greece, Hung., Ireland, Italy, Lat., Lith., Lux., Malta, Neth., Pol., Port., Rom., Slvka., Slvna., Spain, Swed., UK

G8   *Group of 8*
ESTABLISHED: 1994
MEMBERS: Canada, France, Germany, Italy, Japan, Russia, UK, US

IMF   *International Monetary Fund*
(UN agency)
ESTABLISHED: 1945
MEMBERS: 188

NAFTA
North American Free Trade Agreement
ESTABLISHED: 1994
MEMBERS: Canada, Mexico, US

NATO
North Atlantic Treaty Organization
ESTABLISHED: 1949
MEMBERS: Albania, Belg., Bulg., Canada, Croatia, Czech Rep., Denmark, Est., France, Ger., Greece, Hung., Iceland, Italy, Lat., Lith., Lux., Neth., Norway, Poland, Port., Rom., Slovakia, Slovenia, Spain, Turkey, UK, US

OPEC   *Organization of Petroleum Exporting Countries*
ESTABLISHED: 1960
MEMBERS: Algeria, Angola, Ecuador, Iran, Iraq, Kuwait, Libya, Nigeria, Qatar, Saudi Arabia, United Arab Emirates, Venezuela

UN   *United Nations*
ESTABLISHED: 1945
MEMBERS: 193; all nations are represented, except Taiwan and Kosovo. The Vatican City has "observer status" only. In 2012, a UN resolution granted Palestine the status of "non-member observer state."

WTO   *World Trade Organization*
ESTABLISHED: 1995
MEMBERS: 160 *(including EU, Hong Kong, Macao)*

Abbreviations

This glossary provides a comprehensive guide to the abbreviations used in this atlas.

abbrev. abbreviation
Afgh. Afghanistan
Amh. Amharic
anc. ancient
Ar. Arabic
Arm. Armenia/Armenian
Aus. Austria
Aust. Australia
Az. Azerbaijan

Bas. Basque
Bel. Belorussian
Belg. Belgium/Belgian
Bos. & Herz. Bosnia & Herzegovina
Bul. Bulgarian
Bulg. Bulgaria
Bur. Burmese

C Central
C. Cape
Cam. Cambodian
Cast. Castilian
Chin. Chinese
Comm. Commonwealth
Cord. Cordillera (Sp. mts.)
Cz. Czech
Czech Rep. Czech Republic

D.C. District of Columbia
Dan. Danish
Dominican Rep. Dominican Republic

E East
Emb. Embalse
Eng. English
Eq. Guinea Equatorial Guinea
Est. Estonia/Estonian

Faer. Faeroese
Fin. Finland/Finnish
Flem. Flemish

Fr. France/French
Geo. Georgia
Geor. Georgian
Ger. Germany/German
Gk. Greek

Heb. Hebrew
Hung. Hungary/Hungarian

I. Island
Ind. Indonesia, Indonesian
Is. Islands
It. Italian

Kaz. Kazakhstan/Kazakh
Kep. Kepulauan (Ind. island group)
Kir. Kirghiz
Kor. Korean
Kos. Kosovo
Kurd. Kurdish
Kyrgy. Kyrgyzstan

L. Lake, Lago
Lat. Latvia
Latv. Latvian
Leb. Lebanon
Liech. Liechtenstein
Lith. Lithuania/Lithuanian
Lux. Luxembourg

Mac. Macedonia
Med. Sea Mediterranean Sea
Mon. Montenegro
Mold. Moldova
Mt. Mount/Mountain
Mts. Mountains

N North
N. Korea North Korea
Neth. Netherlands
NW Northwest
NZ New Zealand

P. Pulau (Ind. island)
Peg. Pegunungan (Ind. mountain range)
Per. Persian
Pol. Poland/Polish
Port. Portugal, Portuguese

prev. previously
R. River, Rio, Río
Res. Reservoir
Rom. Romania/Romanian
Rus. Russian
Russ. Fed. Russian Federation

S South
S. Korea South Korea
SA South Africa
SCr. Serbian and Croatian
Serb. Serbia
Slvka. Slovakia
Slvna. Slovenia
Som. Somali
Sp. Spanish
St, St. Saint
Str. Strait
Swed. Swedish
Switz. Switzerland

Tajik. Tajikistan
Th. Thai
Turk. Turkish
Turkm. Turkmen
Turkmen. Turkmenistan

U.A.E. United Arab Emirates
UK United Kingdom
Ukr. Ukrainian
Uninhab. Uninhabitable
Unincorp. Unincorporated
Urug. Uruguayan
US United States of America
Uzb. Uzbek
Uzbek. Uzbekistan

var. variant
Vdkhr. Vodokhranilishche (Rus. reservoir)
Vdskh. Vodoskhovyshche (Ukr. reservoir)
Ven. Venezuela

W West
W. Sahara Western Sahara
Wel. Welsh

Yugo. Yugoslavia

Zamb. Zambian

A

Aabenraa Denmark 67 A8
Aachen Germany 76 A4
Aalborg Denmark 67 B7
Aalst Belgium 69 B5
Aba Nigeria 57 G5
Ābādān Iran 102 C4
Abadan Turkmenistan *prev.* Bezmein, Büzmeýin 104 B3
Abashiri Japan 112 D2
Abéché Chad 58 D3
Aberdeen Scotland, UK 70 D3
Aberdeen South Dakota, USA 25 E2
Aberdeen Washington, USA 26 A2
Aberystwyth Wales, UK 71 C6
Abhā Saudi Arabia 103 B6
Abidjan Côte d'Ivoire 56 D5
Abilene Texas, USA 29 F3
Abomey Benin 57 F4
Abu Dhabi *capital of* United Arab Emirates *var.* Abū Ẓaby 103 D5
Abuja *capital of* Nigeria 57 G4
Abū Ẓaby *see* Abu Dhabi
Acapulco Mexico 33 E5
Acarai Mountains *mountain range* Brazil/Guyana 41 F3
Acarigua Venezuela 40 D1
Accra *capital of* Ghana 57 E5
Acklins Island *island* The Bahamas 36 D2
Aconcagua, Cerro *peak* Argentina 46 B4
A Coruña Spain *Cast.* La Coruña 74 C1
ACT *see* Australian Capital Territory
Adalia *see* Antalya
Adalia, Gulf of *see* Antalya Körfezi
'Adan Yemen *Eng.* Aden 103 B7
Adana Turkey *var.* Seyhan 98 D4

Adapazarı Turkey *var.* Sakarya 98 B2
Ad Dahnā' *desert* Saudi Arabia 103 C5
Ad Dakhla Western Sahara 52 A4
Ad Dawḥah *see* Doha
Addis Ababa *capital of* Ethiopia *Amh.* Ādīs Ābeba 55 C5
Adelaide Australia 131 B6
Adélie, Terre d' *territory* Antarctica 136 C4
Aden *see* 'Adan
Aden, Gulf of *sea feature* Indian Ocean 122 A3
Adige *river* Italy 78 C2
Ādīs Ābeba *see* Addis Ababa
Adıyaman Turkey 99 E4
Adriatic Sea Mediterranean Sea 78 D4
Aegean Sea Mediterranean Sea *Gk.* Aigaío Pélagos, *Turk.* Ege Denizi 87 D5
Aeolian Islands *see* Isole Eolie
Afghanistan *country* C Asia 104-105
Africa 50-51
Africa, Horn of *physical region* Ethiopia/Somalia 122 A3
Afyon Turkey *prev.* Afyonkarahisar 98 B3
Afyonkarahisar *see* Afyon
Agadez Niger 57 G3
Agadir Morocco 52 B2
Agassiz Fracture Zone *tectonic feature* Pacific Ocean 135 E4
Agen France 73 B6
Āgra India 116 D3
Agrigento Italy 79 C7
Agrínio Greece 87 B5
Aguarico *river* Ecuador/Peru 40 B4
Aguascalientes Mexico 32 D4
Ahaggar *mountains* Algeria *var.* Hoggar 53 E4
Ahmadābād India 116 C4
Ahvāz Iran 102 C4
Ahvenanmaa *see* Åland
Aigaío Pélagos *see* Aegean Sea
Aintab *see* Gaziantep

Aïr, Massif de l' *region* Niger 57 G2
Aix-en-Provence France 73 D6
Ajaccio Corse, France 73 E7
Ajdābiyā Libya 53 G2
Ajmer India 116 D3
Akaba *see* Al 'Aqabah
Akchâr *desert* Mauritania 56 C2
Akimiski Island *island* Canada 20 C3
Akita Japan 112 D3
Akjoujt Mauritania 56 C2
Akmola *see* Astana
Akmolinsk *see* Astana
Akpatok Island *island* Canada 21 E1
Akra Kanestron *see* Palioúri, Akrotíri
Akron Ohio, USA 22 D3
Aksai Chin *disputed region* China/India 108 B4
Aktau Kazakhstan *prev.* Shevchenko 96 A4
Akureyri Iceland 65 E4
Akyab *see* Sittwe
Alabama *state* USA 30 D3
Alacant *see* Alicante
Alajuela Costa Rica 34 D4
Alamogordo New Mexico, USA 28 D3
Åland *island group* Finland *Fin.* Ahvenanmaa 67 D6
Al 'Aqabah Jordan *var.* Akaba 101 B7
Alaska *state* USA 18
Alaska, Gulf of *sea feature* Pacific Ocean 16 C3
Alaska Range *mountain range* Alaska, USA 18 C3
Albacete Spain 75 E3
Alba Iulia Romania 90 B4
Albania *country* SE Europe 83
Albany Australia 129 B7
Albany Georgia, USA 31 E3
Albany New York, USA 23 F3
Albany Oregon, USA 26 A3
Albany *river* Canada 20 B3

Al Başrah Iraq *var.* Basra 102 C4

Al Baydā' Libya 53 G2

Albert, Lake *lake* Uganda/Dem. Rep. Congo 59 E5

Alberta *province* Canada 19 E4

Albi France 73 C6

Albuquerque New Mexico, USA 28 D2

Alcácer do Sal Portugal 74 C4

Aldabra Group *island group* Seychelles 61 G2

Aleg Mauritania 56 C3

Aleksandriya *see* Oleksandriya

Aleksandropol' *see* Gyumri

Aleksinac Serbia 82 E4

Alençon France 72 B3

Alessandria Italy 78 B2

Ålesund Norway 67 A5

Aleutian Basin *undersea feature* Bering Sea 134 D1

Aleutian Islands *islands* Alaska, USA 18 A3

Aleutian Trench *undersea feature* Pacific Ocean 134 D1

Alexander Island *island* Antarctica 136 A3

Alexandra New Zealand 133 B7

Alexandretta *see* İskenderun

Alexandria *see* Al Iskandariyah

Alexandria Louisiana, USA 30 B3

Alexandroúpoli Greece 86 D3

Al Fāshir *see* El Fasher

Alföld *see* Great Hungarian Plain

Algarve *region* Portugal 74 C4

Algeciras Spain 74 D5

Algeria *country* N Africa 52-53

Alghero Italy 79 A5

Algiers *capital of* Algeria 52 D1

Al Ḥasakah Syria 100 D2

Al Ḥudaydah Yemen 103 B7

Al Ḥufūf Saudi Arabia 103 C5

Alicante Spain *Cat.* Alacant 75 F4

Alice Springs Australia 130 A4

Al Iskandarīyah Egypt *Eng.* Alexandria 54 B1

Al Ismā'īliya Egypt *Eng.* Ismalia 54 B1

Al Jawf Saudi Arabia 102 B4

Al Jazīrah *region* Iraq/Syria 100 E2

Al Jīzah Egypt *var.* El Gîza 54 B1

Al Karak Jordan 101 B6

Al Khalīl *see* Hebron

Al Khārijah Egypt *var.* El Khârga 54 B2

Al Khums Libya 53 F2

Al Khurţūm *see* Khartoum

Alkmaar Netherlands 68 C2

Al Kufrah Libya 53 H4

Al Lādhiqīyah Syria *Eng.* Latakia 100 B3

Allahābād India 117 E4

Allenstein *see* Olsztyn

Allentown Pennsylvania, USA 23 F4

Alma-Ata *capital of* Kazakhstan *Rus./Kaz.* Almaty 96 C5

Al Madīnah Saudi Arabia *Eng.* Medina 102 A5

Al Mafraq Jordan 101 B5

Almalyk Uzbekistan *Uzb.* Olmaliq 105 E2

Al Manāmah *see* Manama

Al Marj Libya 53 G2

Almaty *see* Alma-Ata

Al Mawşil Iraq *Eng.* Mosul 102 B3

Almelo Netherlands 68 E3

Almería Spain 75 E5

Al Minyā Egypt 54 B2

Al Mukallā Yemen 103 C7

Alofi *capital of* Niue 127 F5

Alor, Kepulauan *island group* Indonesia 121 E5

Alps *mountain range* C Europe 62 D4

Al Qāhirah *see* Cairo

Al Qāmishlī Syria *var.* Kamishli 100 E1

Al Qunayţirah Syria 100 B4

Altai Mountains *mountain range* C Asia 108 C2

Altamura Italy 79 E5

Altar, Desierto de *Desert* Mexico/USA *var.* Sonoran Desert 32 A1

Altay China 108 C2

Altay Mongolia 108 D2

Altun Shan *mountain range* China 108 C3

Alturas California, USA 26 B4

Al Uqşur Egypt *Eng.* Luxor 54 B2

Alytus Lithuania *Pol.* Olita 89 B5

Amadeus, Lake *seasonal lake* Australia 129 E5

Amakusa-nada *island group* Japan 113 A6

Amami-Ō-shima *island* Japan 113 A8

Amarillo Texas, USA 29 E2

Amazon *river* South America 38 C3

Amazon Basin *region* C South America 42 D2

Ambanja Madagascar 61 G2

Ambarchik Russian Federation 97 G2

Ambato Ecuador 40 A4

Amboasary Madagascar 61 F4

Ambon Indonesia 121 F4

Ambositra Madagascar 61 G3

Ambriz Angola 60 B1

Amdo China 108 C4

Ameland *island* Netherlands 68 D1

American Falls Reservoir *Reservoir* Idaho, USA 26 E4

American Samoa *unincorporated territory* USA, Pacific Ocean 127 F4

Amersfoort Netherlands 68 D3

Amga *river* Russian Federation 95 F2

Amiens France 72 C3

Amīndivi Islands *island group* India 114 C2

Amirante Islands *island group* Seychelles 61 H1

Amman *capital of* Jordan 101 B5

Ammassalik Greenland *var.* Angmagssalik 64 D4

Ammochostos see Gazimağusa

Âmol Iran 102 C3

Amorgós island Greece 87 D6

Amritsar India 116 D2

Amsterdam capital of Netherlands 68 C3

Amsterdam Island island French Southern and Antarctic Lands 123 C6

Am Timan Chad 58 C3

Amu Darya river C Asia 104 D3

Amundsen Gulf sea feature Canada 19 E2

Amundsen Plain undersea feature Pacific Ocean 136 B4

Amundsen Sea Antarctica 97 G4

Amur river E Asia 97 G4 107 E1

Anabar river Russian Federation 97 E1

Anadolu Dağları see Doğu Karadeniz Dağları

Anadyr' Russian Federation 97 H1

Anápolis Brazil 43 F4

Anatolia region SE Europe 85 G3

Anchorage Alaska, USA 18 C3

Ancona Italy 78 C3

Andalucia region Spain 74 D4

Andaman Islands island group India 115 H2 119 A5

Andaman Sea Indian Ocean 122 D3

Andes mountain range South America 39 B6

Andijon Uzbekistan Rus. Andizhan 105 F2

Andhra Pradesh state India 115 E1

Andizhan see Andijon

Andorra country SW Europe 73 B6

Andorra la Vella capital of Andorra 73 B6

Ándros island Greece 87 D6

Andros Island island The Bahamas 36 C1

Angara river C Asia 95 D3

Ángel de la Guarda, Isla island Mexico 32 B2

Angel Falls see Salto Ángel

Angeles Philippines 121 E1

Ángel, Salto waterfall Venezuela Eng. Angel Falls 41 F2

Ångermanälven river Sweden 66 C4

Angers France 72 B4

Anglesey island Wales, UK 71 C5

Angmagssalik see Ammassalik

Angola country C Africa 60

Angola Basin undersea feature Atlantic Ocean 49 D6

Angora see Ankara

Angoulême France 73 B5

Angren Uzbekistan 105 E2

Anguilla overseas territory UK, West Indies 37

Anhui province China var. Anhwei, Wan 111 C5

Anhwei see Anhui

Anjouan island Comoros 61 F2

Ankara capital of Turkey prev. Angora 98 C3

Annaba Algeria 53 E1

An Nafūd desert region Saudi Arabia 102 B4

An Najaf Iraq var. Najaf 102 B4

Annapolis Maryland, USA 23 F4

Ann Arbor Michigan, USA 22 C3

Annecy France 73 D5

Anshan China 110 A4

Ansongo Mali 57 E3

Antakya Turkey var. Hatay 98 D4

Antalaha Madagascar 61 G2

Antalya Turkey prev. Adalia 98 B4

Antalya, Gulf of see Antalya Körfezi

Antalya Körfezi sea feature Mediterranean Sea Eng. Gulf of Antalya, var. Gulf of Adalia 98 B4

Antananarivo capital of Madagascar prev. Tananarive 61 G3

Antarctica 136

Antarctic Peninsula peninsula Antarctica 136 A2

Antequera Spain 74 D5

Anticosti, Île d' island Canada 21 F3

Antigua island Antigua & Barbuda 37 G3

Antigua & Barbuda country West Indies 37

Anti-Lebanon mountains Lebanon/Syria 100 B4

Antipodes Islands island group New Zealand124 D5

Antofagasta Chile 46 B2

Antsirañana Madagascar 61 G2

Antsohihy Madagascar 61 G2

Antwerp see Antwerpen

Antwerpen Belgium Eng. Antwerp 69 C5

Anyang China 110 C4

Aoga-shima island Japan 113 D6

Aomori Japan 112 D3

Aoraki peak New Zealand var. Cook, Mount 133 B6

Aosta Italy 78 A2

Aoukâr plateau Mauritania 56 D3

Apeldoorn Netherlands 68 D3

Apennines see Appennino

Apia capital of Samoa 127 F4

Appalachian Mountains mountain range E USA 17 D5

Appennino mountain range Italy Eng. Apennines 78 C4

Apure river Venezuela 40 D2

Aqaba see Al 'Aqabah

Aqaba, Gulf of sea feature Red Sea Ar. Khalij al 'Aqabah 101 A8

'Aqabah, Khalij al see Aqaba, Gulf of

Āqchah Afghanistan var. Āqcheh 104 D3

Āqcheh see Āqchah

Arabian Basin undersea feature Indian Ocean 122 B3

Arabian Peninsula peninsula Asia 85 H5 94 B5 103 C5

Arabian Sea Indian Ocean 122 B3

Aracaju Brazil 43 H3

Arad Romania 90 B4

Arafura Sea Asia/Australasia 126 A4

Araguaia *river* Brazil 43 F3

Arāk Iran 102 C3

Araks *see* Aras

Arak's *see* Aras

Aral Sea *inland sea* Kazakhstan/Uzbekistan 94 C3

Araouane Mali 57 E2

Ararat, Mount *peak* Turkey *var.* Great Ararat, *Turk.* Büyükağrı Dağı 94 F3

Aras *river* SW Asia *Arm.* Arak's, *Per.* Rūd-e Aras, *Rus.* Araks, *Turk.* Aras Nehri 99 G3

Aras Nehri *see* Aras

Arauca Colombia 40 C2

Arauca *river* Colombia/Venezuela 40 C2

Arbil Iraq *Kurd.* Hawlēr 102 B3

Arctic Ocean 18-19 137

Arda *river* Bulgaria/Greece 86 C3

Ardabil Iran 102 C3

Ardennes *region* W Europe 69 D7

Arendal Norway 67 A6

Arensburg *see* Kuressaare

Arequipa Peru 42 B4

Arezzo Italy 78 C3

Argentina *country* S South America 46-47

Argentine Basin *undersea feature* Atlantic Ocean 49 B7

Argun *river* China/Russian Federation 95 E3

Århus Denmark 67 A7

Arica Chile 46 B1

Arizona *state* USA 28 B2

Arkansas *state* USA 30 B1

Arkansas *river* C USA 17 C5

Arkhangel'sk Russian Federation 92 C3 96 C2

Arles France 73 D6

Arlington Texas, USA 29 G3

Arlington Virginia, USA 23 E4

Arlon Belgium 69 D8

Armenia *country* SW Asia 99 G2

Armenia Colombia 40 B3

Armidale Australia 131 D5

Arnhem Netherlands 68 D4

Arnhem Land *region* Australia 128 E2

Arno *river* Italy 78 B3

Arran *island* Scotland, UK 70 C4

Ar Raqqah Syria 100 C2

Arras France 72 C3

Ar Riyāḍ *see* Riyadh

Ar Rub 'al Khālī *desert* Asia *Eng.* Empty Quarter, Great Sandy Desert 103 C6

Ar Rustāq Oman *var.* Rostak 103 D5

Artesia New Mexico, USA 28 D3

Artigas Uruguay 44 B4

Aru, Kepulauan *island group* Indonesia 121 G5

Arua Uganda 55 B6

Aruba *constituent country* Netherlands, West Indies 37 E5

Arusha Tanzania 55 C7

Asad, Buḩayrat al *Lake* Syria *Eng.* Lake Assad 100 C2

Asadābād Afghanistan 105 E4

Asahikawa Japan 112 D2

Asamankese Ghana 57 E5

Ascension Island *overseas territory* UK, Atlantic Ocean 49 C5

Ascoli Piceno Italy 78 C4

'Aseb Eritrea *var.* Assab 54 D4

Ashburton New Zealand 133 C6

Asheville North Carolina, USA 31 E1

Aşgabat *capital of* Turkmenistan *prev.* Ashkhabad, Poltoratsk 104 C3

Ashkhabad *see* Aşgabat

Ashmore and Cartier Islands *Australian external territory* Indian Ocean 124 A3

Ash Shāriqah United Arab Emirates *Eng.* Sharjah 103 D5

Asia 94-95 106-107

Asmara *capital of* Eritrea *Amh.* Asmera 54 C4

Asmera *see* Asmara

Assab *see* 'Aseb

As Salţ Jordan *var.* Salt 101 B5

Assamakka Niger 57 F2

Assen Netherlands 68 E2

Assad, Lake *see* Asad, Buḩayrat al

As Sulayyil Saudi Arabia 103 B6

As Suwaydā' Syria 101 B5

As Suways Egypt *Eng.* Suez 54 B1

Astana *country capital* Kazakhstan *prev.* Akmola, Akmolinsk, Tselinograd, Kaz. Aqmola. 96 C4

Astoria Oregon, USA 26 A2

Astrakhan' Russian Federation 93 B7

Astypálaia *island* Greece 87 D6

Asunción *capital of* Paraguay 44 B3

Aswān Egypt 54 B2

Asyūţ Egypt 54 B2

Atacama Desert *desert* Chile 46 B2

Atamyrat *prev.* Kerki. Turkmenistan 104 D3

Aţār Mauritania 56 C2

Atbara Sudan 54 C3

Athabasca, Lake *lake* Canada 19 F4

Athens *capital of* Greece *Gk.* Athína, *prev.* Athínai 87 C5

Athens Georgia, USA 31 E2

Athína *see* Athens

Athínai *see* Athens

Athlone Ireland 71 B5

Ati Chad 58 C3

Atlanta Georgia, USA 30 D2

Atlantic City New Jersey, USA 23 F4

Atlantic Ocean 48-49

Atlantic-Indian Basin *undersea feature* Indian Ocean 136 B1

Atlantic-Indian Ridge *undersea feature* Atlantic Ocean 49 D7

Atlas Mountains *mountain range* Morocco 52 C2

Aţ Ţalfīlah Jordan 101 B6

At Ta'if — Balabac Strait

At Ta'if Saudi Arabia 102 B6
Attapu Laos 119 E5
Attawapiskat Canada 20 C3
Attawapiskat *river* Canada 20 B3
Attu Island *island* Alaska, USA 18 A2
Auch France 73 B6
Auckland New Zealand 132 D3
Auckland Islands *island group* New Zealand124 D5
Augsburg Germany 77 C6
Augusta Australia 129 B7
Augusta Georgia, USA 31 E2
Augusta Maine, USA 23 G2
Aurillac France 73 C5
Aurora Colorado, USA 24 D4
Aurora Illinois, USA 22 B3
Aussig *see* Ústí nad Labem
Austin Texas, USA 29 G4
Australasia 124-125
Australes, Îles *island group* French Polynesia 125 F4
Austral Fracture Zone *tectonic feature* Pacific Ocean 125 H4
Australia *country* Pacific Ocean 124
Australian Alps Australia 131 D7
Australian Capital Territory *territory* Australia *abbrev.* A.C.T. *131* D6
Austria *country* C Europe 77
Auxerre France 72 C4
Avarua *capital of* Cook Islands 127 G5
Aveiro Portugal 74 C2
Avignon France 73 D6
Ávila Spain 74 D2
Avilés Spain 74 D1
Awbārī Libya 53 F3
Axel Heiberg Island *island* Canada 19 F1
Axios *see* Vardar
Ayacucho Peru 42 B4
Aydarko'l Ko'li *lake* Uzbekistan *var.* Aydarkül 104 D2
Aydarkül *see* Aydarko'l Ko'li
Aydın Turkey 98 A3

Ayer's Rock *see* Uluru
Ayr Scotland, UK 70 C4
Ayutthaya Thailand 119 C5
Ayvalık Turkey 98 A3
Azaouâd *desert* Mali 57 E2
A'zāz Syria 100 B2
Azerbaijan *country* SW Asia 99 G2
Azores *islands* Portugal, Atlantic Ocean 48 C3
Azov, Sea of Black Sea *Ukr.* Azovs'ke More, *Rus.* Azovskoye More 93 A6 91 G4
Azovs'ke More *see* Azov, Sea of
Azovskoye More *see* Azov, Sea of
Azul Argentina 46 D4
Azur, Côte d' *coastal region* France 73 E6
Az Zarqā' Jordan 101 B5
Az Zāwiyah Libya 53 F2

B

Baalbek Lebanon *var.* Ba'labakk 100 B4
Babeldaob *island* Palau 124 B2
Babruysk/Bobruysk Belarus *Rus.* Bobruysk 89 D6
Babuyan Channel *channel* Philippines 121 E1
Bacan, Pulau *island* Indonesia 121 F4
Bačka Topola Serbia 82 D3
Bacău Romania 90 C4
Badajoz Spain 74 C4
Baden Switzerland 77 E6
Bādiyat ash Shām *see* Syrian Desert
Baffin Bay *sea feature* Atlantic Ocean 48 B1
Baffin Island *island* Canada 19 G2
Bafing *river* Africa 56 C3
Bafoussam Cameroon 58 B4
Bagdad *see* Baghdad
Bagé Brazil 44 C4

Baghdad *capital of* Iraq *var.* Bagdad, *Ar.* Baghdād 102 B3
Baghdād *see* Baghdad
Baghlān Afghanistan 105 E3
Bago Myanmar *prev.* Pegu 118 B4
Bagoé *river* Côte d'Ivoire/Mali 56 D4
Baguio Philippines 121 E1
Bahamas, The *country* West Indies, Atlantic Ocean 36
Baharden *see* Baharly
Baharly Turkmenistan *prev.* Baharden, Bäherden, Bakharden, Bakherden 104 B3
Bahāwalpur Pakistan 116 C3
Bäherden *see* Baharly
Bahía Blanca Argentina 47 C5
Bahia, Islas de la *islands* Honduras 34 D2
Bahir Dar Ethiopia 54 C4
Bahrain *country* SW Asia 103 C5
Baia Mare Romania 90 B3
Baikal, Lake *see* Baykal, Ozero
Bairiki *capital of* Kiribati 127 E2
Baishan China 110 E3
Baja Hungary 81 C7
Baja California *peninsula* Mexico *Eng.* Lower California 32 B2
Bajo Nuevo *island* Colombia 35 F2
Baker Oregon, USA 26 C3
Baker & Howland Islands *unincorporated territory* USA, Pacific Ocean 125 E2
Bakersfield California, USA 27 C7
Bakharden *see* Baharly
Bakherden *see* Baharly
Bākhtarān *see* Kermānshāh
Bakı *see* Baku
Baku *capital of* Azerbaijan *Az.* Bakı, *var.* Baky 99 H2
Baky *see* Baku
Balabac Strait *sea feature* South China Sea/Sulu Sea 120 D2

Ba'labakk see Baalbek

Balakovo Russian Federation 93 C6

Bālā Murghāb Afghanistan 104 D4

Balaton lake Hungary var. Lake Balaton, Ger. Plattensee 81 C7

Balaton, Lake see Balaton

Balbina, Represa Reservoir Brazil 42 D2

Baleares, Islas island group Spain Eng. Balearic Islands 75 H3

Balearic Islands see Baleares, Islas

Bali island Indonesia 120 D5

Balıkesir Turkey 98 A3

Balikpapan Indonesia 120 D4

Balkanabat Turkmenistan prev. Nebitdag 104 B2

Balkan Mountains mountain range Bulgaria Bul. Stara Planina 86 C2

Balkhash Kazakhstan 96 C5

Balkhash, Lake see Balkhash, Ozero

Balkhash, Ozero lake Kazakhstan Eng. Lake Balkhash 94 C3

Ballarat Australia 131 C7

Balsas river Mexico 33 E5

Bălţi Moldova 90 D3

Baltic Port see Paldiski

Baltic Sea Atlantic Ocean 67 C7

Baltimore Maryland, USA 23 F4

Baltischport see Paldiski

Baltiski see Paldiski

Bamako capital of Mali 56 D3

Bambari Central African Republic 58 D4

Bamenda Cameroon 58 B4

Banaba island Kiribati prev. Ocean Island 127 E2

Bandaaceh Indonesia 120 A3

Banda, Laut see Banda Sea

Banda Sea sea feature Pacific Ocean Ind. Laut Banda 121 F4

Bandar-e 'Abbās Iran 102 D4

Bandar-e Büshehr Iran 102 C4

Bandar Lampung Indonesia prev. Tanjungkarang 120 C4

Bandar Seri Begawan capital of Brunei 120 D3

Bandon Oregon, USA 26 A3

Bandundu Dem. Rep. Congo 59 C6

Bandung Indonesia 120 C5

Bangalore India 114 D2

Banggai, Kepulauan island group Indonesia 121 E4

Banghāzī Libya Eng. Benghazi 53 G2

Bangka, Palau island Indonesia 120 C4

Bangkok capital of Thailand Th. Krung Thep 119 C5

Bangladesh country S Asia 117

Bangor Northern Ireland, UK 71 B5

Bangor Maine, USA 23 G2

Bangui capital of Central African Republic 58 C4

Bani river Mali 56 D3

Bani Suwayf Egypt var. Beni Suef 54 B1

Banja Luka Bosnia & Herzegovina 82 B3

Banjarmasin Indonesia 120 D4

Banjul capital of Gambia 56 B3

Banks Island island Canada 19 E2

Banks Islands island group Vanuatu, Pacific Ocean 126 D4

Banks Peninsula peninsula New Zealand 133 C6

Banks Strait sea feature Tasman Sea 131 C7

Banská Bystrica Slovakia Ger. Neusohl, Hung. Besztercebánya 81 C6

Bantry Bay sea feature Ireland 71 A6

Banyo Cameroon 58 B4

Banzare Seamounts undersea feature Indian Ocean 123 C7

Baotou China 109 F3

Baranavichy/Baranovichi Belarus Rus. Baranovichi, Pol. Baranowicze 89 C6

Baranovichi see Baranavichy/ Baranovichi

Baranowicze see Baranavichy/ Baranovichi

Barbados country West Indies 37 H4

Barbuda island Antigua & Barbuda 37 G3

Barcaldine Australia 130 C4

Barcelona Spain 75 G2

Barcelona Venezuela 41 E1

Barcolod City Philippines 121 E2

Bareilly India 117 E3

Barentsburg Svalbard 65 F2

Barentsøya island Svalbard 65 G2

Barents Sea Arctic Ocean 137 H5

Bari Italy 79 E5

Barinas Venezuela 40 D2

Barisan, Pegunungan mountains Indonesia 120 B4

Barkly Tableland plateau Australia 130 B3

Barlavento, Ilhas de island group Cape Verde var. Windward Islands 56 A2

Bar-le-Duc France 72 D3

Barlee, Lake lake Australia 129 B 5

Barlee Range mountain range Australia 128 B4

Barnaul Russian Federation 96 D4

Barnstaple England, UK 71 C7

Barquisimeto Venezuela 40 D1

Barra island Scotland, UK 70 B3

Barranquilla Colombia 40 B1

Barrier Range mountain range Australia 131 C5

Barrow river Ireland 71 B6

Barstow California, USA 27 C7

Bartang river Tajikistan 105 F3

Bartica Guyana 41 G2

Baruun-Urt Mongolia 109 F2

Barwon River river Australia 131 D5

Barysaw Belarus Rus. Borisov 89 D5

Basarabeasca Moldova 90 D4

Basel Switzerland 77 B6

Basra see Al Başrah

Bassein see Pathein

Basse-Terre capital of Guadeloupe 37 G4

Basseterre capital of St Kitts & Nevis 37 G3

Bass Strait sea feature Australia 131 C7

Bastia Corse, France 73 E7

Bastogne Belgium 69 D7

Bata Equatorial Guinea 58 A5

Batangas Philippines 121 E2

Bătdâmbâng Cambodia 119 D5

Bath England, UK 71 D6

Bathurst Canada 21 F4

Bathurst Island island Australia 128 D2

Bathurst Island island Canada 19 F2

Bâtin, Wâdî al dry watercourse Asia 102 C4

Batman Turkey var. İluh 99 E4

Batna Algeria 53 E1

Baton Rouge Louisiana, USA 30 B3

Batticaloa Sri Lanka 115 E3

Batumi Georgia 99 F2

Bauru Brazil 44 D2

Bavarian Alps mountains Austria/Germany 77 C6

Bayamo Cuba 36 C2

Bayan Har Shan mountain range China 108 D4

Bayanhongor Mongolia 108 D2

Bay City Michigan, USA 22 C3

Baydhabo Somalia 55 D6

Baykal, Ozero lake Russian Federation Eng. Lake Baikal 95 E3

Bayonne France 73 A6

Bayramaly Turkmenistan 104 C3

Bayrūt see Beirut

Beaufort Sea Arctic Ocean 137 F2

Beaufort West South Africa 60 D5

Beaumont Texas, USA 29 H4

Beauvais France 72 C3

Béchar Algeria 52 C2

Be'er Sheva' Israel 101 A6

Beijing capital of China var. Peking 110 C4

Beira Mozambique 61 E3

Beirut capital of Lebanon var. Beyrouth, Bayrūt 100 B4

Beja Portugal 74 C4

Béjaïa Algeria 53 E1

Bek-Budi see Karshi

Békéscsaba Hungary 81 D7

Belarus country E Europe var. Belorusia 89

Belau see Palau

Belcher Islands islands Canada 20 C2

Beledweyne Somalia 55 D5

Belém Brazil 42 F1

Belfast Northern Ireland, UK 71 B5

Belfort France 72 E4

Belgaum India 114 C1

Belgium country W Europe 69

Belgorod Russian Federation 93 A5

Belgrade capital of Serbia SCr. Beograd 82 D3

Belitung, Pulau island Indonesia 120 C4

Belize country Central America 34

Belize City Belize 34 C1

Belle Île island France 72 A4

Belle Isle, Strait of sea feature Canada 21 G3

Bellevue Washington, USA 26 B2

Bellingham Washington, USA 26 B1

Bellingshausen Sea Antarctica 136 A3

Bello Colombia 40 B2

Bellville South Africa 60 C5

Belmopan capital of Belize 34 C1

Belo Horizonte Brazil 45 F1

Belorussia see Belarus

Belostok see Białystok

Beloye More Arctic Ocean Eng. White Sea 71

Belyy, Ostrov island Russian Federation 137 H4

Bend Oregon, USA 26 B3

Bendery see Tighina

Bendigo Australia 131 C7

Benevento Italy 79 D5

Bengal, Bay of sea feature Indian Ocean 122 D3

Bengbu China 111 D5

Benghazi see Banghāzī

Bengkulu Indonesia 120 B4

Benguela Angola 60 B2

Beni river Bolivia 42 C4

Benidorm Spain 75 F4

Beni-Mellal Morocco 52 C2

Benin country N Africa prev. Dahomey 57

Benin, Bight of sea feature W Africa 57 F5

Benin City Nigeria 57 F5

Beni Suef see Banī Suwayf

Ben Nevis mountain Scotland, UK 70 C3

Benue river Cameroon/Nigeria 57 G4

Beograd see Belgrade

Berat Albania 83 D6

Berbera Somalia 54 D4

Berbérati Central African Republic 58 C5

Berdyans'k Ukraine 91 G4

Bereket Turkmenistan prev. Gazandzhyk, var. Kazandzhik, Turkm. Gazanjyk 104 B2

Berezina see Byerazino

Bergamo Italy 78 B2

Bergen Norway 67 A5

Bergse Maas river Netherlands 68 D4

Bering Sea Pacific Ocean 134 D1

Bering Strait sea feature Bering Sea/Chukchi Sea 134 D1

Berkeley California, USA 27 B6

Berlin capital of Germany 76 D3

Bermejo river Argentina 46 D2

Bermuda overseas territory UK, Atlantic Ocean 48 B3

Bern capital of Switzerland Fr. Berne 77 B7

Berne *see* Bern
Berner Alpen *mountain range* Switzerland 77 B7
Bertoua Cameroon 59 B5
Besançon France 72 D4
Besztercebánya *see* Banská Bystrica
Bethlehem West Bank 101 A5
Beyrouth *see* Beirut
Béziers France 73 C6
Bezmein *see* Abadan
Bhamo Myanmar 118 B2
Bhāvnagar India 116 C4
Bhōpal India 116 D4
Bhutan *country* S Asia 117
Biak, Pulau *island* Indonesia 121 G4
Białystok Poland *Rus.* Belostok 80 E3
Biel Switzerland 77 B7
Bielefeld Germany 76 B4
Bielitz-Biala *see* Bielsko-Biała
Bielsko-Biała Poland *Ger.* Bielitz-Biala 81 C5
Bié Plateau *upland* Angola 51 C6
Bighorn Mountains *mountains* C USA 24 C2
Bignona Senegal 56 B3
Big Spring Texas, USA 29 E3
Bihać Bosnia & Herzegovina 82 B3
Bihār *state* India 117 F3
Bijelo Polje Montenegro 82 D4
Bikāner India 116 C3
Bila Tserkva Ukraine 91 E2
Bilbao Spain 75 E1
Billings Montana, USA 24 C2
Bilma, Grand Erg de *desert* Niger 57 G3
Biloela Australia 130 D4
Biloxi Mississippi, USA 30 C3
Biltine Chad 58 D3
Binghamton New York, USA 23 F3
Birāk Libya 53 F3
Birātnagar Nepal 117 F3
Birmingham England, UK 71 D6

Birmingham Alabama, USA 30 D2
Bîr Mogreïn Mauritania 56 C1
Birsen *see* Biržai
Biržai Lithuania *Ger.* Birsen 88 C4
Biscay, Bay of *sea feature* Atlantic Ocean 62 C4
Bishkek *capital of* Kyrgyzstan *prev.* Frunze, Pishpek 105 F2
Bishop California, USA 27 C6
Biskra Algeria 53 E2
Bismarck North Dakota, USA 25 E2
Bismarck Archipelago *island group* Papua New Guinea 126 B3
Bismarck Sea *sea* Pacific Ocean 124 B2
Bissau *capital of* Guinea-Bissau 56 B4
Bitola Macedonia 83 E6
Bitterroot Range *mountains* NW USA 26 D2
Biwa-ko *lake* Japan 113 C5
Bizerte Tunisia 53 E1
Bjelovar Croatia 82 B2
Bjørnøya *island* N Norway *Eng.* Bear Island 65 G3
Black Drin *river* Albania/ Macedonia 83 E6
Black Forest *see* Schwarzwald
Black Hills *mountains* USA 24 D3
Blackpool England, UK 71 D5
Black River *river* China/Vietnam 118 D3
Black Sea Asia/Europe 63 F4
Black Volta *river* Ghana/Côte d'Ivoire 57 E4
Blackwater *river* Ireland 71 A6
Blagoevgrad Bulgaria 86 C3
Blagoveshchensk Russian Federation 97 G4
Blanca, Bahía *sea feature* Argentina 39 D5
Blanche, Lake *lake* Australia 131 B5
Blantyre Malawi 61 E2
Blenheim New Zealand 133 D5
Blida Algeria 52 D1

Bloemfontein *financial capital* of South Africa 60 D4
Blois France 72 C4
Bloomington Indiana, USA 22 C4
Bluefields Nicaragua 35 E3
Blue Mountains *mountains* W USA 26 C2
Blue Nile *river* Ethiopia/Sudan 54 C4
Blumenau Brazil 44 D3
Bo Sierra Leone 56 C4
Boa Vista Brazil 42 D1
Boa Vista *island* Cape Verde 56 A3
Bobo-Dioulasso Burkina Faso 56 D4
Bobruysk *see* Babruysk/ Bobruysk
Boca de la Serpiente *see* Serpent's Mouth, The
Bochum Germany 76 B4
Bodø Norway 66 C3
Bodrum Turkey 98 A4
Bogor Indonesia 120 C5
Bogotá *capital of* Colombia 40 B3
Bo Hai *sea feature* Yellow Sea 110 D4
Bohemian Forest *region* Germany 77 D5
Bohol Sea *Sea* Philippines 121 E2
Boise Idaho, USA 26 D3
Boké Guinea 56 C4
Bokhara *see* Buxoro
Bol Chad 58 B3
Bolivia *country* C South America 42-43
Bologna Italy 78 C3
Bolton England, UK 71 D5
Bolzano Italy *Ger.* Bozen 78 C2
Boma Dem. Rep. Congo 59 B7
Bombay *see* Mumbai
Bomu *river* Central African Republic/Dem. Rep. Congo 59 D5
Bonaire *special municipality* Netherlands, West Indies 37 E5

Bongo, Massif des *upland* Central African Republic 58 D4

Bongor Chad 58 C3

Bonn Germany 76 B4

Boosaaso Somalia 54 E4

Borås Sweden 67 B7

Bordeaux France 73 B5

Borger Texas, USA 29 E2

Borisov *see* Barysaw

Borlänge Sweden 67 C6

Borneo *island* SE Asia 120-121

Bornholm *island* Denmark 67 C8

Bosanski Šamac Bosnia & Herzegovina 82 C3

Bosna *river* Bosnia & Herzegovina 82 C3

Bosna I Hercegovina, Federacija Admin. region *republic* Bosnia and Herzegovina 82 C4

Bosnia & Herzegovina *country* SE Europe 82-83

Bosporus *sea feature* Turkey *Turk.* İstanbul Boğazí 98 B2

Bossangoa Central African Republic 58 C4

Bosten Hu *Lake* China 108 C3

Boston Massachusetts, USA 23 G3

Bothnia, Gulf of *sea feature* Baltic Sea 67 C5

Botoşani Romania 90 C3

Botswana *country* southern Africa 60

Bouar Central African Republic 58 C4

Bougainville Island *island* Papua New Guinea 126 C3

Bougouni Mali 56 D4

Boulder Colorado, USA 24 C4

Boulogne-sur-Mer France 72 C2

Bourges France 72 C4

Bourgogne *region* France *Eng.* Burgundy 72 D4

Bourke Australia 131 C5

Bournemouth England, UK 71 D7

Bouvet Island *external territory* Norway, Atlantic Ocean 49 D7

Bowen Australia 130 D3

Bowling Green Kentucky, USA 22 C5

Bozeman Montana, USA 24 B2

Bozen *see* Bolzano

Brač *island* Croatia 82 B4

Bradford England, UK 71 D5

Braga Portugal 74 C2

Bragança Portugal 74 C2

Brahmaputra *river* Asia 117 G3

Brăila Romania 90 D4

Brainerd Minnesota, USA 25 F2

Brandon Canada 19 F5

Brasília *capital of* Brazil 43 F4

Braşov Romania 90 C4

Bratislava *capital of* Slovakia *Ger.* Pressburg, *Hung.* Pozsony 81 C6

Bratsk Russian Federation 97 E4

Braunau am Inn Austria 77 D6

Braunschweig Germany *Eng.* Brunswick 76 C4

Brazil *country* South America 42-43

Brazil Basin *undersea feature* Atlantic Ocean 49 C5

Brazilian Highlands *upland* Brazil 43 G4

Brazos *river* SW USA 29 G3

Brazzaville *capital of* Congo 59 B6

Brecon Beacons *hills* Wales, UK 71 C6

Breda Netherlands 68 C4

Bregenz Austria 77 B7

Bremen Germany 76 B3

Bremerhaven Germany 76 B3

Brescia Italy 78 B2

Breslau *see* Wrocław

Brest Belarus *Pol.* Brześć nad Bugiem, *prev.* Brześć Litewski, *Rus.* Brest-Litovsk 89 B6

Brest France 72 A3

Brest-Litovsk *see* Brest

Bretagne *region* France *Eng.* Brittany 72 A3

Brezhnev *see* Naberezhnyye Chelny

Bria Central African Republic 58 D4

Bridgetown *capital of* Barbados 37 H4

Brig Switzerland 77 B5

Brighton England, UK 71 E7

Brindisi Italy 79 E5

Brisbane Australia 131 E5

Bristol England, UK 71 D6

British Columbia *province* Canada 18-19

British Indian Ocean Territory *overseas territory* UK, Indian Ocean 122 C4

British Isles *islands* W Europe 70-71

British Virgin Islands *overseas territory* UK, West Indies 37

Brittany *see* Bretagne

Brno Czech Republic *Ger.* Brünn 81 B5

Broken Arrow Oklahoma, USA 29 G1

Broken Hill Australia 131 B6

Broken Ridge *undersea feature* Indian Ocean 123 D6

Bromberg *see* Bydgoszcz

Brooks Range *mountains* Alaska, USA 18 D2

Brookton Australia 129 B6

Broome Australia 128 C3

Brownfield Texas, USA 29 E2

Brownsville Texas, USA 29 G5

Bruges *see* Brugge

Brugge Belgium *Fr.* Bruges 69 A5

Brunei *country* E Asia 120 D3

Brünn *see* Brno

Brunswick Georgia, USA 31 E3

Brunswick *see* Braunschweig

Brusa *see* Bursa

Brussel *see* Brussels

Brussels *capital of* Belgium *Fr.* Bruxelles, *Flem.* Brussel 69 C6

Brüx *see* Most

Bruxelles *see* Brussels

Bryan Texas, USA 29 G3

Bryansk Russian Federation
93 A5 96 A2

Brześć Litewski *see* Brest

Brześć nad Bugiem *see* Brest

Bucaramanga Colombia
40 C2

Buchanan Liberia 56 C5

Bucharest *capital of* Romania
90 C5

Budapest *capital of* Hungary
81 C6

Budweis *see* České Budějovice

Buenaventura Colombia 40 B3

Buenos Aires *capital of*
Argentina 46 D4

Buenos Aires, Lago *lake*
Argentina/Chile 47 B6

Buffalo New York, USA 23 E3

Bug *river* E Europe 90 C1

Bujumbura *capital of* Burundi
prev. Usumbura 55 B7

Bukavu Dem. Rep. Congo 59 E6

Bukhara *see* Buxoro

Bulawayo Zimbabwe 60 D3

Bulgan Mongolia 109 E2

Bulgaria *country* E Europe 86

Bumba Dem. Rep. Congo 59 D5

Bunbury Australia 129 B6

Bundaberg Australia 130 E4

Bunia Dem. Rep. Congo 59 E6

Buraydah Saudi Arabia 103 B5

Burē Ethiopia 54 C4

Burgas Bulgaria 86 E2

Burgos Spain 75 E2

Burgundy *see* Bourgogne

Burketown Australia 130 B3

Burkina Faso *country*
W Africa 57

Burlington Iowa, USA 25 G4

Burlington Vermont, USA 23 F2

Burma *see* Myanmar

Burnie Tasmania 131 C8

Burns Oregon, USA 26 C3

Bursa Turkey *prev.* Brusa 98 B3

Būr Sa'īd Egypt *Eng.* Port Said
54 B1

Burtnieku Ezers *lake* Latvia
88 C3

Buru, Pulau *island* Indonesia
121 E4

Burundi *country* C Africa 55

Busan South Korea *prev.*
Pusan110 E4

Busselton Australia 129 B7

Butembo Dem. Rep. Congo
59 E5

Buton, Pulau *island* Indonesia
121 E4

Butte Montana, USA 24 B2

Butuan Philippines 121 F2

Buxoro Uzbekistan *var.*
Bokhara, *Rus.* Bukhara
104 D2

Büyükağrı Dağı *see* Ararat,
Mount

Buzău Romania 90 C4

Büzmeyin *see* Abadan

Byarezina *river* Belarus *Rus.*
Berezina 89 D6

Bydgoszcz Poland *Ger.*
Bromberg 80 C3

Byzantium *see* İstanbul

C

Caazapá Paraguay 44 C3

Cabanatuan Philippines 121 E1

Cabimas Venezuela 40 C1

Cabinda *exclave* Angola 60 B1

Cabot Strait *sea feature*
Atlantic Ocean 21 G4

Čačak Serbia 82 D4

Cáceres Spain 74 D3

Cachoeiro de Itapemirim Brazil
45 F1

Cadiz Philippines 121 E2

Cádiz Spain 74 D5

Caen France 72 B3

Cagayan de Oro Philippines
121 F2

Cagliari Italy 79 A5

Cahors France 73 B5

Cairns Australia 130 D3

Cairo *capital of* Egypt *Ar.* Al
Qāhirah, *var.* Al Qāhira 54 B1

Čakovec Croatia 82 B2

Calabar Nigeria 57 G5

Calabria *region* Italy 79 D6

Calafate *see* El Calafate

Calais France 72 C2

Calais Maine, USA 23 H1

Calama Chile 46 B2

Calbayog Philippines 121 F2

Calcutta *see* Kolkata

Caldas da Rainha Portugal
74 B3

Caldwell Idaho, USA 27 C3

Caleta Olivia Argentina 47 C6

Calgary Canada 19 E5

Cali Colombia 40 B4

Calicut India *see* Kozhikode
114 D2

California *state* USA 26-27

California, Golfo de *sea feature*
Pacific Ocean *Eng.* California,
Gulf of 32 B2 123 F2

Callabonna, Lake *lake*
Australia131 B5

Callao Peru 42 A3

Caltanissetta Italy 79 C7

Camagüey Cuba 36 C2

Cambodia *country* SE Asia
Cam. Kampuchea 119

Cambridge England, UK 71 E6

Cambridge New Zealand132 D2

Cameroon *country* W Africa
58-59

Campbell Plateau *undersea
feature* Pacific Ocean
134 C5

Campeche Mexico 33 G4

Campeche, Bahía de *sea
feature* Mexico *Eng.* Gulf of
Campeche 33 G4

Campina Grande Brazil 43 H3

Campinas Brazil 45 E2

Campo Grande Brazil 44 C1

Campos Brazil 45 F2

Canada *country* North America
16-17

Canada Basin *undersea feature*
Arctic Ocean *var.* Laurentian
Basin 137 F2

Canadian River *river* SW USA
29 E2

Çanakkale Turkey 98 A3

Çanakkale Boğazı *see*
Dardanelles

Canarias, Islas *islands* Spain
Eng. Canary Islands 50 A2

Canary Basin — Ceuta

Cévennes *mountains* France 73 C6

Ceylon *see* Sri Lanka

Ceylon Plain *undersea feature* Indian Ocean 122 C4

Chad *country* C Africa 58

Chad, Lake *lake* C Africa 58 B3

Chāgai Hills *mountains* Pakistan 116 A2

Chagos-Laccadive Plateau *undersea feature* Indian Ocean 122 C4

Chagos Trench *undersea feature* Indian Ocean 122 C4

Chalkída Greece 87 C5

Challenger Deep *undersea feature* Pacific Ocean 134 B3

Châlons-en-Champagne France 72 D3

Chambéry France 73 D5

Champaign Illinois, USA 22 B4

Chañaral Chile 46 B2

Chandīgarh India 116 D2

Chang, Ko *island* Thailand 119 C5

Changchun China 110 D3

Chang Jiang *river* China *var.* Yangtze 111 B6

Changsha China 111 C6

Chaniá Greece 87 C7

Channel Islands *island group* California, USA 27 B8

Channel Islands *island group* UK 71 D8

Channel-Port-aux-Basques Canada 21 G4

Channel Tunnel France/UK 71 E7

Chapala, Lago de *lake* Mexico 32 D4

Chardzhev *see* Türkmenabat

Chardzhou *see* Türkmenabat

Chari *river* C Africa 58 C3

Chārīkār Afghanistan 105 E4

Chärjew *see* Türkmenabat

Charleroi Belgium 69 C6

Charleston South Carolina, USA 31 F2

Charleston West Virginia, USA 31 F1

Charleville Australia 130 C4

Charlotte North Carolina, USA 31 F1

Charlotte Amalie *capital of* Virgin Islands 37 F3

Charlottesville Virginia, USA 23 E5

Charlottetown Canada 21 G4

Charters Towers Australia 130 D3

Chartres France 72 C3

Châteauroux France 72 C4

Chatham Islands *islands* New Zealand 134 C5

Chattanooga Tennessee, USA 30 D1

Chauk Myanmar 118 A3

Chaves Portugal 74 C2

Cheboksary Russian Federation 93 C5

Cheboygan Michigan, USA 22 C2

Chech, Erg *desert* Algeria/Mali 56 D1

Che-chiang *see* Zhejiang

Cheju-do *see* Jeju-do

Cheju Strait *see* Jeju Strait

Chekiang *see* Zhejiang

Cheleken *see* Hazar

Chelyabinsk Russian Federation 96 C3

Chemnitz Germany *prev.* Karl-Marx-Stadt 76 D4

Chenāb *river* Pakistan 116 C2

Chengdu China 111 B5

Chennai India *prev.* Madras 115 E2

Cherbourg France 72 B3

Cherepovets Russian Federation 92 B4

Cherkasy Ukraine 91 E2

Cherkessk Russian Federation 93 A7

Chernigov *see* Chernihiv

Chernihiv Ukraine *Rus.* Chernigov 91 E1

Chernivtsi Ukraine *Rus.* Chernovtsy, *Rom.* Cernăuţi 90 C3

Chernobyl' *see* Chornobyl'

Chernovtsy *see* Chernivtsi

Chernyakhovsk Kaliningrad, Russian Federation 88 B4

Chesapeake Bay *sea feature* USA 23 F5

Chester England, UK 71 D5

Cheyenne Wyoming, USA 24 D4

Chiang-hsi *see* Jiangxi

Chiang Mai Thailand 118 B4

Chiang-su *see* Jiangsu

Chiba Japan 113 D5

Chicago Illinois, USA 22 B3

Chiclayo Peru 42 A3

Chico California, USA 27 B5

Chicoutimi Canada 21 E4

Chifeng China *var.* Ulanhad 109 F2

Chihli *see* Hebei

Chihuahua Mexico 32 C2

Chile *country* S South America 46-47

Chile Basin *undersea feature* Pacific Ocean 135 G4

Chile Chico Chile 47 B6

Chile Rise *undersea feature* Pacific Ocean 135 G4

Chi-lin *see* Jilin

Chillán Chile 46 B4

Chiloé, Isla de *island* Chile 47 B6

Chimborazo *peak* Ecuador 38 A3

Chimbote Peru 42 A3

Chimkent *see* Shymkent

Chimoio Mozambique 61 E3

China *country* E Asia 108-109

Chinandega Nicaragua 34 C3

Chindwinn *river* Myanmar 118 A2

Chinghai *see* Qinghai

Chingola Zambia 60 D2

Chinook Trough *undersea feature* Pacific Ocean 134 D1

Chíos Greece 87 D5

Chíos *island* Greece *prev.* Khíos 87 D5

Chirchik Uzbekistan *Uzb.* Chirchiq 105 E2

Chirchiq *see* Chirchik

Chiriquí, Golfo de *sea feature* Panama 35 E5

Chişinău *capital of* Moldova, *var.* Kishinev 90 D3

Chita Russian Federation 97 F4

Chitré Panama 35 F5

Chittagong Bangladesh 117 G4

Chitungwiza Zimbabwe 60 D3

Choluteca Honduras 34 C3

Choma Zambia 60 D3

Chona *river* Russian Federation 95 E2

Chon Buri Thailand 119 C5

Ch'ŏngjin North Korea 110 E3

Chongqing *province* China *var.* Chungking 111 B5

Chonos, Archipiélago de los *island group* Chile 47 B6

Chornobyl' Ukraine *Rus.* Chernobyl' 91 E1

Choûm Mauritania 56 C2

Choybalsan Mongolia 109 F2

Christchurch New Zealand 133 C6

Christmas Island *external territory* Australia, Indian Ocean 122 D5

Christmas Island *see* Kiritimati

Christmas Ridge *undersea feature* Pacific Ocean 125 F1

Chuan *see* Sichuan

Chubut *river* Argentina 47 B6

Chudskoye Ozero *see* Peipus, Lake

Chuí *see* Chuy

Chukchi Plain *undersea feature* Arctic Ocean 137 G2

Chukchi Sea Arctic Ocean *Rus.* Chukotskoye More 137 F1

Chukotskoye More *see* Chukchi Sea

Chula Vista California, USA 27 C8

Chulym *river* Russian Federation 94 D3

Chumphon Thailand 119 C6

Chungking *see* Chongqing

Chuquicamata Chile 46 B2

Chur Switzerland 77 B7

Churchill Canada 19 G4

Chuuk Islands *island group* Micronesia 126 B1

Chuy Brazil *var.* Chuí 44 C5

Cienfuegos Cuba 36 B2

Cieza Spain 75 F4

Cilacap Indonesia 120 C5

Cincinnati Ohio, USA 22 C4

Ciudad Bolívar Venezuela 41 E2

Ciudad del Este Paraguay 44 C3

Ciudad de México *see* Mexico City

Ciudad Guayana Venezuela 41 E2

Ciudad Juárez Mexico 32 C1

Ciudad Obregón Mexico 32 B2

Ciudad Ojeda Venezuela 40 C1

Ciudad Real Spain 75 E3

Ciudad Valles Mexico 33 E3

Ciudad Victoria Mexico 33 E3

Clarence *river* New Zealand 133 C5

Clarion Fracture Zone *tectonic feature* Pacific Ocean 125 G1

Clarksville Tennessee, USA 30 D1

Clearwater Florida, USA 31 E4

Clermont Australia 130 D4

Clermont-Ferrand France 73 C5

Cleveland Ohio, USA 22 D3

Clipperton Fracture Zone *tectonic feature* Pacific Ocean 125 G2

Clipperton Island *external territory* France, Pacific Ocean 135 F3

Cloncurry Australia 130 C3

Clovis New Mexico, USA 29 E2

Cluj-Napoca Romania 90 B3

Clutha *river* New Zealand 133 B7

Coast Ranges *mountain range* W USA 26 A5

Coats Island *island* Canada 20 C1

Coats Land *physical region* Antarctica 136 B2

Coatzacoalcos Mexico 33 G4

Cobán Guatemala 34 B2

Cochabamba Bolivia 42 C4

Cochin India *see* Kochi 114 D3

Cochrane Canada 20 C4

Cochrane Chile 47 B6

Coco *river* Honduras/Nicaragua 34 D2

Cocos Basin *undersea feature* Indian Ocean 122 D4

Cocos Islands *external territory* Australia, Indian Ocean 122 D5

Cod, Cape *coastal feature* NE USA 23 G3

Coeur d'Alene Idaho, USA 26 C2

Coffs Harbour Australia 131 E6

Coihaique Chile 47 B6

Coimbatore India 114 D3

Coimbra Portugal 74 C3

Colchester England, UK 71 E6

Colmar France 72 E4

Cologne *see* Köln

Colombia *country* N South America 40-41

Colombo *administrative capital of* Sri Lanka 115 E4

Colón Panama 35 F4

Colón, Archipiélago de *see* Galapagos Islands

Colorado *state* USA 24 C4

Colorado *river* USA 16 B5

Colorado *river* Argentina 47 C5

Colorado Plateau *upland region* S USA 28 B1

Colorado Springs Colorado, USA 24 D4

Columbia South Carolina, USA 31 F2

Columbia *river* NW USA 26 C1

Columbus Georgia, USA 30 D3

Columbus Mississippi, USA 30 C2

Columbus Nebraska, USA 25 E4

Columbus Ohio, USA 22 D4

Comayagua Honduras 34 C2

Dumfries Scotland, UK 70 C4
Düna see Western Dvina
Dünaburg see Daugavpils
Dundalk Ireland 71 B5
Dundee Scotland, UK 70 D3
Dunedin New Zealand 133 B7
Dunkerque France *Eng.*
Dunkirk 72 C2
Dunkirk see Dunkerque
Duqm Oman 103 E6
Durango Mexico 32 D3
Durango Colorado, USA 24 C5
Durazno Uruguay 44 C5
Durban South Africa 60 E4
Durham North Carolina, USA
31 F1
Durrës Albania 83 C5
Dushanbe *capital of* Tajikistan
var. Dyushambe, *prev.*
Stalinabad 105 E3
Düsseldorf Germany 76 A4
Dutch Harbor Alaska, USA
18 B3
Dvinsk see Daugavpils
Dynapro see Dnieper
Dyushambe see Dushanbe
Dzaudzhikau see Vladikavkaz
Dzhalal-Abad Kyrgyzstan *Kir.*
Jalal-Abad 105 F2
Dzhambul see Taraz
Dzhezkazgan see Zhezkazgan
Dzvina see Western Dvina

E

Eagle Pass Texas, USA
29 F4
East Antarctica *region*
Antarctica 136 C3
East Cape *coastal feature* New
Zealand 132 E2
East China Sea Pacific Ocean
111 E5
Easter Fracture Zone *tectonic
feature* Pacific Ocean
135 G4
Easter Island *island* Pacific
Ocean 135 F4
Eastern Ghats *mountain range*
India 117 B5

Eastern Sierra Madre see Sierra
Madre Oriental
East Falkland *island* Falkland
Islands 47 D7
East Indiaman Ridge *undersea
feature* Indian Ocean
23 D5
East Indies *island group* Asia
122 E4
East London South Africa
60 D5
Eastmain *river* Canada 20 D3
East Pacific Rise *undersea
feature* Pacific Ocean
135 F4
East Siberian Sea see
Vostochno-Sibirskoye More
East St Louis Illinois, USA
22 B4
East Timor *country* SE Asia
121
East Novaya Zemlya Trench
var. Novaya Zemlya Trench.
Undersea feature Kara Sea
137 H4
Eau Claire Wisconsin, USA
22 A2
Ebolowa Cameroon 59 B5
Ebro *river* Spain 75 F2
Ecuador *country* NW South
America 40
Ede Netherlands 68 D3
Ede Nigeria 57 F4
Edgeøya *island* Svalbard 65 G2
Edinburgh Scotland, UK 70 C4
Edirne Turkey 98 A2
Edmonton Canada 19 E5
Edward, Lake *lake* Uganda/
Dem. Rep. Congo 59 E6
Edwards Plateau *upland* S USA
29 F4
Efate *island* Vanuatu *prev.*
Sandwich Island 124 D4
Effingham Illinois, USA 22 B4
Eforie-Sud Romania 90 D5
Egadi, Isole *island group* Italy
79 B6
Ege Denizi see Aegean Sea
Eger see Ohře
Egypt *country* NE Africa 54
Eighty Mile Beach *beach*
Australia 128 C3

Eindhoven Netherlands 69 D5
Eisenstadt Austria 77 E6
Eivissa see Ibiza
Elat Israel 101 A7
Elazig Turkey 99 E3
Elba, Isola d' *island* Italy 78 B4
Elbasan Albania 83 D6
Elbe *river* Czech Republic/
Germany 81 B5
Elbing see Elbląg
Elbląg Poland *Ger.* Elbing
80 D2
El'brus *peak* Russian
Federation 93 A7
El Calafate Argentina *var.*
Calafate 47 B7
Elche Spain *Cat.* Elx 75 F4
Elda Spain 75 F4
Eldoret Kenya 55 C6
Eleuthera *island* The Bahamas
36 C1
El Fasher Sudan *var.* Al Fāshir
54 A4
El Geneina Sudan 54 A4
Elgin Scotland, UK 70 C3
El Giza see Al Jīzah
El Hank *cliff* Mauritania 56 D1
Elista Russian Federation 93 B6
El Khalīl see Hebron
El Khārga see Al Khārijah
Elko Nevada, USA 27 D5
Ellensburg Washington, USA
26 B2
Ellesmere Island *island* Canada
19 F1
Ellsworth Land *region*
Antarctica 136 A3
Elmira New York, USA 23 E3
El Mreyyé *desert* Mauritania
56 D2
El Obeid Sudan 54 B4
El Paso Texas, USA 28 D3
El Puerto de Santa María Spain
74 D5
El Qâhira see Cairo
El Salvador *country* Central
America 34
Eltanin Fracture Zone *tectonic
feature* Pacific Ocean
135 E5
El Tigre Venezuela 41 E2

Elx see Elche

Ely Nevada USA 27 D5

Emden Germany 76 B3

Emerald Australia 130 D4

Emmen Netherlands 68 E2

Empty Quarter see Ar Rub' al Khali

Ems river Germany/Netherlands 76 B3

Encarnación Paraguay 44 C3

Enderbury Island atoll Kiribati 136 C2

Enderby Land region Antarctica 136 C2

Enderby Plain undersea feature Indian Ocean 123 B7

England national region UK 70-71

English Channel sea feature Atlantic Ocean 71 D7

Enguri river Georgia Rus. Inguri 99 F1

Enid Oklahoma, USA 29 F1

Ennedi plateau Chad 58 D2

Enns river Austria 77 D6

Enschede Netherlands 68 E3

Ensenada Mexico 32 A1

Entebbe Uganda 55 B6

Enugu Nigeria 57 G5

Eolie, Isole island group Italy Eng. Lipari Islands, var. Aeolian Islands 79 D6

Eperies see Prešov

Eperjes see Prešov

Épinal France 72 E4

Equatorial Guinea country W Africa 59

Erdenet Mongolia 109 E2

Erechim Brazil 44 D3

Erenhot China 109 F2

Erevan see Yerevan

Ereğli Turkey 98 C4

Erfurt Germany 76 C4

Erie Pennsylvania, USA 22 D3

Erie, Lake lake Canada/USA 17 D5

Eritrea country E Africa 54

Erivan see Yerevan

Erlangen Germany 77 C5

Ernākulam India 114 D3

Er Rachidia Morocco 52 C2

Erzerum see Erzurum

Erzgebirge mountain range Czech Republic/Germany var. Krušné Hory 77 D5

Erzincan Turkey 99 E3

Erzurum Turkey prev. Erzerum 99 F3

Esbjerg Denmark 67 A7

Esch-sur-Alzette Luxembourg 69 D8

Escuintla Guatemala 34 B2

Eşfahān Iran 102 C3

Esh Sham see Damascus

Eskişehir Turkey 98 B3

Esmeraldas Ecuador 40 A4

Esperance Australia 129 C6

Espíritu Santo Island Vanuatu 124 D3

Espoo Finland 67 D6

Esquel Argentina 47 B6

Essaouira Morocco 52 B2

Essen Germany 76 A4

Essequibo river Guyana 41 G3

Esteli Nicaragua 34 D3

Estevan Canada 19 F5

Estonia country E Europe 88 D2

Ethiopia country E Africa 54-55

Ethiopian Highlands upland E Africa 50 D4

Etna, Mount peak Sicily, Italy 79 D7

Etosha Pan salt basin Namibia 60 C3

Eucla Australia 129 D6

Eugene Oregon, USA 26 A3

Eugene Washington, USA 26 B1

Euphrates river SW Asia 102 C4

Europe 62-63

Evansville Indiana, USA 22 B5

Everest, Mount peak China/Nepal 108 B5

Everett Washington, USA 26 B1

Everglades, The wetlands Florida, USA 31 F5

Évvoia island Greece 87 C5

Exeter England, UK 71 C7

Exmoor region England, UK 71 C7

Exmouth Australia 128 A4

Exmouth Gulf gulf Australia 128 A4

Exmouth Plateau undersea feature Indian Ocean 123 E5

Eyre North, Lake salt lake Australia 131 B5

Eyre Peninsula peninsula Australia 131 A6

Eyre South, Lake salt lake Australia 131 B5

F

Fada-N'gourma Burkina Faso 57 E4

Faroe Islands external territory Denmark, Atlantic Ocean Far. Fóroyar, Dan. Færoerne, var. Faeroe Islands 65 F5

Færoerne see Faroe Islands

Faguibine, Lac lake Mali 57 E3

Fairbanks Alaska, USA 18 D3

Fairlie New Zealand 133 B6

Faisalābād Pakistan 116 C2

Faïzābād Afghanistan prev. Feyzābād 105 E3

Falkland Islands overseas territory UK, Atlantic Ocean 47 D7

Fallon Nevada, USA 27 C5

Falun Sweden 67 C6

Famagusta see Gazimağusa

Farafangana Madagascar 61 G4

Farāh Afghanistan 104 C5

Farasān, Jazā'ir island group Saudi Arabia 103 B6

Farewell, Cape headland New Zealand 132 C4

Farewell, Cape see Nunap Isua

Farghona see Farg'ona

Farg'ona Uzbekistan prev. Novyy Margilan, Uzb. Farghona 105 F2

Fargo North Dakota, USA 25 E2

Farkhor Tajikistan 105 E3

Farmington New Mexico, USA 28 C1

Fraser Island *island* Australia 130 E4

Frauenburg *see* Saldus

Fray Bentos Uruguay 44 B5

Fredericksburg Virginia, USA 23 E4

Fredericton Canada 21 F4

Frederikshavn Denmark 67 B7

Fredrikstad Norway 67 B6

Freeport The Bahamas 36 C1

Freeport Texas, USA 29 G4

Freetown *capital of* Sierra Leone 56 C4

Freiburg im Breisgau Germany 77 B6

Fremantle Australia 129 B6

French Guiana *overseas department* France, N South America 41

French Polynesia *overseas collectivity* France, Pacific Ocean 135 E3

French Southern and Antarctic Lands *French overseas territory* Indian Ocean *Fr.* Terres Australes et Antarctiques Françaises 123 C7

Fresnillo Mexico 32 D1

Fresno California, USA 27 B6

Fobisher Bay *see* Iqaluit

Frome, Lake *salt lake* Australia 131 B5

Frunze *see* Bishkek

Fu-chien *see* Fujian

Fuerte Olimpo Paraguay 44 B1

Fuerteventura *island* Spain 52 A3

Fuhkien *see* Fujian

Fujian *province* China *var.* Fu-chien, Fuhkien, Fukien, Min 111 D6

Fukien *see* Fujian

Fukui Japan 113 C5

Fukuoka Japan 113 A6

Fukushima Japan 112 D4

Fulda Germany 77 C5

Fünfkirchen *see* Pécs

Fushun China 110 D3

Furnas, Represa de *Reservoir* Brazil 45 E1

Fuxin China 110 D3

Fujian China *prev.* Linchuan 111 D6

FYR Macedonia *see* Macedonia

G

Gaalkacyo Somalia 55 E5

Gabès Tunisia 53 E2

Gabon *country* W Africa 59

Gaborone *capital of* Botswana 60 D4

Gabrovo Bulgaria 86 D2

Gadsden Alabama, USA 30 D2

Gaeta, Golfo di *sea feature* Italy 79 C5

Gafsa Tunisia 53 E2

Gagnoa Côte d'Ivoire 56 D5

Gagra Georgia 99 E1

Gairdner, Lake *lake* Australia 131 B6

Galapagos Fracture Zone *tectonic feature* Pacific Ocean 135 F3

Galapagos Islands *islands* Ecuador, Pacific Ocean *var.* Tortoise Islands, *Sp.* Archipiélago de Colón 135 G3

Galapagos Rise *undersea feature* Pacific Ocean 135 G3

Galaţi Romania 90 D4

Galesburg Illinois, USA 22 B4

Galicia *region* Spain 74 C1

Galilee, Sea of *see* Tiberias, Lake

Galle Sri Lanka 115 E4

Gallego Rise *undersea feature* Pacific Ocean 135 F3

Gallipoli Italy 79 E5

Gällivare Sweden 66 D3

Gallup New Mexico, USA 28 C2

Galveston Texas, USA 29 G4

Galway Ireland 71 A5

Gambia *country* W Africa 56

Gambia *River* Africa 56 C3

Gambier, Îles *island group* French Polynesia 135 E4

Gan *see* Gansu

Gan *see* Jiangxi

Gäncä Azerbaijan *Rus.* Gyandzha, *prev.* Kirovabad, Yelisavetpol 99 G2

Gand *see* Gent

Gander Canada 21 H3

Gandia Spain 75 F3

Ganges *river* S Asia 116 F4

Ganges Fan *Undersea feature* Bay of Bengal 122 D3

Ganges, Mouths of the *wetlands* Bangladesh/India 117 G4

Gangtok India 117 G3

Gansu *province* China *var.* Gan, Kansu 111 B5

Gao Mali 57 E3

Gaoual Guinea 56 C4

Gaoxiong Taiwan *prev.* Kaohsiung 111 D7

Gar China *var.* Shiquanhe 108 A4

Garagum Kanaly *canal* Turkmenistan *prev.* Karakumskiy Kanal 104 C3

Garagum *desert* Turkmenistan *var.* Kara Kum, Karakumy 104 C2

Garda, Lago di *lake* Italy 78 B2

Gardēz Afghanistan *prev.* Gardīz 105 E4

Gardīz *see* Gardēz

Garissa Kenya 55 C6

Garmo Peak *see* Communism Peak

Garonne *river* France 73 B5

Garoowe Somalia 55 E5

Garoua Cameroon 58 B4

Gary Indiana, USA 22 B3

Gaspé Canada 21 F4

Gastonia North Carolina, USA 31 E1

Gävle Sweden 67 C5

Gaya India 117 F4

Gaza Gaza Strip 101 A6

Gazandzhyk *see* Bereket

Gazanjyk *see* Bereket

Gaza Strip *disputed territory* SW Asia 101 A6

Gaziantep — Grampian Mountains

Gaziantep Turkey *prev.* Aintab 98 D4

Gazimağusa Cyprus *var.* Famagusta *Gk.* Ammochostos 98 C5

Gdańsk Poland *Ger.* Danzig 80 C2

Gdingen *see* Gdynia

Gdynia Poland *Ger.* Gdingen 80 C2

Gedaref Sudan 54 C4

Geelong Australia 131 C7

Gëkdepe *see* Gökdepe

Gemena Dem. Rep. Congo 59 C5

General Eugenio A. Garay Paraguay 44 A1

General Santos Philippines 121 F3

Geneva *see* Genève

Geneva, Lake *lake* France/ Switzerland *Fr.* Lac Léman, *var.* Le Léman, *Ger.* Genfer See 77 A7

Genève Switzerland *Eng.* Geneva 77 A7

Genfer See *see* Geneva, Lake

Gengen Gol China 109 F1

Genk Belgium 69 D5

Genoa *see* Genova

Genova Italy *Eng.* Genoa 78 B3

Genova, Golfo di *sea feature* Italy 78 B3

Gent Belgium *Fr.* Gand, *Eng.* Ghent 69 B5

Geok-Tepe *see* Gökdepe

George South Africa 60 D5

George V Land *physical region* Antarctica 136 C4

Georgenburg *see* Jurbarkas

George Town *capital of* Cayman Islands 36 B3

Georgetown *capital of* Guyana 41 G2

George Town Malaysia 120 B3

Georgia *country* SW Asia 99 F2

Georgia *state* USA 31 E3

Gera Germany 76 C4

Geraldton Australia 129 A5

Gereshk Afghanistan 104 D5

Germany *country* W Europe 76-77

Gerona *see* Girona

Getafe Spain 75 E3

Gettysburg Pennsylvania, USA 23 E4

Gevgelija Macedonia 83 E6

Ghana *country* W Africa 57

Ghanzi Botswana 60 C3

Ghardaïa Algeria 52 D2

Gharyān Libya 53 F2

Ghaznī Afghanistan 105 E4

Ghent *see* Gent

Gibraltar *overseas territory* UK, SW Europe 74 D5

Gibson Desert *desert region* Australia 128 C4

Gijón Spain *var.* Xixón 74 D1

Gilbert Islands *see* Tungaru

Gilbert River *river* Australia 130 C3

Gillette Wyoming, USA 24 C3

Gingin Australia 129 B6

Girin *see* Jilin

Girne Cyprus *var.* Kyrenia 98 C5

Girona Spain *var.* Gerona 75 G2

Gisborne New Zealand 132 E3

Giurgiu Romania 90 C5

Gjirokastër Albania 83 D6

Gjøvik Norway 67 B5

Glasgow Scotland, UK 70 C4

Gleiwitz *see* Gliwice

Glendale Arizona, USA 28 B2

Glendive Montana, USA 24 D2

Gliwice Poland *Ger.* Gleiwitz 81 C5

Gloucester England, UK 71 D6

Glubokoye *see* Hlybokaye

Gobi *desert* China/Mongolia 108 D3

Godāveri *river* India 106 B3 115 E1

Godoy Cruz Argentina 46 B4

Godthåb *see* Nuuk

Godwin Austin, Mount *see* K2

Goiânia Brazil 43 F4

Gökdepe Turkmenistan *prev.* Geok-Tepe, *prev.* Gëkdepe 104 B3

Golan Heights *disputed territory* SW Asia 100 B4

Gold Coast *coastal region* Australia 131 E5

Goldingen *see* Kuldīga

Golmud China 108 D4

Goma Dem. Rep. Congo 59 E6

Gomel' *see* Homyel'/Gomel

Gómez Palacio Mexico 32 D2

Gonaïves Haiti 36 D3

Gonder Ethiopia 54 C4

Gongola *river* Nigeria 57 G4

Good Hope, Cape of *coastal feature* South Africa 60 C5

Goondiwindi Australia 131 D5

Goose Lake *lake* W USA 26 B4

Goré Chad 58 C4

Gorē Ethiopia 55 C5

Gore New Zealand 133 B7

Gorgān Iran 102 D3

Gorki *see* Horki

Gor'kiy *see* Nizhniy Novgorod

Gorlovka *see* Horlivka

Gorontalo Indonesia 121 E4

Gorzów Wielkopolski Poland *Ger.* Landsberg 80 B3

Gospić Croatia 82 B3

Gosford Australia 131 D6

Gostivar Macedonia 83 D5

Göteborg Sweden 67 B7

Gotel Mountains *mountain range* Nigeria 57 G4

Gotland *island* Sweden 67 C7

Gotō-rettō *island group* Japan 113 A6

Göttingen Germany 76 C4

Gouda Netherlands 68 C4

Gough Island *overseas territory* UK, Atlantic Ocean 49 D7

Gouin, Réservoir *Reservoir* Canada 20 D4

Gouré Niger 57 G3

Governador Valadares Brazil 43 G4 45 F1

Goví Altayn Nuruu *mountain range* Mongolia 109 E3

Gozo *island* Malta 79 C7

Grafton Australia 131 E5

Grampian Mountains *mountains* Scotland, UK 70 C3

Guaviare *river* Colombia 40 D3
Guayaquil Ecuador 40 A4
Guayaquil, Golfo do *sea feature* Ecuador/Peru 40 A5
Guernsey *British Crown Dependency* Channel Islands 71 D8
Güney Dogu Toroslar *mountain range* SE Turkey 99 F3
Guiana Highlands *upland* N South America 38 C2
Guider Cameroon 58 B4
Guimarães Portugal 74 C2
Guinea *country* W Africa 56
Guinea, Gulf of *sea feature* Atlantic Ocean 49 D5
Guinea-Bissau *country* W Africa 56
Guiyang China 111 B6
Guizhou *province* China *var.* Kuei-chou, Kweichow, Qian 111 B6
Gujarāt *state* India 116 C4
Gujrānwāla Pakistan 116 C2
Gujrāt Pakistan 116 C2
Gulf, The *sea feature* Arabian Sea *var.* Persian Gulf 122 B2
Gulfport Mississippi, USA 30 C3
Gulu Uganda 55 B6
Gumbinnen *see* Gusev
Gunnbjørn Fjeld *mountain* Greenland 64 D4
Guri, Embalse de *Reservoir* Venezuela 41 E2
Gusau Nigeria 57 F3
Gusev Kaliningrad, Russian Federation *prev.* Gumbinnen 88 B4
Gushgy *see* Serhetabat
Guwāhāti India 117 G3
Guyana *country* NE South America 41
Gwalior India 116 D3
Gwangju South Korea *prev.* Kwangju 111 E4
Gyandzha *see* Gäncä
Gyangzê China 108 C5
Győr Hungary *Ger.* Raab 81 C6

Gyumri Armenia *Rus.* Kumayri, *prev.* Leninakan, Aleksandropol'99 F2
Gyzylarbat *see* Serdar

H

Ha'apai Group *islands* Tonga 127 F5
Haapsalu Estonia *Ger.* Hapsal 88 C2
Haarlem Netherlands 68 C3
Haast New Zealand 133 B6
Hachijō-jima *island* Japan 113 D5
Hachinohe Japan 112 D3
Hadejia *river* Nigeria 57 G3
Ḥaḍramawt *Mountain range* Yemen 103 C7
Hagåtña Guam 126 B1
Hague, The *see* 's-Gravenhage
Haibowan *see* Wuhai
Haicheng China 110 D4
Haifa *see* Hefa
Ḥā'il Saudi Arabia 102 B4
Hailar *see* Hulun Buir
Hainan *island* China *var.* Hainan Dao 106 D3 111 C8
Hainan *province* China *var.* Qiong 111 C7
Hainan Dao *see* Hainan Dao
Hai Phong Vietnam 118 D3
Haiti *country* West Indies 36
Hajdarken *see* Khaydarkan
Hakodate Japan 112 D3
Ḥalab Syria 100 B2
Hala'ib Triangle *disputed region* NE Africa 54 C3
Ḥalāniyāt, Juzur al *Island group* Oman 103 D6
Halden Norway 67 B6
Halfmoon Bay New Zealand 133 A7
Halifax Canada 21 F4
Halle Germany 76 C4
Hallein Austria 77 D7
Halls Creek Australia 128 D3

Halmahera, Pulau *island* Indonesia 121 F3
Halmahera Sea *Sea* Indonesia 121 F4
Halmstad Sweden 67 B7
Ha Long Vietnam *prev.* Hông Gai 118 E3
Hamada Japan 113 B5
Hamadān Iran 102 C3
Ḥamāh Syria 100 B2
Hamamatsu Japan 113 C5
Hamar Norway 67 B5
Hamburg Germany 76 B3
Hämeenlinna Finland 67 D5
HaMelah, Yam *see* Dead Sea
Hamersley Range *mountain range* Australia 128 B4
Hamhŭng North Korea 110 E4
Hami China 108 C3
Hamilton Canada 20 D5
Hamilton New Zealand 132 D3
Hamm Germany 76 B4
Hammerfest Norway 66 D2
Handan China 110 C4
HaNegev *desert region* Israel *Eng.* Negev 101 A6
Hangayn Nuruu *mountain range* Mongolia 108 D2
Hangzhou China 111 D5
Hannover Germany *Eng.* Hanover 76 B4
Hanoi *capital of* Vietnam 118 D3
Hanover *see* Hannover
Hanzhong China 111 B5
Hapsal *see* Haapsalu
Ḥaraḍ Yemen 103 C5
Harare *capital of* Zimbabwe 61 E3
Harbin China 110 E3
Hargeysa Somalia 55 D5
Hari *river* Indonesia 120 B4
Harīrūd *river* C Asia 104 D4
Harper Liberia 56 D5
Harrisburg Pennsylvania, USA 23 E4
Harstad Norway 66 C2
Hartford Connecticut, USA 23 G3
Har Us Nuur *lake* Mongolia 108 C2

Hasselt Belgium 69 D5
Hastings New Zealand 132 E4
Hastings Nebraska, USA 24 E4
Hatay *see* Antakya
Hatteras, Cape *coastal feature* North Carolina, USA 31 G1
Hattiesburg Mississippi, USA 30 C3
Hat Yai Thailand 119 C7
Haugesund Norway 67 A6
Hauraki Gulf *gulf* New Zealand 132 D2
Havana *capital of* Cuba *Sp.* La Habana 36 B2
Havelock North Carolina, USA 31 G1
Havre Montana, USA 24 C1
Havre-Saint-Pierre Canada 21 F3
Hawaii *state* USA 135 E2
Hawai'ian Islands *islands* USA 125 F1
Hawai'ian Ridge *undersea feature* Pacific Ocean 134 D2
Hawera New Zealand 132 D4
Hawke Bay *bay* New Zealand 132 E4
Hawlêr *see* Arbīl
Hawthorne Nevada, USA 27 C6
Hay River Canada 19 E4
Hays Kansas, USA 25 E4
Hazar Turkmenistan *prev.* Cheleken 104 A2
Heard & McDonald Islands *islands* Indian Ocean 123 C7
Hebei *province* China *var.* Hopeh, Hopei, Ji; *prev.* Chihli 110 C4
Hebron West Bank *var.* Al Khalīl, El Khalil, *Heb.* Ḥevron 101 D7
Heerenveen Netherlands 68 D2
Heerlen Netherlands 69 D6
Hefa Israel *prev.* Haifa 101 A5
Hefei China 111 D5
Hei *see* Heilongjiang
Heidelberg Germany 77 B5
Heilbronn Germany 77 B5
Heilongjiang *province* China *var.* Hei, Hei-lung-chiang 110 E3

Hei-lung-chiang *see* Heilongjiang
Helena Montana, USA 24 B2
Hells Canyon *valley* Idaho/Oregon USA 26 C3
Helmand *river* Afghanistan 104 C5
Helmond Netherlands 69 D5
Helsingborg Sweden 67 B7
Helsinki *capital of* Finland 67 D6
Henan *province* China *var.* Honan, Yu 111 C5
Hengduan Shan *mountain range* China 111 A6
Hengelo Netherlands 68 E3
Hengyang China 111 C6
Henzada *see* Hinthada
Herāt Afghanistan 104 C4
Hermansverk Norway 67 A5
Hermosillo Mexico 32 B2
Herning Denmark 67 A7
Heywood Islands *island group* Australia 128 C3
Hiiumaa *island* Estonia *Ger.* Dagden, *Swed.* Dagö 88 C2
Hildesheim Germany 76 C4
Hilversum Netherlands 68 C3
Himalayas *mountain range* S Asia 106 B2
Himora Ethiopia 54 C4
Ḥimṣ Syria 100 B3
Hinchinbrook Island *island* Australia 130 D3
Hindu Kush *mountain range* C Asia 105 E4
Hinthada Myanmar *prev.* Henzada 118 A4
Hiroshima Japan 113 B5
Hitachi Japan 112 D4
Hjørring Denmark 67 A7
Hlybokaye Belarus *Rus.* Glubokoye 89 D5
Hobart Tasmania 131 C8
Hobbs New Mexico, USA 29 E3
Hồ Chí Minh Vietnam *var.* Ho Chi Minh City, *prev.* Saigon 119 E6
Ho Chi Minh City *see* Hồ Chí Minh
Hodeida *see* Al Ḥudaydah

Hoek van Holland Netherlands 68 B4
Hoggar *see* Ahaggar
Hohe Tauern *mountain range* Austria 77 C7
Hohhot China 109 F3
Hokitika New Zealand 133 B5
Hokkaidō *island* Japan 112 D2
Holguín Cuba 36 C2
Holland *see* Netherlands
Hollabrunn Austria 77 E6
Holon Israel 101 A5
Holyhead Wales, UK 71 C5
Hombori Mopti, Mali 57 E3
Homyel'/Gomel' Belarus *Rus.* Gomel' 89 E7
Honan *see* Henan
Honduras *country* Central America 34-35
Honduras, Gulf of *sea feature* Caribbean Sea 34 C2
Hønefoss Norway 67 B6
Hông Gai *see* Ha Long
Hong Kong *special administrative region* China, E Asia 111 C6
Honiara *capital of* Solomon Islands 126 C3
Honshū *island* Japan 112 D3
Hoorn Netherlands 68 C2
Hopa Turkey 99 E2
Hopedale Canada 21 F2
Hopeh *see* Hebei
Hopei *see* Hebei
Hopkinsville Kentucky, USA 22 B5
Horki Belarus *Rus.* Gorki 89 E5
Horlivka Ukraine *Rus.* Gorlovka 90 G3
Horn, Cape *see* Hornos, Cabo
Hornos, Cabo *Eng* Cape Horn *coastal feature* Chile 47 C8
Horsham Australia 131 C6
Hospitalet *see* L'Hospitalet de Llobregat
Hot Springs Arkansas, USA 30 B2
Houston Texas, USA 29 G4
Hovd Mongolia 108 C2
Hövsgöl Nuur *lake* Mongolia 108 D1

Inverness Scotland, UK 70 C3
Investigator Ridge *undersea feature* Indian Ocean 122 D4
Ioánnina Greece 86 A4
Ióniol Nisiá *island group* Greece *Eng.* Ionian Islands 87 A5
Ionian Islands *see* Ióniol Nisiá
Ionian Sea Mediterranean Sea 87 A6
Íos *island* Greece 87 D6
Iowa *state* USA 25 F3
Ipoh Malaysia 120 B3
Ipswich England, UK 71 E6
Iqaluit Canada *prev.* Frobisher Bay 19 H3
Iquique Chile 46 B1
Iquitos Peru 42 B2
Irákleio Greece 87 D7
Iran *country* SW Asia 102-103
Iranian Plateau *upland* Iran 102 D4
Iraq *country* SW Asia 102
Irbid Jordan 101 B5
Ireland *country* W Europe 70-71
Irian Jaya *see* Papua
Irish Sea British Isles 71 C5
Irkutsk Russian Federation 97 E4
Iron Mountain Michigan, USA 22 B2
Ironwood Michigan, USA 22 B1
Irrawaddy *river* Myanmar 118 B2
Irrawaddy, Mouths of the *wetlands* Myanmar 118 A4
Irtysh *River* Asia 94 C3
Iruña *see* Pamplona
Ishim *River* Kazakhstan/Russian Federation 94 C3
Isiro Dem. Rep. Congo 59 E5
İskenderun Turkey *Eng.* Alexandretta 98 D4
Iskǔr *river* Bulgaria 86 C1
Iskǔr, Yazovir *Reservoir* Bulgaria 86 C2
Islay Scotland, UK 70 B4
Islāmābād *capital of* Pakistan 116 C1
Ismaila *see* Al Ismā'īlīya
Isna Egypt 54 B2

Ísparta Turkey 98 B4
Israel *country* SW Asia 100-101
Issyk-Kul, Ozero *lake* Kyrgyzstan 105 G2
İstanbul Turkey *var.* Stambul, *prev.* Constantinople, Byzantium, *Bul.* Tsarigrad 98 B2
İstanbul Boğazi *see* Bosporus
Itabuna Brazil 43 G4
Itagüí Colombia 40 B2
Italy *country* S Europe 78-79
Ittoqqortoormiit Greenland 65 E3
Iturup *island* Japan/Russian Federation (disputed) 112 E1
Ivanhoe Australia 131 C6
Ivano-Frankivs'k Ukraine 90 C2
Ivanovo Russian Federation 92 B4
Ivittuut Greenland 64 B4
Ivory Coast *see* Côte d'Ivoire
Ivujivik Canada 20 D1
Iwaki Japan 112 D4
Izabal, Lago de *lake* Guatemala 34 C2
Izhevsk Russian Federation 93 C5 96 B3
İzmir Turkey *prev.* Smyrna 98 A3
İzmit Turkey *var.* Kocaeli 98 B2
Izu-shotō *island group* Japan 113 D6

J

Jabal ash Shifá *desert* Saudi Arabia 102 A4
Jabalpur India 116 E4
Jackson Mississippi, USA 30 C2
Jacksonville Florida, USA 31 E3
Jacksonville Texas, USA 29 G3
Jacmel Haiti 36 D3
Jaén Spain 75 E4
Jaffna Sri Lanka 115 E3
Jagdaqi China 109 G1
Jiangxi *province* China 111 C6
Jaipur India 116 D3

Jajce Bosnia & Herzegovina 82 C4
Jakarta *capital of* Indonesia 120 C5
Jakobstad Finland 66 D4
Jakobstad *see* Jēkabpils
Jalālābād Afghanistan 105 E4
Jalal-Abad *see* Dzhalal-Abad
Jalandhar India 116 D2
Jalapa *see* Xalapa
Jamaame Somalia 55 D6
Jamaica *country* West Indies 36
Jamālpur Bangladesh 117 G4
Jambi Indonesia 120 B4
James Bay *sea feature* Canada 20 C4
Jammu & Kashmir *disputed region* India/Pakistan 116 D2
Jāmnagar India 116 B4
Jan Mayen *external territory* Norway, Arctic Ocean 65 F3
Japan *country* E Asia 112-113
Japan, Sea of Pacific Ocean 112 B3
Jarvis Island *unincorporated territory* USA, Pacific Ocean 125 F2
Java *see* Jawa
Java Sea Pacific Ocean *var.* Laut Jawa 122 D4
Java Trench *undersea feature* Indian Ocean 122 D4
Jawa *island* Indonesia *var.* Java 120 C5
Jawa, Laut *see* Java Sea
Jayapura Indonesia 121 H4
Jaz Mūriān, Hāmūn-e *lake* Iran 102 E4
Jedda *see* Jiddah
Jefferson City Missouri, USA 25 G4
Jeju-do *island* South Korea *prev.* Cheju-do 111 E5
Jeju Strait *sea feature* South Korea *prev.* Cheju Strait 111 E5
Jēkabpils Latvia *Ger.* Jakobstadt 88 C4
Jelgava Latvia *Ger.* Mitau 88 C3
Jember Indonesia 120 D5
Jena Germany 76 C4

Jenīn *var.* Janin, Jinīn; *anc.* Engannim. West Bank 101 D6

Jérémie Haiti 36 D3

Jerevan *see* Yerevan

Jericho West Bank 101 B5

Jerid, Chott el *salt lake* Africa 84 D4

Jersey *British Crown Dependency* Channel Islands 71 D8

Jerusalem *capital of* Israel 101 B5

Jhelum Pakistan 116 C2

Ji *see* Hebei

Ji *see* Jilin

Jiangsu *province* China *var.* Chiang-su, Kiangsu, Su 111 D5

Jiangxi *province* China *var.* Chiang-hsi, Gan, Kiangsi 111 C6

Jiaxing Zhejiang, China 111 D5

Jibuti *see* Djibouti

Jiddah Saudi Arabia *Eng.* Jedda 103 A5

Jiftlik Post West Bank 101 D7

Jihlava Czech Republic *Ger.* Iglau 81 B5

Jilin *province* China *var.* Chi-lin, Girin, Ji, Kirin 110 E3

Jilin China 110 E3

Jīma Ethiopia 55 C5

Jin *see* Shanxi

Jinan China 111 C4

Jingdezhen China 111 D5

Jinhua China 111 D5

Jining *see* Ulan Qab

Jinotega Nicaragua 34 D3

Jinsha Jiang *river* China 108 D5

Jinzhou China 110 D4

Jīzān Saudi Arabia 103 B6

João Pessoa Brazil 43 H3

Jodhpur India 116 C3

Joensuu Finland 67 E5

Johannesburg South Africa 60 D4

Johnston Atoll *US unincorporated territory* Pacific Ocean 125 E1

Johor Bahru Malaysia 120 C3

Joinville Brazil 44 D3

Joliet Illinois, USA 22 B3

Jönköping Sweden 67 B7

Jonquière Canada 21 E4

Jordan *country* SW Asia 100–101

Jordan *river* SW Asia 101 B5

Joseph Bonaparte Gulf *gulf* Australia 128 D2

Jos Plateau *upland* Nigeria 57 G4

Juan Fernandez, Islas *islands* Chile 46 A4

Juàzeiro Brazil 43 G3

Juàzeiro do Norte Brazil 43 G3

Juba *capital of* South Sudan 55 B5

Júcar *river* Spain 75 E3

Judenburg Austria 77 D7

Juigalpa Nicaragua 34 D3

Juiz de Fora Brazil 43 G5 45 F2

Juneau Alaska, USA 18 D4

Junggar Pendi *desert* China 108 C2

Junín Argentina 46 D4

Jura *mountains* France/Switzerland 77 A7

Jura *island* Scotland, UK 70 B4

Jurbarkas Lithuania *Ger.* Jurburg, *var.* Georgenburg 88 B4

Jurburg *see* Jurbarkas

Juruá *river* Brazil/Peru 42 C2

Juticalpa Honduras 34 D2

Jutland *see* Jylland

Juventud, Isla de la *island* Cuba 36 B2

Jylland *peninsula* Denmark *Eng.* Jutland 67 A7

Jyväskylä Finland 67 D5

K

K2 *peak* China/Pakistan *Eng.* Mount Godwin Austen 116 D1

Kaachka *see* Kaka

Kaakhka *see* Kaka

Kabale Uganda 55 B6

Kabinda Dem. Rep. Congo 59 D7

Kābol *see* Kabul

Kabul *capital of* Afghanistan *Per.* Kabol 105 E4

Kachch, Gulf of *sea feature* Arabian Sea 116 B4

Kachch, Rann of *wetland* India/Pakistan *var.* Rann of Kutch 116 B4

Kadugli Sudan 54 B4

Kaduna Nigeria 57 G4

Kaédi Mauritania 56 C3

Kâghet *Physical region* Mauritania 56 D1

Kagoshima Japan 113 A6

Kahramanmaraş Turkey *var.* Marash, Maraş 98 D4

Kai, Kepulauan *island group* Indonesia 121 G4

Kaifeng China 111 C5

Kaikohe New Zealand 132 C2

Kaikoura New Zealand 133 C5

Kainji Reservoir *Reservoir* Nigeria 57 F4

Kairouan Tunisia 53 E1

Kaiserslautern Germany 77 B5

Kaitaia New Zealand 132 C2

Kajaani Finland 66 E4

Kaka Turkmenistan *prev.* Kaakhka, *var.* Kaachka 104 C3

Kakhovka Ukraine 91 F4

Kakhovs'ka Vodoskhovyshche *Reservoir* Ukraine 91 F3

Kalahari Desert *desert* southern Africa 60 C4

Kalamariá Greece 86 B4

Kalámata Greece 87 B6

Kalât *see* Qalāt

Kalbarri Australia 129 A5

Kalemie Dem. Rep. Congo 59 E7

Kalgoorlie Australia 129 C6

Kalimantan *geopolitical region* Indonesia *Eng.* Indonesian Borneo 120 D4

Kaliningrad *external territory* Russian Federation 96 A2

Kaliningrad Kaliningrad, Russian Federation *prev.* Königsberg 88 A4

Kalinkavichy Belarus *Rus.* Kalinkovichi 89 D7

Kalinkovichi *see* Kalinkavichy
Kalisch *see* Kalisz
Kalispell Montana, USA 24 B1
Kalisz Poland *Ger.* Kalisch
80 C4
Kalmar Sweden 67 C7
Kalpeni Island *island* India
114 C3
Kama *river* Russian Federation
92 D4
Kamchatka *peninsula* Russian
Federation 97 H3
Kamchiya *river* Bulgaria 86 E2
Kamina Dem. Rep. Congo
59 D7
Kamishli *see* Al Qāmishlī
Kamloops Canada 19 E5
Kampala *capital of* Uganda
55 B6
Kâmpóng Cham Cambodia
119 D6
Kâmpóng Chhnăng Cambodia
119 D5
Kâmpóng Saôm *see*
Sihanoukville
Kâmpôt Cambodia 119 D6
Kampuchea *see* Cambodia
Kam"yanets'-Podil's'kyy
Ukraine 90 C3
Kananga Dem. Rep. Congo
59 D7
Kanazawa Japan 112 C4
Kandahār Afghanistan
var. Qandahār 104 D5
Kandi Benin 57 F4
Kaniv's'ke Vodoskhovyshche
Reservoir Ukraine 91 E2
Kandy Sri Lanka 115 E3
Kanestron, Ákra *see* Palioúri,
Akrotírio
Kangaroo Island *island*
Australia 131 B7
Kangertittivaq *region*
Greenland 64 E3
Kangikajik *headland*
Greenland 65 E4
Kanjiža Serbia 82 D2
Kankan Guinea 56 D4
Kano Nigeria 57 G4
Kānpur India *prev.* Cawnpore
117 E3

Kansas *state* USA 24-25
Kansas City Kansas, USA 25 F4
Kansas City Missouri, USA 25 F4
Kansk Russian Federation 97 E4
Kansu *see* Gansu
Kaohsiung *see* Gaoxiong
Kaolack Senegal 56 B3
Kapfenberg Austria 77 E7
Kaposvár Hungary 81 C7
Kapsukas *see* Marijampolė
Kapuas *river* Indonesia 120 D4
Kara-Balta Kyrgyzstan 105 F2
Karabük Turkey 98 C2
Karāchi Pakistan 116 B4
Karaganda *see* Karagandy
Karagandy Kazakhstan *prev.*
Karaganda 96 C4
Karakol Kyrgyzstan *prev.*
Przheval'sk 105 G2
Kara Kum *see* Garagum
Karakumskiy Kanal *see*
Garagum Kanaly
Karakumy *see* Garagum
Karamay China 108 C2
Karamea Bight *gulf* New
Zealand 133 C5
Karasburg Namibia 60 C4
Kara Sea *see* Karskoye More
Karditsa Greece 86 B4
Kariba, Lake *lake* Zambia/
Zimbabwe 60 D3
Karimata, Selat *strait* Indonesia
120 C4
Karkinits'ka Zatoka *sea feature*
Black Sea 91 E4
Karl-Marx-Stadt *see* Chemnitz
Karlovac Croatia 82 B3
Karlovy Vary Czech Republic
Ger. Karlsbad 81 A5
Karlsbad *see* Karlovy Vary
Karlskrona Sweden 67 C7
Karlsruhe Germany 77 B5
Karlstad Sweden 67 B6
Karnātaka *state* India 114 D1
Kárpathos *island* Greece 87 E7
Kars Turkey 99 F2
Karshi Uzbekistan *prev.* Bek-
Budi, *Uzb.* Qarshi 104 D3
Karskoye More Arctic Ocean
Eng. Kara Sea 137 H3

Kasai *river* Dem. Rep. Congo
59 C6
Kasama Zambia 61 E2
Kaschau *see* Košice
Kāshān Iran 102 C3
Kashi China 108 A3
Kasongo Dem. Rep. Congo
59 E6
Kassa *see* Košice
Kassala Sudan 54 C4
Kassel Germany 76 B4
Kastamonu Turkey 98 C2
Katanning Australia 129 B6
Katerini Greece 86 B4
Katha Myanmar 118 B2
Katherine Australia 128 E2
Kathmandu *capital of* Nepal
117 F3
Katsina Nigeria 57 G3
Katowice Poland 81 C5
Kauen *see* Kaunas
Kaunas Lithuania *Ger.* Kauen,
Pol. Kowno, *Rus.* Kovno
88 B4
Kavadarci Macedonia 82 E5
Kavála Greece 86 C3
Kavaratti Island *island* India
114 C3
Kavīr, Dasht-e *Salt pan* Iran
102 C3
Kawasaki Japan 113 D5
Kayan *river* Indonesia 120 D3
Kayes Mali 56 C3
Kayseri Turkey 98 C3
Kazakhstan *country* C Asia 96
Kazan' Russian Federation
96 B3
Kazandzhik *see* Bereket
Kazanlŭk Bulgaria 86 D2
Kecskemét Hungary 81 D7
Kediri Indonesia 120 D5
Keetmanshoop Namibia 60 C4
Kefallonía *island* Greece *Eng.*
Cephalonia 87 A5
Keá *see* Tziá
Kelang *see* Klang
Kelmė Lithuania 88 B4
Kelowna Canada 19 E5
Kemerovo Russian Federation
96 D4
Kemi Finland 66 D4

Kemi — Kizyl-Arvat

Kladno Czech Republic 81 A5
Klagenfurt Austria 77 D7
Klaipėda Lithuania *Ger.* Memel 88 B4
Klamath Falls Oregon, USA 26 B4
Khang Malaysia *var.* Kelang 120 B2
Ključ Bosnia & Herzegovina 82 B3
Knin Croatia 82 B4
Knoxville Tennessee, USA 31 E1
Knud Rasmussen Land *region* Greenland 64 D1
Kōbe Japan 113 C5
Koblenz Germany 77 B5
Kobryn Belarus 89 B6
Kocaeli *see* Izmit
Kočani Macedonia 83 E5
Kōchi Japan 113 B6
Kochi India *see* Cochin 114 D3
Kodiak Alaska, USA 18 C3
Kodiak Island *island* Alaska, USA 18 C3
Koedoes *see* Kudus
Kohīma India 117 H3
Kohtla-Järve Estonia 88 D2
Kokand *see* Qo'qon
Kokchetav Kazakhstan 96 C4
Kokkola Finland 66 D4
Koko Nor *see* Qinghai
Koko Nor *see* Qinghai Hu
Kokshaal-Tau *mountain range* Kyrgyzstan 105 G2
Kola Peninsula *see* Kol'skiy Poluostrov
Kolguyev, Ostrov *island* Russian Federation 92 D2
Kolhumadulu Atoll *island* Maldives 114 C5
Kolka Latvia 88 C3
Kolkata India *var.* Calcutta 117 F4
Köln Germany *Eng.* Cologne 76 B4
Kol'skiy Poluostrov *peninsula* Russian Federation *Eng.* Kola Peninsula 63 F1 92 C2
Kolwezi Dem. Rep. Congo 59 D8

Kolyma *river* Russian Federation 95 G2
Kommunizma, Pik *see* Communism Peak
Komoé *river* Côte d'Ivoire 57 E4
Komotiní Greece 86 D3
Komsomol'sk-na-Amure Russian Federation 97 G4
Kondoz *see* Kunduz
Konduz *see* Kunduz
Köneürgench Turkmenistan *prev.* Kunya-Urgench, *prev.* Këneurgench 104 C2
Kong Christian IX Land *region* Greenland 64 D4
Kong Christian X Land *region* Greenland 64 E3
Kong Frederik VI Kyst *region* Greenland 64 C4
Kong Frederik VIII Land *region* Greenland 64 E2
Kong Frederik IX Land *region* Greenland 64 C3
Kong Karls Land *island group* Svalbard 65 G2
Kong Oscar Fjord *fjord* Greenland 65 E3
Konia *see* Konya
Königgrätz *see* Hradec Králové
Königsberg *see* Kaliningrad
Konispol Albania 83 D7
Konjic Bosnia & Herzegovina 82 C4
Konya Turkey *prev.* Konia 98 C4
Kopaonik *mountains* Serbia 83 D4
Koper Slovenia 77 D8
Koprivnica Croatia 82 B2
Korçë Albania 83 D6
Korčula *island* Croatia 82 B4
Korea Bay *bay* China/North Korea 110 D4
Korea Strait *sea feature* Japan/South Korea 110-111 E5
Korinthiakós Kólpos *sea feature* Greece *Eng.* Gulf of Corinth 87 B5
Kórinthos Greece *Eng.* Corinth 87 B5

Kōriyama Japan 113 D4
Korla China 108 C3
Korosten' Ukraine 90 D1
Kortrijk Belgium 69 A6
Kos *island* Greece 87 E6
Kosciusko, Mount *peak* Australia 131 D7
Košice Slovakia *Ger.* Kaschau, *Hung.* Kassa 81 D6
Köslin *see* Koszalin
Kosovo *country* SE Europe 83 D5
Kosovska Mitrovica *see* Mitrovicë/Mitrovica
Kosrae *island* Micronesia 126 C2
Kossou, Lac de *lake* Côte d'Ivoire 56 D4
Kostanay Kazakhstan *var.* Kustanay 96 C4
Kostyantynivka Ukraine 91 G3
Koszalin Poland *Ger.* Köslin 80 B2
Kota India 116 D4
Kota Bharu Malaysia 120 B3
Kota Kinabalu Malaysia 120 D3
Kotka Finland 67 E5
Kotlas NW Russia 92 C4
Kotuy *river* Russian Federation 95 E2
Koudougou Burkina Faso 57 E4
Kourou French Guiana 41 H2
Kousséri Cameroon 58 B3
Kouvola Finland 67 E5
Kovel' Ukraine 90 C1
Kovno *see* Kaunas
Kowno *see* Kaunas
Kozáni Greece 86 B4
Kozhikode India *see* Calicut 114 D2
Kra, Isthmus of *coastal feature* Myanmar/Thailand 119 B6
Kragujevac Serbia 82 D4
Krakau *see* Kraków
Kraków Poland *Eng.* Cracow, *Ger.* Krakau 81 D5
Kralendijk Bonaire 37 E5
Kraljevo Serbia 82 D4
Kranj Slovenia 77 D7
Krasnodar Russian Federation 93 A6

La Crosse Wisconsin, USA
22 A2
Ladoga, Lake *see* Ladozhskoye
Ozero
Ladozhskoye Ozero *lake*
Russian Federation *Eng.* Lake
Ladoga 92 B3
Ladysmith Wisconsin, USA
22 A2
Lae Papua New Guinea 126 B3
La Esperanza Honduras 34 C2
Lafayette Louisiana, USA 30 B3
Laghouat Algeria 52 D2
Lagos Nigeria 57 F5
Lagos Portugal 74 C4
Lagouira Western Sahara 52 A4
La Grande Oregon, USA 26 C3
La Habana *see* Havana
Lahore Pakistan 116 C2
Laï Chad 58 C4
Laila *see* Laylá
Lajes Brazil 44 D3
Lake Charles Louisiana, USA
30 B3
Lake District *region* England,
UK 71 C5
Lakewood Colorado, USA
24 D4
Lakshadweep *island group*
India *Eng.* Laccadive Islands
114 B2
La Ligua Chile 46 B4
La Louvière Belgium 69 B6
Lambaré Paraguay 44 B3
Lambaréné Gabon 59 B6
Lamia Greece 86 B4
Lancaster England, UK 71 D5
Lancaster California, USA 27 C7
Lancaster Sound *sea feature*
Canada 19 F2
Landsberg *see* Gorzów
Wielkopolski
Land's End *coastal feature*
England, UK 71 C7
Landshut Germany 77 D6
Lang Son Vietnam 118 D3
Länkäran Azerbaijan *Rus.*
Lenkoran' 99 H3
Lansing Michigan, USA 22 C3
Lanzarote *island* Spain 52 B3

Lanzhou China 110 B4
Laon France 72 D3
La Oroya Peru 42 B3
Laos *country* SE Asia 118
La Palma *island* Spain 52 A3
La Paz *legislative &
administrative capital of*
Bolivia 42 C4
La Paz Mexico 32 B3
La Pérouse Strait *sea feature*
Japan 112 D1
Lapland *region* N Europe 66 C3
La Plata Argentina 46 D4
Lappeenranta Finland 67 E5
Laptev Sea *see*
Laptevykh, More
Laptevykh, More Arctic Ocean
Eng. Laptev Sea 97 F2
L'Aquila Italy 78 D4
Laramie Wyoming, USA 24 C4
Laredo Texas, USA 29 F5
La Rioja Argentina 46 C3
Lárisa Greece 86 B4
Lārkāna Pakistan 116 B3
Larnaca Cyprus *var.* Larnaka,
Larnax 98 C5
Larnaka *see* Larnaca
Larnax *see* Larnaca
La Rochelle France 72 B4
La Roche-sur-Yon France 72 B4
La Romana Dominican Republic
36 E3
Las Cruces New Mexico, USA
28 D3
Las Piedras Uruguay 44 C5
La Serena Chile 46 B3
La Spezia Italy 78 B3
Las Tablas Panama 35 F5
Las Vegas Nevada, USA 27 D7
Latakia *see* Al Lādhiqīyah
Latvia *country* NE Europe 88
Launceston Tasmania 131 C8
Laurentian Basin *see* Canada
Basin
Laurentian Mountains *upland*
Canada 16 D4
Lausanne Switzerland 77 A7
Laut, Pulau *prev.* Laoet. *Island*
Indonesia 120 D4
Laval France 72 B4
Lawton Oklahoma, USA 29 F2

Laylá Saudi Arabia 103 C5
Lazarev Sea *sea* Antarctica
136 B2
Lebanon *country* SW Asia
100-101
Lebu Chile 47 B5
Lecce Italy 79 E5
Leduc Canada 19 E5
Leeds England, UK 71 D5
Leeuwarden Netherlands 68 D1
Leeward Islands *see* Sotavento,
Ilhas de
Lefkáda *island* Greece *prev.*
Levkás 87 A5
Lefkoşa *see* Nicosia
Lefkosia *see* Nicosia
Legaspi *see* Legazpi City
Legazpi City Philippines *var.*
Legaspi 120 E2
Legnica Poland *Ger.* Liegnitz
80 B4
Le Havre France 72 B3
Leicester England, UK 71 D6
Leiden Netherlands 68 C3
Leipzig Germany 76 D4
Lek *river* Netherlands 68 C4
Le Léman *see* Geneva, Lake
Lelystad Netherlands 68 D3
Léman, Lac *see* Geneva, Lake
Le Mans France 72 B4
Lemesos *see* Limassol
Lemnos *see* Límnos
Lena *river* Russian Federation
97 F3
Leninabad *see* Khŭjand
Leninakan *see* Gyumri
Leningrad *see* St Petersburg
Leninsk *see* Türkmenabat
Lenkoran' *see* Länkäran
León Mexico 33 E4
León Nicaragua 34 C3
León Spain 74 D1
Léopoldville *see* Kinshasa
Lepel' *see* Lyepyel'
Le Puy France 73 C5
Lérida *see* Lleida
Lerwick Scotland, UK 70 D1
Lesbos *see* Lésvos
Leshan China 111 B5

Leskovac Serbia 82 E4

Lesotho *country* southern Africa 60

Lesser Antilles *island group* West Indies 37 G4

Lésvos *island* Greece *Eng.* Lesbos 86 D4

Lethbridge Canada 19 E5

Leti, Kepulauan *island group* Indonesia 121 F5

Leuven Belgium 69 C6

Leverkusen Germany 76 A4

Levin New Zealand 132 D4

Levkás *see* Lefkáda

Lewis *island* Scotland, UK 70 B2

Lewiston Idaho, USA 26 C2

Lewiston Maine, USA 23 G2

Lexington Kentucky, USA 22 C5

Lezhë Albania 83 D5

Lhasa China 108 C5

Lhazê China 108 C4

L'Hospitalet de Llobregat *var.* Hospitalet. Spain 75 G2

Liao *see* Liaoning

Liaoning *province* China *var.* Liao, Shengking; *hist.* Fengtien, Shenking. Admin. region 110 D3

Libau *see* Liepāja

Liberec Czech Republic *Ger.* Reichenberg 80 B4

Liberia *country* W Africa 56

Liberia Costa Rica 34 D4

Libreville *capital* of Gabon 59 A5

Libya *country* N Africa 53

Libyan Desert *desert* N Africa 50 C3

Lichuan China 111 B5

Liechtenstein *country* C Europe 77 B7

Liège Belgium 69 D6

Liegnitz *see* Legnica

Lienz Austria 77 D7

Linz Austria 77 D7

Liepāja Latvia *Ger.* Libau 88 B3

Liffey *river* Ireland 71 B5

Ligurian Sea Mediterranean Sea 78 A3

Likasi Dem. Rep. Congo 59 E8

Lille France 72 D2

Lillehammer Norway 67 B5

Lilongwe *capital* of Malawi 61 E2

Lima *capital* of Peru 42 B4

Limassol Cyprus *var.* Lemesos 98 C5

Limerick Ireland 71 A6

Limnos *island* Greece *var.* Lemnos 86 D4

Limoges France 72 C5

Limón Costa Rica 35 E4

Limpopo *river* southern Africa 60 D3

Linares Chile 46 B4

Linares Spain 75 E4

Linchuan *see* Fuzhou

Lincoln England, UK 71 D5

Lincoln Nebraska, USA 25 F4

Lincoln Sea Arctic Ocean 64 E1

Linden Guyana 41 G2

Lindi Tanzania 55 C8

Line Islands *island group* Kiribati 127 C5

Linköping Sweden 67 C6

Linz Austria 77 D6

Lion, Golfe du *sea feature* Mediterranean Sea 73 D6

Lipari, Isola *island* Italy 79 D6

Lipari Islands *see* Isole Eolie

Lira Uganda 55 B6

Lisbon *capital* of Portugal *Port.* Lisboa 74 B3

Litani *river* SW Asia 91 B4

Lithuania *country* E Europe 88-89

Little Andaman *island* India 115 G2

Little Minch *sea feature* Scotland, UK 70 B3

Little Rock Arkansas, USA 30 B2

Liuzhou China 111 C6

Liverpool England, UK 71 D5

Livingstone Zambia 60 D3

Livno Bosnia & Herzegovina 82 B4

Livorno Italy 78 B3

Ljubljana *capital* of Slovenia 77 D7

Ljusnan *river* Sweden 67 B5

Llanos *region* Colombia/ Venezuela 41 E2

Lleida Spain *Cast.* Lérida 75 F2

Lobatse Botswana 60 D4

Lobito Angola 60 B2

Locarno Switzerland 77 B7

Lodja Dem. Rep. Congo 59 D6

Łódź Poland *Rus.* Lodz 80 D4

Lofoten *island group* Norway 66 B3

Logroño Spain 75 E2

Loire *river* France 72 B4

Loja Ecuador 40 A5

Lokitaung Kenya 55 C5

Loksa Estonia *Ger.* Loxa 88 D2

Lombok, Pulau *island* Indonesia 120 D5

Lomé *capital* of Togo 57 E5

Lomond, Loch *lake* Scotland, UK 70 C4

London Canada 20 C5

London *capital* of UK 71 E6

Londonderry Northern Ireland, UK 70 B4

Londonderry, Cape *coastal feature* Australia 128 D2

Londrina Brazil 44 D2

Long Beach California, USA 27 C8

Long Island *island* The Bahamas 34 D2

Long Island *island* NE USA 23 G3

Longreach Australia 130 C4

Long Strait *Strait* Russian Federation 95 H2

Longview Texas, USA 29 G3

Longview Washington, USA 26 B2

Longyearbyen Svalbard 65 F2

Lop Nur *lake* China 108 C3

Lorca Spain 75 E4

Lord Howe Island *island* Australia 124 C4

Lord Howe Rise *undersea feature* Pacific Ocean 124 D4

Maicao — Marías, Islas

Maicao Colombia 40 C1
Maiduguri Nigeria 57 H4
Maïmanah Afghanistan *prev.*
Meymaneh 104 D4
Maine *state* USA 23 G1
Maine, Gulf of *gulf* USA 23 G2
Mainz Germany 77 B5
Maio *Island* Cape Verde 56 A3
Maiz, Islas del *islands*
Nicaragua 35 E3
Majorca *see* Mallorca
Majuro *island* Marshall Islands
126 D1
Makarska Croatia 82 B4
Makarov Basin *undersea
feature* Arctic Ocean 137 G3
Makassar Indonesia *prev.*
Ujungpandang 121 E4
Makassar Strait *strait* Indonesia
120 D4
Makeyevka *see* Makiyivka
Makhachkala Russian
Federation 93 B7 96 A4
Makiyivka Ukraine *Rus.*
Makeyevka 91 G5
Makkah Saudi Arabia *Eng.*
Mecca 103 A5
Makkovik Canada 21 F2
Malabo *capital of* Equatorial
Guinea 59 A5
Malacca, Strait of *sea feature*
Indonesia/ Malaysia
106 C4 119 C8 120 B3
Maladzyechna Belarus *Rus.*
Molodechno, *Pol.*
Molodechno 89 C5
Málaga Spain 74 D5
Malakal South Sudan 55 B5
Malang Indonesia 120 D5
Malanje Angola 60 C2
Malatya Turkey 99 E3
Malawi *country* southern
Africa 61
Malay Peninsula *peninsula*
Malaysia/Thailand 119 D8
Malaysia *country* Asia 120
Malden Island *atoll* Kiribati
125 F2
Maldives *country* Indian Ocean
114 C4
Male' *capital of* Maldives
114 C4

Malekula *island* Vanuatu 124 D3
Mali *country* W Africa 57
Malindi Kenya 55 C7
Mallorca *island* Spain *Eng.*
Majorca 75 H3
Malmö Sweden 67 B7
Malta *country* Mediterranean
Sea 79 C8
Malta Montana, USA 24 C1
Malta Channel *sea feature*
Mediterranean Sea 79 C7
Maluku *island group*
Indonesia *var.* Moluccas
107 E4 121 F4
Maluku, Laut Pacific Ocean
Eng. Molucca Sea 121 F4
Mamberamo *river* Indonesia
121 H4
Mamoudzou *capital of*
Mayotte 61 G2
Man, Isle of *British Crown
Dependency* UK 71 C5
Manado Indonesia 121 F3
Managua *capital of Nicaragua*
34 D3
Manama *capital of Bahrain Ar.*
Al Manāmah 103 C5
Mananjary Madagascar 61 G3
Manaus Brazil 42 D2
Manchester England, UK
71 D5
Manchester New Hampshire,
USA 23 G2
Manchurian Plain *plain* E Asia
107 F1
Mandalay Myanmar 118 B3
Mangalia Romania 90 D5
Mangalore India 114 C2
Manicouagan, Réservoir
Reservoir Canada 21 E3
Manihiki *atoll* Cook Islands
125 F3
Maniitsoq Greenland 64 D3
Manila *capital of Philippines*
121 E1
Manisa Turkey *prev.* Saruhan
98 A3
Manitoba *province* Canada
19 G4
Manizales Colombia 40 B3
Manjimup Australia 129 B7
Mannar Sri Lanka 115 E3

Mannar, Gulf of *sea feature*
Indian Ocean 114 D3
Mannheim Germany 77 B5
Manono Dem. Rep. Congo
59 E7
Mansel Island *island* Canada
20 C1
Mansfield Ohio, USA 22 D4
Manta Ecuador 40 A4
Mantes-la-Jolie France 72 C3
Mantova Italy *Eng.* Mantua
78 B2
Mantua *see* Mantova
Manurewa New Zealand
132 D3
Manzhouli China 109 F1
Mao Chad 58 B3
Maoke, Pegunungan
mountains Indonesia
121 H4
Maputo *capital of*
Mozambique 61 E4
Mar, Serra do *mountains* Brazil
38 D4
Maracaibo Venezuela 40 C1
Maracaibo, Lago de *inlet*
Venezuela 40 C1
Maracay Venezuela 40 D1
Maradi Niger 57 F3
Marāgheh Iran 102 C3
Marajó, Ilha de *island* Brazil
43 F2
Marañón *river* Peru 42 B2
Maraş *see* Kahramanmaraş
Marash *see* Kahramanmaraş
Marbella Spain 74 D5
Marble Bar Australia 128 B4
Mar Chiquita, Laguna *salt lake*
Argentina 46 C3
Mardān Pakistan 116 C1
Mar del Plata Argentina 47 D5
Mardin Turkey 99 E4
Margarita, Isla de *island*
Venezuela 41 E1
Mārgow, Dasht-e- *desert*
Afghanistan 104 C5
Mariana Trench *undersea
feature* Pacific Ocean
124 B1 126 B1
Marías, Islas *islands* Mexico
32 C4

Melbourne Australia 131 C7

Melbourne Florida, USA 31 F4

Melghir, Chott *Salt lake* Algeria 53 E2

Melilla *external territory* Spain, N Africa 52 C1

Melitopol' Ukraine 91 F4

Melo Uruguay 44 C4

Melville Island *island* Australia 128 E2

Melville Island *island* Canada 19 E2

Memel *see* Klaipėda

Memel *see* Neman

Memphis Tennessee, USA 30 C1

Mendaña Fracture Zone *tectonic feature* Pacific Ocean 135 G3

Mende France 73 C6

Mendeleyev Ridge *undersea feature* Arctic Ocean 137 G2

Mendocino Fracture Zone *tectonic feature* Pacific Ocean 134 D2

Mendoza Argentina 46 B4

Menengiyn Tal *plain* Mongolia 109 F3

Menongue Angola 60 C2

Menorca *island* Spain *Eng.* Minorca 75 H3

Metairie Louisiana, USA 30 C3

Mentawai, Kepulauan *island group* Indonesia 120 B4

Meppel Netherlands 68 D2

Merced California, USA 27 B6

Mercedes Uruguay 44 B5

Mergui *see* Myeik

Mergui Archipelago *island chain* Myanmar 119 B6

Mérida Mexico 33 H3

Mérida Spain 74 D3

Mérida Venezuela 40 C2

Meridian Mississippi, USA 30 C2

Merredin Australia 129 B6

Mersin Turkey *var.* İçel 98 C4

Meru Kenya 55 C6

Merv *see* Mary

Mesa Arizona, USA 28 B2

Meshed *see* Mashhad

Messina Italy 79 D6

Messina, Stretto di *sea feature* Ionian Sea/Tyrrhenian Sea 79 D7

Mestre Italy 78 C2

Meta *river* Colombia/Venezuela 40 C2

Metković Croatia 82 C4

Metz France 72 E3

Meuse *river* W Europe *var.* Maas 72 D3

Mexicali Mexico 32 A1

Mexico *country* North America 32-33

México, Golfo de *see* Mexico, Gulf of

Mexico, Gulf of *sea feature* Atlantic Ocean/Caribbean Sea 48 A4

Mexico City *capital of* Mexico *Sp.* Ciudad de México 33 E4

Meymaneh *see* Maīmanah

Mezen' *river* Russian Federation 92 D3

Miami Florida, USA 31 F5

Miami Beach Florida, USA 31 F5

Mianyang China 111 B5

Michigan *state* USA 22 C2

Michigan, Lake *lake* USA 17 C5

Micronesia *country* Pacific Ocean 126 B2

Micronesia *region* Pacific Ocean 126

Mid Atlantic Ridge *undersea feature* Atlantic Ocean 48 B4

Middelburg South Africa 60 D5

Middle Andaman *island* India 115 G2

Middlesbrough England, UK 71 D5

Mid-Indian Basin *undersea feature* Indian Ocean 122 C4

Mid-Indian Ridge *undersea feature* Indian Ocean 123 C5

Midland Texas, USA 29 E3

Mid-Pacific Mountains *var.* Mid-Pacific Seamounts. *Undersea feature* Pacific Ocean 124 C1

Mid-Pacific Seamounts *see* Mid-Pacific Mountains

Midway Islands *US territory* Pacific Ocean 134 D2

Mikhaylovka Russian Federation 93 B6

Milagro Ecuador 40 A4

Milan *see* Milano

Milano Italy *Eng.* Milan 78 B2

Mildura Australia 131 C6

Millennium Island *island* Kiribati *prev.* Caroline Island 127 H3

Miles Australia 131 D5

Miles City Montana, USA 24 C2

Milford Haven Wales, UK 71 C6

Milford Sound New Zealand 133 E6

Milford Sound *inlet* New Zealand 133 A6

Milos *island* Greece 87 C6

Milwaukee Wisconsin, USA 22 B3

Min *see* Fujian

Minatitlán Mexico 33 G4

Minch, The *strait* Scotland, UK 70 C3

Mindanao *island* Philippines 121 F2

Mindoro *island* Philippines 121 E2

Mindoro Strait *sea feature* South China Sea/Sulu Sea 121 E2

Mingäçevir Azerbaijan *Rus.* Mingechaur 99 G2

Mingechaur *see* Mingäçevir

Minho *river* Portugal/Spain *Sp.* Miño 74 C2

Minicoy Island *island* India 114 C3

Minneapolis Minnesota, USA 23 F2

Minnesota *state* USA 25 F2

Miño *river* Portugal/Spain *Port.* Minho 74 C1

Minorca *see* Menorca

Minot North Dakota, USA 24 D1

Mīnā' Qābūs Oman 122 B3

Minsk *capital of* Belarus 89 C5

Minto, Lake *lake* Canada 20 D2

Moselle *river* W Europe *Ger.* Mosel 72 E4

Mosgiel New Zealand 133 B7

Moshi Tanzania 55 C7

Moskva *see* Moscow

Mosquito Coast *coastal region* Nicaragua 35 E3

Moss Norway 67 B6

Mossendjo Congo 59 B6

Mossoró Brazil 43 H2

Most Czech Republic *Ger.* Brüx 80 A4

Mostaganem Algeria 52 D1

Mostar Bosnia & Herz. 82 C4

Mosul *see* Al Mawşil

Motril Spain 75 E5

Motueka New Zealand 133 C5

Moulins France 72 C4

Moulmein *see* Mawlamyine

Moundou Chad 58 C4

Mount Gambier Australia 131 B7

Mount Isa Australia 130 B4

Mount Magnet Australia 129 B5

Mount Vernon Illinois, USA 22 B5

Mouscron Belgium 69 A6

Moyobamba Peru 42 B2

Moyu China 108 B2

Mozambique *country* SE Africa 61

Mozambique Channel *sea feature* Indian Ocean 61 F3

Mozyr' *see* Mazyr

Mpika Zambia 61 E2

Mtwara Tanzania 55 C8

Muang Không Laos 119 D5

Muang Xaignabouri *see* Xaignabouri

Mudanjiang China 110 E3

Mufulira Zambia 60 D2

Muğla Turkey 98 A4

Mulhouse France 72 E4

Mull *island* Scotland, UK 70 B3

Muller, Pegunungan *mountains* Indonesia 120 C3

Multán Pakistan 116 C2

Mumbai India *var.* Bombay 117 C5

München Germany *Eng.* Munich 77 C6

Muncie Indiana, USA 22 C4

Munich *see* München

Münster Germany 76 B4

Muqdisho *see* Mogadishu

Mur *river* C Europe 77 E7

Murchison River *river* Australia 129 B5

Murcia Spain 75 F4

Mures *river* Hungary/Romania 81 D7

Murfreesboro Tennessee, USA 30 D1

Murgab Tajikistan 105 F3

Murgap *river* Turkmenistan *var.* Murghab 104 C3

Murghab *see* Murgap

Müritz *lake* Germany 76 D3

Murmansk Russian Federation 92 C2 96 C1

Murray *river* Australia 131 B6

Murray Fracture Zone *tectonic feature* Pacific Ocean 135 E2

Murray Ridge *Undersea feature* Arabian Sea 122 B3

Murwillumbah Australia 131 E5

Murzuq Libya 53 F3

Muş Turkey 99 F3

Muscat *capital of* Oman *Ar.* Masqaţ 103 E5

Musgrave Ranges *mountain range* Australia 129 D5

Musters, Lago *lake* Argentina 46 C6

Mu Us Shadi *Desert* China 109 E3

Mvonioälv *river* Finland/ Sweden 66 D3

Mwali *island* Comoros 61 F2

Mwanza Tanzania 55 B6

Mwene-Ditu Dem. Rep. Congo 59 D7

Mweru, Lake *lake* Dem. Rep. Congo/Zambia 59 E7

Myanmar *country* SE Asia *var.* Myanmar 118-119

Myeik Myanmar *prev.* Mergui 119 B5

Mykolayiv Ukraine *Rus.* Nikolayev 91 E4

Mykonos *island* Greece 87 D5

Mysore India 114 D2

Mzuzu Malawi 61 E2

N

Naberezhnyye Chelny Russian Federation *prev.* Brezhnev 93 C5

Nablus West Bank *var.* Nābulus, *Heb.* Shekhem 101 D6

Nābulus *see* Nablus

Nacala Mozambique 61 F2

Naga Philippines 120 E2

Nagano Japan 112 C4

Nagasaki Japan 113 A6

Nāgercoil India 114 D3

Nagorno-Karabakh *region* Azerbaijan 99 G2

Nagoya Japan 113 C5

Nāgpur India 116 D4

Nagqu China 108 C5

Nagykanizsa Hungary *Ger.* Grosskanizsa 83 C7

Nagyszombat *see* Trnava

Naha Japan 113 A8

Nain Canada 21 F2

Nairobi *capital of* Kenya 55 C6

Najaf *see* An Najaf

Najrān Saudi Arabia 103 B6

Nakamura Japan 113 B6

Nakhichevan' *see* Naxçıvan

Nakhon Ratchasima Thailand 119 C5

Nakhon Sawan Thailand 119 C5

Nakhon Si Thammarat Thailand 119 C6

Nakuru Kenya 55 C6

Nal'chik Russian Federation 96 A4

Namangan Uzbekistan 105 E2

Nam Co *lake* China 108 C4

Nam Đinh Vietnam 118 D3

Namib Desert *desert* Namibia 60 B3

Namibe Angola 60 B2

Namibia *country* southern Africa 60

Nampa Idaho, USA 26 C3

Namp'o North Korea 110 E4
Nampula Mozambique 61 F2
Namur Belgium 69 C6
Nanchang China 111 C5
Nancy France 72 D3
Nänded India 116 D5 114 D1
Nanjing China 111 D5
Nanning China 111 B6
Nanortalik Greenland 64 C5
Nansen Basin *undersea feature*
   Arctic Ocean 137 G4
Nantes France 72 B4
Napier New Zealand 132 E4
Naples *see* Napoli
Napo *river* Ecuador/Peru 42 B2
Napoli Italy *Eng.* Naples 79 D5
Narbonne France 73 C6
Nares Strait *sea feature*
   Canada/Greenland 64 C1
Narew *river* Poland 80 E3
Narmada *river* India 116 D4
Narva Estonia 88 E2
Narva *river* Estonia/Russian
   Federation 88 E2
Narva Bay *sea feature* Gulf of
   Finland *Est.* Narva Laht, *Rus.*
   Narvskiy Zaliv 88 E2
Narva Laht *see* Narva Bay
Narvik Norway 66 C3
Narvskiy Zaliv *see* Narva Bay
Naryn Kyrgyzstan 105 G2
Nāshik India 116 C5
Nashville Tennessee, USA 30 D1
Nâşir, Buheiret *see* Nasser, Lake
Nassau *capital of* The Bahamas
   36 C1
Nasser, Lake *reservoir* Egypt
   var. Nâşir, Buheiret 54 B2
Natal Brazil 43 H3
Natal Basin *Undersea feature*
   Indian Ocean 123 A5
Natitingou Benin 57 E4
Naturaliste Plateau *undersea*
   feature Indian Ocean 123 E6
Natzrat Israel *Eng.* Nazareth
   101 A5
Nauru *country* Pacific Ocean
   126 D3
Navapolatsk/Novopolotsk
   Belarus *Rus.* Novopolotsk
   89 D5

Navassa Island *unincorporated*
   territory USA, West Indies
   36 D3
Navoiy Uzbekistan
   Uzb. Nawoly 104 D2
Nawābshāh Pakistan 116 B3
Nawoly *see* Navoiy
Naxçıvan Azerbaijan *Rus.*
   Nakhichevan' 99 G3
Náxos *island* Greece 87 D6
Nay Pyi Taw *capital of*
   Myanmar 118 B3
Nazareth *see* Natzrat
Nazca Peru 42 B4
Nazrēt Ethiopia 55 C5
Nazwá Oman 103 E5
N'Dalatando Angola 60 B2
Ndélé Central African Republic
   58 C4
N'Djaména *capital of* Chad
   58 B3
Ndola Zambia 60 D2
Nebitdag *see* Balkanabat
Nebraska *state* USA 24-25 E3
Neches *river* S USA 29 H3
Neckar *river* Germany
   77 B5
Necochea Argentina 47 D5
Neftezavodsk *see* Seýdi
Negēlē Ethiopia 55 C5
Negev *see* HaNegev
Negro, Río *river* Argentina
   47 C5
Negro, Río *river* Brazil/Uruguay
   44 C4
Negro, Río *river* N South
   America 40 C1
Neiva Colombia 40 B3
Nellore India 115 E2
Neman *river* NE Europe *Bel.*
   Nyoman, *Lith.* Nemunas, *Ger.*
   Memel, *Pol.* Niemen 88 B4
Nemunas *see* Neman
Nemuro Japan 112 E2
Nepal *country* S Asia 117
Neris *river* Belarus/Lithuania
   Bel. Viliya, *Pol.* Wilja 88 C4
Ness, Loch *lake* Scotland, UK
   70 C3

Netherlands *country* W Europe
   var. Holland 68-69
Netze *see* Noteć
Neubrandenburg Germany
   76 D3
Neuchâtel, Lac de *lake*
   Switzerland 77 A7
Neumünster Germany 76 C2
Neuquén Argentina 47 C5
Neusiedler See *lake* Austria/
   Hungary 77 E6
Neusohl *see* Banská Bystrica
Neutra *see* Nitra
Nevada *state* USA 26-27
Nevers France 72 C4
Nevşehir Turkey 98 C3
New Amsterdam Guyana 41 G2
Newark New Jersey, USA 23 F3
New Britain *island* Papua New
   Guinea 126 B3
New Brunswick *province*
   Canada 21 F4
New Caledonia *special*
   collectivity France, Pacific
   Ocean 126 C5
New Caledonia *island* Pacific
   Ocean 124 D3
New Caledonia Basin *undersea*
   feature Pacific Ocean
   124 D4
Newcastle Australia 131 D6
Newcastle upon Tyne
   England, UK 70 D4
New Delhi *capital of* India
   116 D3
Newfoundland & Labrador
   province Canada 21 F2
Newfoundland *island* Canada
   21 G3
Newfoundland Basin *undersea*
   feature Atlantic Ocean
   48 B3
New Georgia Islands *island*
   group Solomon Is 126 C3
New Guinea *island* Pacific
   Ocean 126 B3
New Hampshire *state* USA
   23 G2
New Haven Connecticut, USA
   23 G3

New Ireland — North Korea

New Ireland *island* Papua New Guinea 126 C3
New Jersey *state* USA 23 F4
Newman Australia 128 B4
New Mexico *state* USA 28-29
New Orleans Louisiana, USA 30 C3
New Plymouth New Zealand 132 D3
Newport Oregon, USA 26 A3
Newport News Virginia, USA 23 F5
New Providence *island* The Bahamas 36 C1
Newry Northern Ireland, UK 71 B5
New Siberian Islands *see* Novosibirskiye Ostrova
New South Wales *state* Australia 131 C6
New York *state* USA 23 F3
New York New York, USA 23 F3
New Zealand *country* Pacific Ocean 132-133
Neyshābūr Iran 102 D3
Ngaoundéré Cameroon 58 B4
Ngerulmud *capital of* Palau 126 A1
N'Giva Angola 60 C3
N'Guigmi Niger 57 H3
Nha Trang Vietnam 119 E5
Niagara Falls *waterfall* Canada/ USA 23 E3
Niamey *capital of* Niger 57 F3
Niangay, Lac *lake* Mali 56 E3
Nias, Pulau *island* Indonesia 120 B3
Nicaragua *country* Central America 34-35
Nicaragua, Lago de *lake* Nicaragua 34 D3
Nice France 73 E6
Nicobar Islands *island group* India 115 H3
Nicosia *capital of* Cyprus *var.* Lefkosia, *Turk.* Lefkoşa 98 C5
Nicoya, Península de *peninsula* Costa Rica 34 D4
Niemen *see* Neman

Nieuw Amsterdam Suriname 41 H2
Niğde Turkey 98 D4
Niger *country* W Africa 57
Niger *river* W Africa 56-57 D3
Niger, Mouths of the *delta* Nigeria 57 F5
Nigeria *country* W Africa 57
Niigata Japan 112 C4
Nijmegen Netherlands 68 D4
Nikolayev *see* Mykolayiv
Nikopol' Ukraine 91 F3
Nile *river* N Africa 54 B3
Nile Delta *wetlands* Egypt 54 B1
Nîmes France 73 D6
Ninetyeast Ridge *undersea feature* Indian Ocean 123 C5
Ningbo China 111 D5
Ningxia *autonomous region* China 110-111 B4
Nioro Mali 56 D3
Nipigon, Lake *lake* Canada 20 B4
Niš Serbia 82 E4
Nitra Slovakia *Ger.* Neutra, *Hung.* Nyitra 81 C6
Nitra *river* Slovakia *Ger.* Neutra, *Hung.* Nyitra 81 C6
Niue *asociated territory* New Zealand, Pacific Ocean 127 F4
Nizāmābād India 114 D1
Nizhnevartovsk Russian Federation 96 D3
Nizhniy Novgorod Russian Federation *prev.* Gor'kiy 93 C5 96 B3
Nkongsamba Cameroon 58 B4
Norak Tajikistan 105 E3
Nord Greenland 65 E2
Nordaustlandet *island* Svalbard 65 G1
Norfolk Virginia, USA 23 F5
Norfolk Island *external territory* Australia, Pacific Ocean 124 D4
Noril'sk Russian Federation 96 D3

Norfolk Ridge *undersea feature* Pacific Ocean 124 D4
Norman Oklahoma, USA 28 F2
Normandie *region* France *Eng.* Normandy 72 B3
Normandy *see* Normandie
Normanton Australia 130 C3
Norrköping Sweden 67 C6
Norseman Australia 129 C6
North Albanian Alps *mountains* Albania/ Montenegro 83 D5
North America 16-17
North Andaman *island* India 115 G2
North Atlantic Ocean 64-65
North Australian Basin *undersea feature* Indian Ocean 124 A2 128 A2
North Bay Canada 20 D4
North Cape *coastal feature* New Zealand 132 C1
North Cape *coastal feature* Norway 66 D2
North Carolina *state* USA 31 F1
North Dakota *state* USA 24-25 D2
North Fiji Basin *undersea feature* Coral Sea 124 D3
Northern Cook Islands *islands* Cook Islands 127 G4
Northern Cyprus, Turkish Republic of *disputed region* Cyprus 98 C5
Northern Dvina *river* Russian Federation *see* Severnaya Dvina 63 G2
Northern Ireland *province* UK 70-71
Northern Mariana Islands *commonwealth territory* USA, Pacific Ocean 124 B1
Northern Sporades *see* Vóreies Sporádes
Northern Territory *territory* Australia 130 A3
North European Plain *region* N Europe 62 E3
North Frisian Islands *islands* Denmark/Germany 76 B2
North Island *island* New Zealand 132 G2
North Korea *country* E Asia 110

North Little Rock Arkansas, USA 30 B1

North Platte Nebraska, USA 25 E4

North Platte *river* C USA 24 D3

North Pole *ice feature* Arctic Ocean 137 G3

North Sea Atlantic Ocean 70 E2

North Siberian Lowland *lowlands* Russian Federation 94-95

North Taranaki Bight *gulf* New Zealand 132 D3

North Uist *island* Scotland, UK 70 B3

Northwest Territories *territory* Canada 19 E3

Norway *country* N Europe 66-67

Norwegian Sea Arctic Ocean 137 G5

Norwich England, UK 71 E6

Noteć *river* Poland *Ger.* Netze 80 C3

Nottingham England, UK 71 D6

Nottingham Island *island* Hudson Strait 20 D1

Nouâdhibou Mauritania 56 B2

Nouakchott *capital of* Mauritania 56 B2

Nouméa *capital of* New Caledonia 126 D5

Nova Gradiška Croatia 82 C3

Nova Iguaçu Brazil 43 F5 45 F2

Novara Italy 78 B2

Nova Scotia *province* Canada 21 F4

Novaya Zemlya *islands* Russian Federation 137 H4

Novaya Zemlya Trench *see* East Novaya Zemlya Trench

Novi Sad Serbia 82 D3

Novokuznetsk Russian Federation *prev.* Stalinsk 96 D4

Novopolotsk *see* Navapolatsk/ Novopolotsk

Novosibirsk Russian Federation 96 D4

Novosibirskiye Ostrova *islands* Russian Federation *Eng.* New Siberian Islands 95 F1

Novo Urgench *see* Urgench

Novyy Margilan *see* Farg'ona

Nsanje Malawi 61 E3

Nsawam Ghana 57 E5

Nubian Desert *desert* Sudan 54 B3

Nu'eima West Bank 101 D7

Nuevo Laredo Mexico 33 E2

Nuku'alofa *capital of* Tonga 127 F5

Nukus Uzbekistan 104 C2

Nullarbor Plain *region* Australia 129 D6

Nunap Isua Island *coastal region* Greenland *var.* Uummannaruaq *Dan.* Kap Farvel 64 C5

Nunavut *Territory* Canada 19 F3

Nunivak Island *island* Alaska, USA 18 B2

Nuoro Italy 79 A5

Nuremberg *see* Nürnberg

Nürnberg Germany *Eng.* Nuremberg 77 C5

Nusa Tenggara *islands* East Timor / Indonesia 120 E5

Nuuk Greenland *var.* Godthåb 64 C4

Nyainqêntanglha Shan *mountain range* China 108 D5

Nyala Sudan 54 A4

Nyasa, Lake *lake* E Africa 51 D5

Nyeri Kenya 55 C6

Nyima China 108 C4

Nyíregyháza Hungary 81 E6

Nyitra *see* Nitra

Nykøbing Denmark 67 B8

Nyköping Sweden 67 C6

Nyngan Australia 131 D6

Nyoman *see* Neman

O

Oakland California, USA 27 B6

Oakley Kansas, USA 25 E4

Oamaru New Zealand 133 B7

Oaxaca Mexico 33 F5

Ob' *river* Russian Federation 96 D4

Oban Scotland, UK 70 C4

Obihiro Japan 112 D2

Obo Central African Republic 58 D7

Oceania 124-125

Ocean Island *see* Banaba

Oceanside California, USA 27 C8

Ochamchira *see* Och'amch'ire

Ochamchire Georgia *prev.* Och'amch'ire, *Rus.* Ochamchira 99 E1

Och'amch'ire *see* Ochamchire

Ödenburg *see* Sopron

Odense Denmark 67 B7

Oder *river* C Europe 80 C4

Odesa Ukraine *Rus.* Odessa 91 E4

Odessa *see* Odesa

Odessa Texas, USA 29 E3

Odienné Côte d'Ivoire 56 D4

Oesel *see* Saaremaa

Ofanto *river* Italy 79 D5

Offenbach Germany 77 B5

Ogaden *plateau* Ethiopia 55 D5

Ogallala Nebraska, USA 24 D4

Ogbomosho Nigeria 57 F4

Ogden Utah, USA 24 B3

Ogdensburg New York, USA 23 F2

Oger *see* Ogre

Ogre Latvia *Ger.* Oger 88 C3

Ogulin Croatia 82 B3

Ohio *state* USA 22 D4

Ohio *river* N USA 22 B5

Ohrid Macedonia 83 D6

Ohrid, Lake *lake* Albania/ Macedonia 83 D6

Ohře *river* Czech Republic/ Germany *Ger.* Eger 81 A5

Ôita Japan 113 B6

Okavango *river var.* Cubango southern Africa 60 C3

Okavango Delta *wetland* Botswana 60 C3

Okayama Japan 113 B5

Okazaki Japan 113 C5

Okeechobee, Lake *lake* Florida, USA 31 F4

Okhotsk Russian Federation 97 G3

Okhotsk, Sea of Pacific Ocean 134 C1

Okinawa *island* Japan 113 A8

Oki-shotō *island group* Japan 113 B5

Oklahoma *state* USA 29 F1

Oklahoma City Oklahoma, USA 29 F2

Okushiri-tō *island* Japan 112 C2

Okāra Pakistan 116 C2

Öland *island* Sweden 67 C7

Olavarría Argentina 46 D4

Olbia Italy 79 B5

Oldenburg Germany 76 B3

Oleksandriya Ukraine *Rus.* Aleksandriya 91 F3

Olenëk Russian Federation 97 E3

Ölgiy Mongolia 108 C2

Olhão Portugal 74 C4

Olita *see* Alytus

Olmaliq *see* Almalyk

Olmütz *see* Olomouc

Olomouc Czech Republic *Ger.* Olmütz 81 C5

Olsztyn Poland *Ger.* Allenstein 80 D2

Olt *river* Romania 90 B5

Olympia Washington, USA 26 B2

Omaha Nebraska, USA 25 F4

Oman *country* SW Asia 103 D6

Oman, Gulf of *sea feature* Indian Ocean 103 E5, 122 B3

Omdurman Sudan 54 B4

Omsk Russian Federation 96 C4

Onega *river* Russian Federation 92 C4

Onega, Lake *see* Onezhskoye Ozero

Onezhskoye Ozero *lake* Russian Federation *Eng.* Lake Onega 92 B3

Ongole India 115 E2

Onitsha Nigeria 57 F5

Onslow Australia 128 A4

Ontario *province* Canada 18 B3

Ontario, Lake *lake* Canada/USA 17 D5

Oostende Belgium *Eng.* Ostend 69 A5

Opole Poland *Ger.* Oppeln 80 C4

Oporto *see* Porto

Oppeln *see* Opole

Oradea Romania 90 B3

Oran Algeria 52 D1

Orange River *river* southern Africa 60 C4

Oranjestad Aruba 37 E5

Orantes *River* Asia 100 B3

Ordu Turkey 98 D2

Ordzhonikidze *see* Vladikavkaz

Örebro Sweden 67 C6

Oregon *state* USA 26

Orël Russian Federation 83 A5

Orem Utah, USA 24 B4

Orenburg Russian Federation 93 C6 96 B4

Orense *see* Ourense

Orestiáda Greece 86 D3

Orinoco *river* Colombia/Venezuela 41 E3

Oristano Italy 79 A5

Orkney *islands* Scotland, UK 70 C2

Orlando Florida, USA 31 E4

Orléans France 72 C4

Örnsköldsvik Sweden 67 C5

Orontes *river* SW Asia 100 B3

Orosirá Rodópis *see* Rhodope Mountains

Orsha Belarus 89 E5

Orsk Russian Federation 93 D6 96 B4

Oruro Bolivia 42 C4

Ōsaka Japan 113 C5

Osborn Plateau *undersea feature* Indian Ocean 123 C5

Ösel *see* Saaremaa

Osh Kyrgyzstan 105 F2

Oshawa Canada 20 D5

Oshkosh Wisconsin, USA 22 B2

Osijek Croatia 82 C3

Oslo *capital of* Norway 67 B6

Osmaniye Turkey 98 D4

Osnabrück Germany 76 B3

Osorno Chile 47 B5

Oss Netherlands 68 D4

Ossora Russian Federation 97 H2

Ostend *see* Oostende

Östersund Sweden 67 C5

Ostrava Czech Republic *Ger.* Mährisch-Ostrau, *prev.* Moravská Ostrava 81 C5

Ostrołęka Poland 80 D3

Ostrowiec Świętokrzyski Poland 80 D4

Ōsumi-shotō *island group* Japan 113 A7

Otago Peninsula *peninsula* New Zealand 133 B7

Otaru Japan 112 D2

Oti *river* Africa 57 E4

Otranto, Strait of *sea feature* Albania/Italy 79 E5

Ottawa *capital of* Canada 20 D4

Ottawa *river* Canada 20 D4

Ou *river* Laos 118 C3

Ouachita *river* SE USA 30 B2

Ouagadougou *capital of* Burkina Faso 57 E3

Ouarâne *desert* Mauritania 56 D2

Ouargla Algeria 53 E2

Ouessant, Île d' *island* France 72 A3

Ouésso Congo 59 C5

Oujda Morocco 52 D2

Oulu Finland 66 D4

Oulu *river* Finland 66 D4

Oulujärvi *lake* Finland 66 E4

Ounasjoki *river* Finland 66 D3

Our *river* W Europe 69 E7

Ourense Spain *Cast.* Orense 74 C2

Ourinhos Brazil 44 D2

Ourthe *river* Belgium 69 D6

Outer Hebrides *island group* UK *var.* Western Isles 70 B3

Outer Islands *island group* Seychelles 61 H2

Ouyen Australia 131 C6

Oviedo Spain 74 D1

Owando Congo 59 C6

Owen Fracture Zone *tectonic feature* Arabian Sea 122 B3

Owensboro Kentucky, USA 22 B5

Oxford England, UK 71 D6
Oxnard California, USA 29 C7
Oyem Gabon 59 B5
Oyo Nigeria 57 F4
Ozark Plateau *plain* Arkansas/
   Missouri, USA 25 G5
Ózd Hungary 81 D6

P

Paamiut Greenland 64 B4
Pachuca Mexico 33 E4
Pacific-Antarctic Ridge
   undersea feature Pacific
   Ocean 136 B5
Pacific Ocean 134-135
Padang Indonesia 120 B4
Paderborn Germany 76 B4
Padova Italy *Eng.* Padua 78 C2
Padre Island *island* Texas, USA
   29 G5
Padua *see* Padova
Paducah Kentucky, USA 22 B5
Paeroa Waikato, New Zealand
   132 D3
Pafos *see* Paphos
Pag *island* Croatia 82 A3
Pago Pago *capital of* American
   Samoa 127 F4
Paide Estonia *Ger.* Weissenstein
   88 D2
Paihia New Zealand 132 D2
Painted Desert *desert* SW USA
   28 C1
País Valenciano *cultural region*
   Spain 75 F3
Pakistan *country* S Asia 116
Pakokku Myanmar 118 A3
Palagruža *island* Croatia 83 B5
Palau *country* Pacific Ocean
   var. Belau 124 B2 126
Palawan *island* Philippines
   121 E2
Palawan Passage *passage*
   Philippines 121 E2
Paldiski Estonia *prev.* Baltiski,
   Eng. Baltic Port, *Ger.*
   Baltischport 88 C2
Palembang Indonesia 120 C4

Palencia Spain 74 D2
Palermo Italy 79 C6
Palikir *capital of* Micronesia
   126 C2
Palioúri, Akrotírio *coastal
   feature* Greece *var.* Akra
   Kanestron 86 C4
Palk Strait *sea feature*
   India/Sri Lanka 115 E3
Palliser, Cape *headland* New
   Zealand 133 D5
Palm Springs California, USA
   27 D8
Palma Spain 75 G3
Palmer Land *physical region*
   Antarctica 136 A3
Palmerston North New Zealand
   132 D4
Palmyra *see* Tudmur
Palmyra Atoll *incorporated
   territory* USA, Pacific Ocean
   125 F2
Palu Indonesia 121 E4
Pamir *river* Afghanistan/
   Tajikistan 105 F3
Pamirs *mountains* Tajikistan
   105 F3
Pampa Texas, USA 29 E2
Pampas *region* South America
   46 C4
Pamplona Spain *var.* Iruña 75 F1
Pānāji India 114 C2
Panama *country* Central
   America 35
Panamá, Golfo de *sea feature*
   Panama 35 F5
Panama Canal *canal* Panama
   35 F4
Panama City *capital of* Panama
   35 F5
Panama City Florida, USA
   30 D3
Pančevo Serbia 82 D3
Panevėžys Lithuania 88 C4
Pantanal *region* Brazil 38 C4
Pantelleria *island* Italy 79 B7
Papeete *capital of* French
   Polynesia 127 H4
Paphos Cyprus *var.* Pafos 98 C5
Papua *province* Indonesia *prev.*
   Irian Jaya 121 H4

Papua New Guinea *country*
   Pacific Ocean 126
Paracel Islands *disputed
   territory* Asia 120 D1
Paragua *river* Venezuela 41 E3
Paraguay *country* South
   America 44
Paraguay *river* C South
   America 38 C4 44 B2
Parakou Benin 57 F4
Paramaribo *capital of* Suriname
   41 G2
Paraná Argentina 46 D4
Paraná *river* C South America
   46 D3
Paranaíba Brazil 43 G2
Paraparaumu New Zealand
   132 D4
Pardubice Czech Republic *Ger.*
   Pardubitz 81 B5
Pardubitz *see* Pardubice
Parepare Indonesia 121 E4
Paris *capital of* France 72 C3
Paris Texas, USA 29 G2
Parma Italy 78 B3
Pärnu Estonia *Rus.* Pyarnu,
   prev. Pernov, *Ger.* Pernau
   88 C2
Páros *island* Greece 87 D6
Pasadena California, USA 27 C7
Pasadena Texas, USA 29 G4
Passo Fundo Brazil 44 D3
Pasto Colombia 40 B4
Patagonia *region* S South
   America 47 C6
Pathein Myanmar *prev.* Bassein
   118 A4
Patna India 117 F3
Patos, Lagoa dos *lagoon* Brazil
   44 D4
Pátra Greece 87 B5
Pattani Thailand 119 C7
Pattaya Thailand 119 C5
Patuca *river* Honduras 34 D2
Pau France 73 B6
Pavlodar Kazakhstan 96 C4
Pavlograd *see* Pavlohrad
Pavlohrad Ukraine *Rus.*
   Pavlograd 91 G3
Paysandú Uruguay 44 B4

Pisa Italy 78 B3
Pisco Peru 42 B4
Pishpek *see* Bishkek
Pistyan *see* Piešťany
Pitcairn, Henderson, Ducie & Oeno Islands *overseas territory* UK, Pacific Ocean 125 G4
Piteå Sweden 66 D4
Pitești Romania 90 C4
Pittsburgh Pennsylvania, USA 23 E4
Piura Peru 42 A2
Pivdennyy Bug *river* Ukraine 91 E3
Plasencia Spain 74 D3
Plata, Río de la *river* Argentina/Uruguay *see.* River Plate 44 B5 46 D4
Plate, River *see* Plata, Río de la
Platte *river* C USA 25 E4
Plattensee *see* Balaton
Plenty, Bay of *bay* New Zealand 132 E3
Pleven Bulgaria 86 C1
Płock Poland 80 D3
Ploiești Romania 90 C4
Plovdiv Bulgaria *Gk.* Philippopolis 86 C2
Plungė Lithuania 88 B4
Plymouth *capital of* Montserrat 37 G3
Plymouth England, UK 71 C7
Plzeň Czech Republic *Ger.* Pilsen 81 A5
Po *river* Italy 78 B2
Pocatello Idaho, USA 26 E4
Po Delta *wetland* Italy 78 C3
Podgorica *capital of* Montenegro 83 C5
Pohnpei Island *island* Micronesia 126 C2
Pointe-Noire Congo 59 B6
Poitiers France 72 B4
Poland *country* E Europe 80-81
Polatsk Belarus 89 D5
Pol-e Khomri *see* Pul-e Khumrī
Poltava Ukraine 91 F2
Poltoratsk *see* Aşgabat
Polynesia *region* Pacific Ocean 127

Pomeranian Bay *bay* Germany/Poland 80 B2
Pompano Beach Florida, USA 31 F5
Ponca City Oklahoma, USA 29 G1
Pondicherry India 115 E2
Ponta Grossa Brazil 44 D2
Pontevedra Spain 74 C1
Pontianak Indonesia 120 C4
Poona *see* Pune
Poopó, Lake *lake* Bolivia 42 C5
Popayán Colombia 40 B3
Poprad Slovakia *Ger.* Deutschendorf 81 D5
Porbandar India 116 B4
Pori Finland 67 D5
Porsgrunn Norway 67 B6
Portalegre Portugal 74 C3
Port Angeles Washington, USA 26 A1
Port Arthur Texas, USA 29 H4
Port Augusta Australia 131 B6
Port-au-Prince *capital of* Haiti 36 D3
Port Blair India 115 G2
Port Douglas Australia 130 D3
Port Elizabeth South Africa 60 D5
Port-Gentil Gabon 59 A6
Port Harcourt Nigeria 57 F5
Port Hardy Canada 18 D5
Port Harrison *see* Inukjuak
Port Hedland Australia 128 B4
Portland Australia 131 B7
Portland Maine, USA 23 G2
Portland Oregon, USA 26 B2
Port Lincoln Australia 131 A6
Port Louis *capital of* Mauritius 61 H4
Port Macquarie Australia 131 E6
Port Moresby *capital of* Papua New Guinea 126 B3
Porto Portugal *Eng.* Oporto 74 C2
Porto Alegre Sao Tome and Principe 44 D4

Port-of-Spain *capital of* Trinidad & Tobago 37 G5
Porto-Novo *capital of* Benin 57 F5
Porto Velho Brazil 42 C3
Portoviejo Ecuador 40 A4
Port Said *see* Būr Sa'īd
Portsmouth England, UK 71 D7
Port Sudan Sudan 54 C3
Portugal *country* SW Europe 74
Port-Vila *capital of* Vanuatu 126 D5
Porvenir Chile 47 B7
Posadas Argentina 46 E3
Posen *see* Poznań
Poste-de-la-Baleine *see* Kuujjuarapik
Pöstyén *see* Piešťany
Potenza S Italy 79 D5
Poti Georgia 99 E2
Potosí Bolivia 42 C5
Potsdam Germany 76 D4
Póvoa de Varzim Portugal 74 C2
Powder *river* N USA 24 C2
Powell, Lake *lake* SW USA 24 B5
Poza Rica Mexico 33 F4
Poznań Poland *Ger.* Posen 80 C3
Pozo Colorado Paraguay 44 B2
Pozsony *see* Bratislava
Prag *see* Prague
Prague *capital of* Czech Republic *Cz.* Praha, *Ger.* Prag 81 B5
Praha *see* Prague
Praia *capital of* Cape Verde 56 A3
Prato Italy 78 B3
Pratt Kansas, USA 25 E5
Preschau *see* Prešov
Prescott Arizona, USA 28 B2
Presidente Prudente Brazil 44 D2
Prešov Slovakia *Ger.* Eperies, *var.* Preschau, *Hung.* Eperjes 81 D5
Prespa, Lake *lake* SE Europe 83 D6 86 A3
Presque Isle Maine, USA 23 G1

Pressburg see Bratislava
Preston England, UK 71 D5
Pretoria judicial capital of South Africa 60 D4
Préveza Greece 86 A4
Prijedor Bosnia & Herzegovina 82 B3
Prilep Macedonia 83 E5
Prince Albert Canada 19 F5
Prince Edward Island province Canada 21 F4
Prince Edward Islands island group South Africa 123 A7
Prince George Canada 19 E5
Prince of Wales Island island Canada 19 F2
Prince Rupert Canada 18 D4
Princess Charlotte Bay bay Australia 126 C1
Princess Elizabeth Land region Antarctica 136 C3
Principe island Sao Tome & Principe 59 A5
Pripet river Belarus/Ukraine 90 C1
Pripet Marshes wetlands Belarus/Ukraine 90 C1
Prishtinë capital of Kosovo 83 D5
Prizren Kosovo 83 D5
Prome see Pyay
Prossnitz see Prostějov
Prostějov Czech Republic Ger. Prossnitz 81 C5
Provence region France 73 D6
Providence Rhode Island, USA 23 G3
Providencia, Isla de island Colombia 35 E3
Provo Utah, USA 24 B4
Prudhoe Bay Alaska, USA 18 D2
Przheval'sk see Karakol
Pskov Russian Federation 92 A4
Pskov, Lake lake Estonia/ Russian Federation Est. Pihkva Järv, Rus. Pskovskoye Ozero 88 D3
Pskovskoye Ozero see Pskov, Lake

Ptich' see Ptsich
Ptsich river Belarus Rus. Ptich' 89 D6
Pucallpa Peru 42 B3
Puebla Mexico 33 F4
Pueblo Colorado, USA 22 D4
Puerto Aisén Chile 47 B6
Puerto Barrios Guatemala 34 C2
Puerto Carreño Colombia 40 D2
Puerto Cortés Honduras 34 C2
Puerto Deseado Argentina 47 C6
Puerto Maldonado Peru 42 C4
Puerto Montt Chile 47 B5
Puerto Natales Chile 47 B7
Puerto Plata Dominican Republic 37 E3
Puerto Princesa Philippines 120 E2
Puerto Rico commonwealth territory USA, West Indies 37 F3
Puerto San Julián Argentina 47 C7
Puerto Suárez Bolivia 42 D4
Puerto Vallarta Mexico 32 D4
Pula Croatia 82 A3
Pul-e Khumri Afghanistan prev. Pol-e Khomrī 105 E4
Pune India prev. Poona 114 C1
Puno Peru 42 C4
Punta Arenas Chile prev. Magallanes 47 B7
Puntarenas Costa Rica 34 D4
Purmerend Netherlands 68 C3
Purus river Brazil/Peru 42 C3
Pusan see Busan
Putrajaya administrative capital of Malaysia 120 B3
Putumayo river NW South America 38 B3
Pyapon Myanmar 118 B4
Pyarnu see Pärnu
Pyay Myanmar prev. Prome 118 A4
Pyongyang capital of North Korea 110 E4

Pyramid Lake lake Nevada, USA 27 C5
Pyrenees mountain range SW Europe 62 C4

Q

Qaanaaq Greenland var. Thule 64 D1
Qābatiya West Bank 101 D7
Qaidam Pendi basin China 108 D4
Qalāt Afghanistan prev. Kalāt 104 D5
Qalqīlya West Bank 101 D7
Qamdo China 108 D5
Qandahār see Kandahār
Qaqortoq Greenland 64 C4
Qara Qum see Karakumy
Qarshi see Karshi
Qasigiannguit Greenland 64 C3
Qatar country SW Asia 105 D5
Qattara Depression see Qaṭṭārah, Munkhafaḍ al
Qaṭṭārah, Munkhafaḍ al desert basin Egypt Eng. Qattara Depression 54 A1
Qausuittuq see Resolute
Qeqertarsuaq Greenland 64 B3
Qeqertarsuaq island Greenland 64 B3
Qian see Guizhou
Qilian Shan mountain range China 108 D4
Qimusseriarsuaq bay Greenland 64 C2
Qinā Egypt 54 B2
Qingdao China 110 D4
Qinghai province China var. Chinghai, Koko Nor, Qing, Tsinghai 108 D4
Qinghai Hu lake China var. Koko Nor 108 D4
Qingzang Gaoyuan plateau China Eng. Plateau of Tibet 110 A4
Qiong see Hainan
Qiqihar China 110 D3
Qira China 108 B4
Qitai China 108 C3

Qom Iran *var.* Kum 102 C3
Qondūz *river* Afghanistan 105 E4
Qondūz *see* Kunduz
Qo'qon Uzbekistan prev. Kokand, *var.* Khokand, 105 E2
Quba Azerbaijan *Rus.* Kuba 99 H2
Québec Canada 21 E4
Québec *province* Canada 20 D3
Queen Charlotte Islands *islands* Canada 18 D4
Queen Charlotte Sound *sea feature* Canada 18 D5
Queen Elizabeth Islands *islands* Canada 19 F1
Queensland *state* Australia 130 C4
Queenstown New Zealand 133 B6
Quelimane Mozambique 61 E3
Querétaro Mexico 33 E4
Quetta Pakistan 116 B2
Quezaltenango Guatemala 34 B2
Quibdó Colombia 40 B2
Quimper France 72 A3
Quy Nhơn Vietnam 119 E5
Qing *see* Qinghai
Quito *capital of* Ecuador 40 A4
Qürghonteppa Tajikistan *Rus.* Kurgan–Tynbe 105 E3
Qyteti Stalin *see* Kuçovë

R

Raab *see* Győr
Raab *see* Rába
Rába *river* Austria/Hungary *Ger.* Raab 81 C7
Rabat *capital of* Morocco 52 C2
Race, Cape *coastal feature* Canada 21 H4
Rach Gia Vietnam 119 D6
Radom Poland 80 D4

Radviliškis Lithuania 88 C4
Ragusa Italy 79 D7
Rahīmyär Khān Pakistan 116 C3
Raipur India 117 E5
Rājahmundry India 115 E1
Rājasthān *state* India 116 C3
Rājkot India 116 C4
Rājshāhi Bangladesh 117 G4
Rakaia *river* New Zealand 133 C6
Rakvere Estonia *Ger.* Wesenberg 88 D2
Raleigh North Carolina, USA 31 F1
Ralik Chain *islands* Marshall Islands 126 D1
Râmnicu Vâlcea Romania prev. Rîmnicu Vîlcea 90 B4
Ramallah West Bank 101 D7
Ramree Island *island* Myanmar 118 A3
Rancagua Chile 46 B4
Rānchi India 117 F4
Randers Denmark 67 A7
Rangiora New Zealand 133 C6
Rangitikei *river* New Zealand 132 D4
Rangoon *see* Yangon
Rankin Inlet Canada 19 G3
Rapid City South Dakota, USA 24 D3
Rarotonga *island* Cook Islands 127 G5
Rasht Iran 102 C3
Ratak Chain *islands* Marshall Islands 126 D1
Ratchaburi Thailand 119 C5
Rat Islands *island group* Alaska, USA 18 A2
Raukumara Range *mountain range* New Zealand 132 E3
Rauma Finland 67 D5
Ravenna Italy 78 C3
Rāwalpindi Pakistan 116 C1
Rawson Argentina 47 C6
Razgrad Bulgaria 86 D1
Reading England, UK 71 D6
Rebecca, Lake *lake* Australia 129 C6
Rebun-tō *island* Japan 112 D1

Rechytsa Belarus 89 D7
Recife Brazil 43 H3
Recklinghausen Germany 76 G4
Red Deer Canada 19 E5
Redding California, USA 27 B5
Red River *river* S USA 30 B3
Red River *river* China/ Vietnam 118
Red Sea Indian Ocean 122 A3
Reefton New Zealand 133 C5
Regensburg Germany 77 C5
Reggane Algeria 52 D3
Reggio di Calabria Italy 79 D6
Reggio nell' Emilia Italy 78 B3
Regina Canada 19 F5
Rehoboth Namibia 60 C4
Reichenberg *see* Liberec
Reid Australia 129 D6
Reims France *Eng.* Rheims 72 D3
Reindeer Lake *lake* Canada 17 C4
Reni Ukraine 90 D4
Rennes France 72 B3
Reno Nevada, USA 27 B5
Resistencia Argentina 46 D3
Reşiţa Romania 90 B4
Resolute Canada *Var.* Qausuittuq 19 F2
Réunion *overseas department* France, Indian Ocean 123 B5
Reus Spain 75 G2
Reutlingen Germany 77 B6
Reval *see* Tallinn
Revel *see* Tallinn
Revillagigedo, Islas *island* Mexico 32 B4
Rey, Isla del *island* Panama 35 F5
Reykjavík *capital of* Iceland 65 E5
Reynosa Mexico 33 E2
Rēzekne Latvia *Ger.* Rositten, *Rus.* Rezhitsa 88 D4
Rezhitsa *see* Rēzekne
Rheims *see* Reims
Rhine *river* W Europe 62 D3
Rhode Island *state* USA 23 G3

Rhodes see Ródos

Rhodope Mountains mountain range Bulgaria/Greece Gk. Orosirá Rodópis, Bul. Despoto Planina 86 C3

Rhône river France/Switzerland 62 C4

Ribeirão Preto Brazil 45 E1

Riberalta Bolivia 42 C3

Ribnița Moldova 90 D3

Richfield Utah, USA 24 B4

Richland Washington, USA 24 C2

Richmond Kentucky, USA 22 C5

Richmond New Zealand 133 C5

Richmond Virginia, USA 23 E5

Richmond Range mountain range New Zealand 133 C5

Ricobayo, Embalse de reservoir Spain 74 D2

Riga capital of Latvia Latv. Rīga 88 C3

Riga, Gulf of sea feature Baltic Sea 88 C3

Riihimäki Finland 67 D5

Rijeka Croatia It. Fiume 82 A3

Rimah, Wādī ar dry watercourse Saudi Arabia 103 B5

Rimini Italy 78 C3

Râmnicu Vilcea see Râmnicu Vâlcea

Riobamba Ecuador 40 A4

Rio Branco Brazil 42 C3

Rio Cuarto Argentina 46 C4

Rio de Janeiro Brazil 45 F2

Rio Gallegos Argentina 47 C7

Rio Grande Brazil 44 D4

Rio Grande river N America 16 B6

Rio Grande Rise undersea feature Atlantic Ocean 49 C6

Rio Verde Mexico 33 E4

Rishiri-tō island Japan 112 D1

Rivas Nicaragua 34 D3

Rivera Uruguay 44 C4

Riverside California, USA 27 C8

Riverton New Zealand 133 A7

Rivne Ukraine Rus. Rovno 90 C2

Riyadh capital of Saudi Arabia Ar. Ar Riyāḍ 103 C5

Rize Turkey 99 E2

Rkiz Mauritania 56 C3

Road Town capital of British Virgin Islands 37 F3

Roanne France 73 D5

Roanoke Virginia, USA 23 E5

Roanoke river SE USA 31 G1

Robinson Range mountain range Australia 129 B5

Rochester Minnesota, USA 25 F3

Rochester New York, USA 23 E3

Rockford Illinois, USA 22 B3

Rockhampton Australia 130 D4

Rock Island Illinois, USA 22 B3

Rock Springs Wyoming, USA 24 C3

Rockstone Guyana 41 G2

Rocky Mountains mountain range Canada/USA 18-19 D4

Rodez France 73 C6

Ródhos see Ródos

Ródos island Greece var. Ródhos, Eng. Rhodes 87 E6

Ródos Greece Eng. Rhodes 87 E6

Rodosto see Tekirdağ

Roeselare Belgium 69 A5

Roma Australia 131 D5

Roma see Rome

Romania country SE Europe 90

Rome capital of Italy It. Roma 78 C4

Rome Georgia, USA 30 D2

Rønne Denmark 67 B8

Ronne Ice Shelf ice feature Antarctica 136 B3

Roosendaal Netherlands 68 C4

Rosario Argentina 46 D4

Roseau capital of Dominica 37 G4

Rosenau see Rožňava

Rositten see Rēzekne

Ross Ice Shelf ice feature Antarctica 136 B4

Ross Sea Antarctica 136 B4

Rostak see Ar Rustāq

Rostock Germany 76 C2

Rostov-na-Donu Russian Federation 96 A3

Roswell New Mexico, USA 28 D2

Rotorua New Zealand 132 D3

Rotorua, Lake lake New Zealand 132 D3

Rotterdam Netherlands 68 C4

Rouen France 72 C3

Rovaniemi Finland 66 D3

Rovno see Rivne

Rovuma river Mozambique/Tanzania 61 F2

Roxas City Philippines 121 E2

Rožňava Slovakia Ger. Rosenau, Hung. Rozsnyó 81 D6

Rozsnyó see Rožňava

Ruatoria New Zealand 132 E3

Ruawai New Zealand 132 D2

Rudnyy Kazakhstan 96 C4

Rudolf, Lake see Lake Turkana

Rügen headland Germany 76 D2

Rukwa, Lake lake Tanzania 55 B7

Rumbek South Sudan 55 B5

Rundu Namibia 60 C3

Ruoqiang China 108 C3

Ruse Bulgaria 86 D1

Russian Federation country Europe/Asia 92-93 96-97

Rustavi Georgia 99 F2

Rutland Vermont, USA 23 F2

Rutog China 108 B4

Rwanda country C Africa 55

Ryazan' Russian Federation 93 B5 96 B3

Rybinskoye Vodokhranilishche Reservoir Russian Federation Eng. Rybinsk Reservoir 92 B3

Rybnik Poland 81 C5

Ryūkyū-rettō island group Japan 113 A8

Samui, Ko *island group* Thailand 119 C6
San *river* Poland 81 E5
Saña Peru 42 A3
Sana *capital of* Yemen *var.* Şan'ā' 103 B7
Sanandaj Sinneh. Iran 102 C3
San Andrés, Isla de *island* Colombia 35 E3
San Angelo Texas, USA 29 F3
San Antonio Chile 46 B4
San Antonio Texas, USA 29 F4
San Antonio *river* S USA 29 G4
San Antonio Oeste Argentina 47 C5
Sanāw Yemen 103 C6
San Bernardino California, USA 27 C7
San Carlos Uruguay 44 C5
San Carlos de Bariloche Argentina 47 B5
San Clemente Island *island* W USA 27 C8
San Cristóbal Venezuela 40 C2
San Diego California, USA 27 C8
Sandwich Island *see* Efate
San Fernando Trinidad & Tobago 37 G5
San Fernando Venezuela 40 D2
San Fernando de Noronha *island* Brazil 43 H2
San Francisco California, USA 27 B6
Sangir, Kepulauan *island group* Indonesia 121 F3
San Ignacio Belize 34 C1
San Joaquin Valley *valley* W USA 27 B6
San José *capital of* Costa Rica 34 D4
San Jose California, USA 27 B6
San José del Guaviare Colombia 40 C3
San Juan Argentina 46 B3
San Juan *river* Costa Rica/ Nicaragua 34 D4
San Juan *capital of* Puerto Rico 37 F3
San Juan Bautista Paraguay 44 B3

San Juan de los Morros Venezuela 40 D1
Sankt Martin *see* Martin
Sankt-Peterburg *see* St Petersburg
Sankt Pölten Austria 77 E6
Şanlıurfa Turkey *prev.* Urfa 98 E4
San Lorenzo Honduras 34 C3
San Luis Potosí Mexico 33 E3
San Marino *country* S Europe 78 C3
San Matías, Golfo *sea feature* Argentina 39 C6
San Miguel El Salvador 34 C3
San Miguel de Tucumán Argentina 46 C3
San Nicolas Island *island* W USA 27 B8
San Pedro Sula Honduras 34 C2
San Remo Italy 78 A3
San Salvador *capital of* El Salvador 34 C3
San Salvador de Jujuy Argentina 46 C2
San Sebastián *see* Donostia/ San Sebastián
Santa Ana El Salvador 34 B2
Santa Ana California, USA 27 C8
Santa Barbara California, USA 27 B7
Santa Catalina Island *island* W USA 27 C8
Santa Clara Cuba 36 B2
Santa Cruz Bolivia 42 D4
Santa Cruz California, USA 27 B6
Santa Cruz Islands *island group* Solomon Islands 126 C4
Santa Fe Argentina 46 D3
Santa Fe New Mexico, USA 28 D2
Santa Maria Brazil 44 C4
Santa Marta Colombia 40 C1
Santander Spain 75 E1
Santanilla, Islas *islands* Honduras 35 E1
Santarém Brazil 43 E2
Santarém Portugal 74 C3
Santaren Channel *Channel* The Bahamas 36 C2

Santa Rosa Argentina 47 C4
Santa Rosa California, USA 27 A6
Santa Rosa de Copán Honduras 34 C2
Santa Rosa Island *island* W USA 27 B8
Santiago *island* Cape Verde 56 A3
Santiago *capital of* Chile 46 B4
Santiago Dominican Republic 37 E3
Santiago Panama 35 F5
Santiago de Compostela Spain 74 C1
Santiago de Cuba Cuba 36 C3
Santiago del Estero Argentina 46 C3
Santo Antão *island* Cape Verde 56 A2
Santo Domingo *capital of* Dominican Republic 37 E3
Santo Domingo de los Colorados Ecuador 40 A4
Santoríni *island* Greece 87 D6
Santos Brazil 45 E2
São Borja Brazil 44 C3
São Francisco *river* Brazil 43 G3
São José do Rio Preto Brazil 44 D1
São Luís Brazil 43 G2
São Nicolau *island* Cape Verde 56 A2
Saône *river* France 72 D4
São Paulo Brazil 43 F5 45 E2
São Tomé *capital of* Sao Tome & Principe 59 A5
São Tomé *island* Sao Tome & Principe 59 A5
Sao Tome & Principe *country* W Africa 59
São Vincente *island* Cape Verde 56 A2
São Vicente, Cabo de *coastal feature* Portugal *Eng.* Cape St Vincent 74 B4
Sapele Nigeria 57 F5
Sapporo Japan 112 D2
Saragossa *see* Zaragoza

Sarajevo *capital of* Bosnia & Herzegovina 82 C4
Sarandë Albania 83 D6
Saransk Russian Federation 93 B5
Saratov Russian Federation 93 B6
Sarawak *state* Malaysia 120 D3
Sardegna *island* Italy *Eng.* Sardinia 79 A5
Sardinia *see* Sardegna
Sarema *see* Saaremaa
Sargasso Sea Atlantic Ocean 48 B4
Sargodha Pakistan 116 C2
Sarh Chad 58 C4
Sārī Iran 102 D3
Saruhan *see* Manisa
Sasebo Japan 113 A6
Saskatchewan *province* Canada 19 F5
Saskatchewan *river* Canada 19 F5
Saskatoon Canada 19 F5
Sassandra *River* Côte d'Ivoire 56 D5
Sassari Italy 79 A5
Satu Mare Romania 90 B3
Saudi Arabia *country* SW Asia 102-103
Sault Sainte Marie Canada 20 C4
Sault Sainte Marie Michigan, USA 22 C1
Saurimo Angola 60 C2
Sava *river* SE Europe 82 C3
Savannah Georgia, USA 31 F3
Savannah *river* SE USA 31 E2
Savissivik Greenland 64 C2
Savona Italy 78 A3
Savu Sea *sea* Indonesia 120 E5
Sawhāj Egypt *var.* Sohâg 54 B2
Şawqirah Oman 103 D6
Saýat Turkmenistan 104 D3
Sayhūt Yemen 103 D7
Saynshand Mongolia 109 E2
Say 'ūn Yemen 103 C6
Scandinavia *geophysical region* Europe 48 D2
Schaffhausen Switzerland 77 B6

Schaulen *see* Šiauliai
Schefferville Canada 21 E2
Scheldt *river* W Europe 69 B5
Schiermonnikoog *island* Netherlands 68 D1
Schneidemühl *see* Piła
Schwäbische Alb *mountains* Germany 77 B6
Schwarzwald *Forested mountain region* Germany *Eng.* Black Forest 77 B6
Schwerin Germany 76 C3
Scilly, Isles of *islands* UK 71 B7
Scotia Sea Atlantic Ocean 136 A1
Scotland *national region* UK 70
Scottsbluff Nebraska, USA 24 D3
Scottsdale Arizona, USA 28 B2
Scranton Pennsylvania, USA 23 F3
Scutari, Lake *lake* Albania/ Montenegro 83 C5
Seddon New Zealand 133 C5
Seattle Washington, USA 26 B2
Ségou Mali 56 D3
Segovia Spain 75 E2
Segura *river* Spain 75 E4
Seikan Tunnel *tunnel* Japan 112 D3
Seinäjoki Finland 67 D5
Seine *river* France 72 C3
Sejong City *administrative capital of* South Korea 110 E4
Selfoss Iceland 65 E5
Semara *see* Smara
Semarang Indonesia 120 D4
Semey Kazakhstan *prev.* Semipalatinsk 96 D4
Semipalatinsk *see* Semey
Sendai Japan 112 D4
Senegal *country* W Africa 56
Senegal *river* Africa 56 C3
Sēn, Stœng *river* Cambodia 119 D5
Seoul *capital of* South Korea *Kor.* Sŏul 110 E4
Sept-Îles Canada 21 F3
Seraing Belgium 69 D6

Seram, Pulau *island* Indonesia 121 F4
Serbia *country* SE Europe 82 D3
Serdar Turkmenistan *prev.* Gyzylarbat, *prev.* Kizyl-Arvat 104 B2
Serhetabat Turkmenistan *prev.* Gushgy, Kushka 104 C4
Serov Russian Federation 96 C3
Serpent's Mouth, The *sea feature* Trinidad & Tobago/ Venezuela *Sp.* Boca de la Serpiente 41 F1
Serra do Mar *mountains* Brazil 44 D3
Sérres Greece 86 C3
Setesdal *valley* Norway 67 A6
Sétif Algeria 53 E1
Setúbal Portugal 74 C4
Seul, Lake *lake* Canada 20 A3
Sevana Lich *lake* Armenia 99 G2
Sevastopol' Ukraine 91 F5
Severn *river* Canada 20 B3
Severn *river* England/Wales, UK 71 D6
Severnaya Dvina *river* Russian Federation *Eng.* Northern Dvina 92 C3
Severnaya Zemlya *island group* Russian Federation 137 H3
Sevilla Spain *Eng.* Seville 74 D4
Seville *see* Sevilla
Seychelles *country* Indian Ocean 61 122 B4
Seyðisfjörður Iceland 65 E4
Seýdi Turkmenistan *prev.* Neftezavodsk 104 D2
Seyhan *see* Adana
Sfax Tunisia 53 F2
's-Gravenhage *capital of* Netherlands *Eng.* The Hague 68 B3
Shaan *see* Shaanxi
Shaanxi *province* China *var.* Shaan, Shan-hsi, Shaanxi Sheng, Shenshi, Shensi 111 C5

Sunyani Ghana 57 E4

Superior Wisconsin, USA 22 A1

Superior, Lake *lake* Canada/ USA 16 C5

Suquţrá *island* Yemen *var.* Socotra 103 D7 122 B3

Şūr Oman 103 E5

Surabaya Indonesia 120 D5

Surakarta Indonesia 120 D5

Sūrat India 116 C5

Surat Thani Thailand 119 C6

Sûre *river* W Europe 69 D7

Surfers Paradise Australia 131 E5

Surinam *see* Suriname

Suriname *country* NE South America *var.* Surinam 41

Surkhob *river* Tajikistan 105 E3

Surt Libya *var.* Sidra 53 G2

Surt, Khalij *sea feature* Mediterranean Sea *Eng.* Gulf of Sirte, Gulf of Sidra 85 E4

Surtsey Island S Iceland 65 E5

Susanville California, USA 27 B5

Suways, Qanât as *see* Suez Canal

Suva *capital* of Fiji 127 E4

Svalbard *external territory* Norway, Arctic Ocean 65 G2

Svay Riêng Cambodia 119 D6

Sverdlovsk *see* Yekaterinburg

Svetlogorsk *see* Svyatlahorsk/ Svetlogorsk

Svyataya Anna Trough *undersea feature* Kara Sea 137 H4

Svyetlahorsk/Svetlogorsk Belarus *Rus.* Svetlogorsk 89 D6

Swakopmund Namibia 60 B3

Swansea Wales, UK 71 C6

Swaziland *country* southern Africa 61

Sweden *country* N Europe 66-67

Sweetwater Texas, USA 29 F3

Swindon England, UK 71 D6

Switzerland *country* C Europe 77

Sydney Australia 131 D6

Sydney Canada 21 G4

Syeverodonets'k Ukraine 91 G1

Syktyvkar Russian Federation 92 D4 96 C3

Sylhet Bangladesh 117 G4

Syracuse *see* Siracusa

Syracuse New York, USA 23 E3

Syr Darya *river* C Asia 104 D1

Syria *country* SW Asia 100-101

Syrian Desert *desert* SW Asia *Ar.* Bādiyat ash Shām 101 C5

Szczecin Poland *Ger.* Stettin 80 B3

Szczeciński, Zalew *bay* Germany/Poland 80 A2

Szechwan *see* Sichuan

Szeged Hungary *Ger.* Szegedin 81 D7

Szegedin *see* Szeged

Székesfehérvár Hungary *Ger.* Stuhlweissenburg 81 C6

Szekszárd Hungary 81 C7

Szolnok Hungary 81 D6

Szombathely Hungary *Ger.* Steinamanger 81 B6

T

Tabariya, Bahrat *see* Tiberius, Lake

Tábor Czech Republic 81 B5

Tabora Tanzania 55 B7

Tabrīz Iran 102 C2

Tabuaeran *island* Kiribati 127 G2

Tabūk Saudi Arabia 102 A4

Tacloban Philippines 120 F2

Tacna Peru 42 C4

Tacoma Washington, USA 26 B2

Tacuarembó Uruguay 44 C4

Tadmur *see* Tudmur

Taegu *see* Daegu

Taejŏn *see* Daejeon

Tafassâsset, Ténéré du *desert* Niger 57 G2

Taguatinga Brazil 43 F3

Tagus *river* Portugal/Spain *Port.* Tejo, *Sp.* Tajo 74 C3

Tahiti *island* French Polynesia 127 H5

Tahoe, Lake *lake* W USA 27 B5

Tahoua Niger 57 F3

Taibei *capital* of Taiwan *var.* Taipei111 D6

T'aichung *see* Taizhong

Taieri *129* New Zealand 133 B7

Taihape New Zealand 132 D4

T'ainan *see* Tainan

Tainan Taiwan *prev.* T'ainan111 D6

Taipei *see* Taibei

Taiping Malaysia 120 B3

Taiwan *country* E Asia *prev.* Formosa 111

Taiwan Strait *sea feature* East China Sea/South China Sea *var.* Formosa Strait 111 D7

Taiyuan China 110 C4

Taizhong Taiwan *prev.* T'aichung 111D6

Ta'izz Yemen 103 B7

Tajikistan *country* C Asia 105

Tajo *see* Tagus

Takapuna New Zealand 132 D2

Takla Makan *see* Taklimakan Shamo

Taklimakan Shamo *desert region* China *var.* Takla Makan 108 B3

Talamanca, Cordillera de *mountains* Costa Rica 35 E4

Talas Kyrgyzstan 105 F2

Talaud, Kepulauan *island group* Indonesia 121 F3

Talca Chile 46 B4

Talcahuano Chile 46 B4

Taldykoigan Kazakhstan 96 C5

Tallahassee Florida, USA 30 D3

Tallinn *capital* of Estonia *prev.* Revel, *Ger.* Reval, *Rus.* Tallin 88 D2

Talsen *see* Talsi

Talsi Latvia *Ger.* Talsen 88 B3

Tamale Ghana 57 E4

Tamanrasset Algeria 53 E4

Tambo Australia 130 C4

Tambov Russian Federation 93 B5

Tamil Nādu *state* India 114 C2

Tampa Florida, USA 31 E4

Tampere Finland 67 D5

Ternopil' Ukraine *Rus.* Ternopol' 90 C2

Ternopol' *see* Ternopil'

Terrassa Spain 75 G2

Terre Haute Indiana, USA 22 B4

Terres Australes et Antarctiques Françaises *see* French Southern and Antarctic Lands

Terschelling *island* Netherlands 68 C1

Teruel Spain 75 F3

Teseney Eritrea 54 C4

Tessalit Mali 57 E2

Tete Mozambique 61 E3

Tétouan Morocco 52 C1

Tetovo Macedonia 83 D5

Tetschen *see* Děčín

Tevere *river* Italy 78 C4

Texas *state* USA 28-29 F3

Texarkana Arkansas, USA 30 A2

Texas City Texas, USA 29 G4

Texel *island* Netherlands 68 C2

Thailand *country* SE Asia 118-119

Thailand, Gulf of *sea feature* South China Sea 119 C6

Thames *river* England, UK 71 D6

Thar Desert *desert* India/Pakistan 116 C3

Tharthār, Buḩayrat ath *lake* Iraq 102 B3

Thásos *island* Greece 86 C3

Thaton Myanmar 118 B4

Theiss *see* Tisza

Thermaic Gulf *see* Thermaïkós Kólpos

Thermaïkós Kólpos *sea feature* Greece *Eng.* Thermaic Gulf 86 B4

Thessaloníki Greece *var.* Salonica 86 B3

The Valley *dependent territory capital* Anguilla 37 G5

Thimphu *capital* of Bhutan 117 G3

Thionville France 72 E3

Thiruvananthapuram India *see* Trivandrum 114 D3

Thompson Canada 19 F4

Thorn *see* Toruń

Thorshavn *see* Tórshavn

Thracian Sea Greece *Gk.* Thrakikó Pélagos 86 D3

Thrakikó Pélagos *see* Thracian Sea

Three Kings Islands *island group* New Zealand 132 C1

Thule *see* Qaanaaq

Thunder Bay Canada 20 B4

Thuner See *lake* Switzerland 77 B7

Thurso Scotland, UK 70 C2

Tianjin China *var.* Tientsin 110 D4

Tiberias, Lake *lake* Israel *var.* Sea of Galilee, *Heb.* Yam Kinneret, *Ar.* Bahrat Tabariya 101 B5

Tibesti *mountains* Chad/Libya 50 C3

Tibet *autonomous region* China *Chin.* Xizang 108 C5

Tibet, Plateau of *see* Qingzang Gaoyuan

Tienen Belgium 69 C6

Tien Shan *mountain range* C Asia 105 G2

Tientsin *see* Tianjin

Tierra del Fuego *island* Argentina/Chile 47 C8

Tiflis *see* Tbilisi

Tighina Moldova *prev.* Bendery 90 D4

Tigris *river* SW Asia 94 B4

Tijuana Mexico 32 A1

Tiki Basin *undersea feature* Pacific Ocean 135 E3

Tiksi Russian Federation 97 F2

Tilburg Netherlands 68 C4

Timaru New Zealand 133 B6

Timişoara Romania 90 A4

Timmins Canada 20 C4

Timor *island* Indonesia 121 F5

Timor Sea Indian Ocean 121 F5

Tindouf Algeria 52 B3

Tinos Greece 87 D5

Tirana *capital* of Albania 83 D6

Tiraspol Moldova 90 D4

Tîrgovişte *see* Târgovişte

Tîrgu Mureş *see* Târgu Mureş

Tirol *region* Austria *var.* Tyrol 77 C7

Tiruchchirāppalli India 114 D3

Tisa *see* Tisza

Tisza *river* E Europe *Ger.* Theiss, *Cz./Rom./SCr.* Tisa 81 D6

Titicaca, Lake *lake* Bolivia/Peru 42 C4

Tlemcen Algeria 52 D2

Toamasina Madagascar 61 G3

Toba, Danau *lake* Indonesia 120 B3

Tobago *island* Trinidad and Tobago 37 G5

Toba Kākar Range *mountains* Pakistan 116 B2

Tobruk *see* Ṭubruq

Tocantins *river* Brazil 43 F3

Tocopilla Chile 46 B2

Togo *country* W Africa 57 E4

Tokat Turkey 98 D3

Tokelau *dependent territory* New Zealand, Pacific Ocean 127 F3

Tokmak Kyrgyzstan 105 F2

Tokuno-shima *island* Japan 113 A8

Tokushima Japan 113 B5

Tokyo *capital* of Japan 113 D5

Toledo Spain 75 E3

Toledo Ohio, USA 22 C3

Toledo Bend Reservoir *Reservoir* S USA 29 H3

Toliara Madagascar 61 E3

Tol'yatti *prev.* Stavropol' Russian Federation 93 C5

Tomakomai Japan 112 D2

Tombouctou Mali 57 E3

Tombua Angola 60 B2

Tomini, Gul of *sea feature* Indonesia 121 E4

Tomsk Russian Federation 96 D4

Tonga *country* Pacific Ocean 127 E5

Tongatapu *island* Tonga 125 E3

Tongking, Gulf of *see* Tonkin, Gulf of

Tongliao China 109 G2

Tongtian He *river* China 108 C4

Tonkin, Gulf of *sea feature* South China Sea *var.* Gulf of Tongking 111 B7

Tônle Kông *river* Cambodia/ Vietnam 118 E5

Tônlé Sap *lake* Cambodia 119 D5

Tonopah Nevada, USA 27 C6

Toowoomba Australia 131 D5

Topeka Kansas, USA 25 F4

Top Springs Australia 130 A3

Torino Italy *Eng.* Turin 78 A2

Tornio Finland 66 D4

Tornionjoki *river* Finland/ Sweden 66 D3

Toronto Canada 20 D5

Toros Dağları *mountain range* Turkey *Eng.* Taurus Mountains 98 C4

Torre del Greco Italy 79 D5

Torrens, Lake *lake* Australia 131 B5

Torreón Mexico 32 D2

Torres Strait *sea feature* Arafura Sea/Coral Sea 126 B4

Torrington Wyoming, USA 24 D3

Tórshavn *capital of* Faroe Islands *Dan.* Thorshavn 65 F5

To'rtko'l Uzbekistan *prev.* Petroaleksandrovsk, *prev.* Turtkul', *Uzb.* Türtkül 104 C2

Tortoise Islands *see* Galapagos Islands

Tortosa Spain 75 F2

Toruń Poland *Ger.* Thorn 80 C3

Toscana *region* Italy *Eng.* Tuscany 78 B3

Toscano, Archipelago *island group* Italy 78 B4

Toshkent *see* Tashkent

Tottori Japan 113 B5

Touggourt Algeria 53 E2

Toulon France 73 D6

Toulouse France 73 B6

Toungoo Myanmar 118 B4

Tournai Belgium 69 B6

Tours France 72 C4

Townsville Australia 130 D3

Toyama Japan 112 C4

Tozeur Tunisia 53 E2

Trâblous *see* Tripoli, Lebanon

Trabzon Turkey *Eng.* Trebizond 99 E2

Tralee Ireland 71 A6

Trang Thailand 119 C7

Transantarctic Mountains *mountain range* Antarctica 136 B3

Transnistria *region* Moldova 90 D3

Transylvania *region* Romania 90 B3

Transylvanian Alps *see* Carpaţii Meridionali

Trapani Italy 79 C6

Traralgon Australia 131 C7

Trasimeno, Lago *Lake* Italy 78 C4

Traverse City Michigan, USA 22 C2

Travis, Lake *lake* Texas, USA 29 F4

Trebinje Bosnia & Herzegovina 83 C5

Trebizond *see* Trabzon

Trelew Argentina 47 C6

Trenčín Slovakia *Ger.* Trentschin *Hung.* Trencsén 81 C6

Trencsén *see* Trenčín

Trento Italy *Ger.* Trient 78 C2

Trenton New Jersey, USA 23 F4

Trentschin *see* Trenčín

Tres Arroyos Argentina 47 D5

Treviso Italy 78 C2

Trient *see* Trento

Trieste Italy 78 C2

Trikala Greece 86 B4

Trincomalee Sri Lanka 115 E3

Trindade *external territory* Brazil, Atlantic Ocean 49 C6

Trinidad Bolivia 42 C4

Trinidad Uruguay 44 B5

Trinidad *island* Trinidad & Tobago 38 C2

Trinidad & Tobago *country* West Indies 37 G5

Trípoli Greece 87 B5

Tripoli Lebanon *var.* Trâblous, Ṭarābulus 100 B4

Tripoli *capital of* Libya *Ar.* Ṭarābulus al-Gharb 53 F2

Tristan da Cunha *overseas territory* UK, Atlantic Ocean 49 D6

Trivandrum India *see* Thiruvananthapuram 114 D3

Trnava Slovakia *Ger.* Tyrnau, *Hung.* Nagyszombat 81 C6

Trois-Rivières Canada 21 E4

Trollhättan Sweden 67 B6

Tromsø Norway 66 C2

Trondheim Norway 66 B4

Trondheimsfjorden *inlet* Norway 66 B4

Troyes France 72 D4

Trujillo Honduras 34 D2

Trujillo Peru 42 A3

Tsarigrad *see* İstanbul

Tschenstochau *see* Częstochowa

Tselinograd *see* Astana

Tsetserleg Mongolia 108 D2

Tshikapa Dem. Rep. Congo 59 C7

Tsinghai *see* Qinghai

Tsumeb Namibia 60 C3

Tsushima *island* Japan 113 A5

Tuamotu Fracture Zone *tectonic feature* Pacific Ocean 125 H3

Tuamotu Islands *island group* French Polynesia 125 G3

Tubmanburg Liberia 56 C4

Ṭubruq Libya *Eng.* Tobruk 53 H2

Tucson Arizona, USA 28 B3

Tucupita Venezuela 41 F1

Tucurui, Represa de *Reservoir* Brazil 43 F2

Tudmur Syria *var.* Tadmur, *Eng.* Palmyra 100 C3

Tuguegarao Philippines 121 E1

Verkhoyansk Range *see* Verkhoyanskiy Khrebet
Vermont *state* USA 23 F2
Vernon Texas, USA 29 F2
Véroia Greece 86 B3
Verona Italy 78 C2
Versailles France 72 C3
Verviers Belgium 69 D6
Vesoul France 72 E4
Veszprém Hungary *Ger.* Veszprim 81 C7
Veszprim *see* Veszprém
Viana do Castelo Portugal 74 C2
Viareggio Italy 78 B3
Vicenza Italy 78 C2
Vichy France 73 C5
Victoria *state* Australia 131 C7
Victoria Canada 18 D5
Victoria *capital of* Seychelles 61 H1
Victoria Texas, USA 29 G4
Victoria *river* Australia 128 D3
Victoria, Lake *lake* E Africa *var.* Victoria Nyanza 55 B6
Victoria Falls *waterfall* Zambia/ Zimbabwe 51 C6
Victoria Island *island* Canada 19 F2
Victoria Land *region* Antarctica 137 C4
Victoria Nyanza *see* Victoria, Lake
Vidin Bulgaria 86 B1
Viedma Argentina 47 C5
Vienna *capital of* Austria *Ger.* Wien 77 E6
Vientiane *capital of* Laos 118 C4
Vietnam *country* SE Asia 118-119
Vigo Spain 74 C2
Vijayawāda India 115 E1
Vila Nova de Gaia Portugal 74 C2
Vila Real Portugal 74 C2
Viliya *see* Neris
Viljandi Estonia *Ger.* Fellin 88 D2
Villach Austria 77 D7
Villahermosa Mexico 33 G4

Villa Mercedes Argentina 46 C4
Villarrica *peak* Chile 39 B6
Villavicencio Colombia 40 C3
Villeurbanne France 73 D5
Vilna *see* Vilnius
Vilnius *capital of* Lithuania *Pol.* Wilno, *Ger.* Wilna, *Rus.* Vilna 89 C5
Viña del Mar Chile 46 B4
Vinh Vietnam 118 D4
Vinnitsa *see* Vinnytsya
Vinnytsya Ukraine *Rus.* Vinnitsa 90 D2
Virgin Islands *unincorporated territory* USA, West Indies 37 F3
Virginia Minnesota, USA 25 F2
Virginia USA 22-23
Virovitica Croatia 82 C3
Virtsu Estonia *Ger.* Werder 88 C2
Visākhapatnam India 117 E5
Visalia California, USA 27 C7
Visby Sweden 67 C7
Viscount Melville Sound *sea feature* Arctic Ocean 19 F2
Viseu Portugal 74 C3
Vistula *see* Wisła
Vitebsk *see* Vitsyebsk/Vitebsk
Viterbo Italy 78 C4
Viti Levu *island* Fiji 127 E4
Vitim *river* Russian Federation 95 E3
Vitória Brazil 43 G5 45 G1
Vitória da Conquista Brazil 43 G4
Vitoria-Gasteiz Spain 75 E1
Vitsyebsk/Vitebsk Belarus *Rus.* Vitebsk 88 E5
Vjosës, Lumi i *river* Albania 83 D6
Vladikavkaz Russian Federation *prev.* Ordzhonikidze, Dzaudzhikau 93 B7
Vladimir Russian Federation 93 B5
Vladimirovka *see* Yuzhno-Sakhalinsk
Vladivostok Russian Federation 97 G5
Vlieland *island* Netherlands 68 C1

Vlissingen Netherlands *Eng.* Flushing 69 B5
Vlorë Albania 83 D6
Vojvodina *region* Serbia 82 D3
Volga *river* Russian Federation 96 A3
Volgograd Russian Federation *prev.* Stalingrad 93 B6, 96 A3
Volkovysk *see* Vawkavysk
Vologda Russian Federation 96 B2
Vólos Greece 86 B4
Volta *river* Ghana 57 E4
Volta, Lake *lake* Ghana 57 E4
Volta Redonda Brazil 45 E2
Vóreies Sporádes *island group* Greece *Eng.* Northern Sporades 86 C4
Vorkuta Russian Federation 92 E3 96 C2
Vormsi *island* Estonia *Ger.* Worms, *Swed.* Ormsö 88 C2
Voronezh Russian Federation 93 B5
Võru Estonia *Ger.* Werro 88 D3
Vosges *mountain range* France 72 E4
Vostochno-Sibirskoye More Arctic Ocean *Eng.* East Siberian Sea 137 G2
Vostok Island *island* Kiribati 127 H4
Vrangel'ya, Ostrov *island* Russian Federation *Eng.* Wrangel Island 97 G1
Vratsa Bulgaria 86 C2
Vršac Serbia 82 D3
Vukovar Croatia 82 C3
Vulcano, Isola *island* Italy 79 D6
Vyatka *river* Russian Federation 93 C5

W

Wa Ghana 57 E4
Waag *see* Váh
Waal *river* Netherlands 68 D4
Wabash *river* C USA 22 B4

Waco — White Sea

Waco Texas, USA 29 G3
Waddeneilanden *island group* Netherlands *Eng.* West Frisian Islands 68 C1
Waddenzee *sea feature* Netherlands 68 D1
Wadi Halfa Sudan 54 B3
Wādī Mūsā Jordan *var.* Petra 101 B6
Wad Medani Sudan 54 B4
Wagga Wagga Australia 131 C6
Wagin Australia 129 B6
Wahai Indonesia 121 F4
Wahībah, Ramlat Āl *Desert* Oman 103 E5
Waiau *river* New Zealand 133 A7
Waipara New Zealand 132 E4
Wairau *river* New Zealand 133 C5
Wairoa New Zealand 132 E3
Waitaki *river* New Zealand 133 B6
Waiuku New Zealand 132 D3
Wakatipu, Lake *lake* New Zealand 133 D7
Wakayama Japan 113 C5
Wake Island *atoll* Pacific Ocean 124 D1
Wake Island *US unincorporated territory* Pacific Ocean 134 C2
Wakkanai Japan 112 D1
Wałbrzych Poland *Ger.* Waldenburg 80 B4
Waldenburg *see* Wałbrzych
Wales *national region* UK *Wel.* Cymru 71
Walgett Australia 131 D5
Walk *see* Valga
Walla Walla Washington, USA 26 C2
Wallis & Futuna *overseas collectivity* France, Pacific Ocean 127 E4
Walnut Ridge Arkansas, USA 30 B1
Walvis Bay Namibia 60 B4
Walvis Ridge *undersea feature* Atlantic Ocean 49 D6
Wan *see* Anhui

Wanaka New Zealand 133 B6
Wanaka, Lake *lake* New Zealand 133 B6
Wandel Sea Arctic Ocean 137 G4
Wanganui New Zealand 132 D4
Wanlaweyn Somalia 55 D6
Warangal India 117 E5
Warkworth New Zealand D2
Warrnambool Australia 131 C7
Warsaw *capital of* Poland *Pol.* Warszawa, *Ger.* Warschau 80 D3
Warschau *see* Warsaw
Warszawa *see* Warsaw
Warta *river* Poland *Ger.* Warthe 80 C4
Warthe *see* Warta
Wash, The *inlet* England, UK 71 E5
Washington *state* USA 26
Washington, D.C. *capital of* USA 23 E4
Waterford Ireland 71 B6
Watertown New York, USA 23 E2
Watertown South Dakota, USA 25 E2
Wau South Sudan 55 B5
Waukegan Illinois, USA 22 B3
Wawa Canada 20 C4
Weddell Plain *undersea feature* Atlantic Ocean 136 B2
Weddell Sea Antarctica 136 A2
Weichsel *see* Wisła
Weissenstein *see* Paide
Wellesley Islands *island group* Australia 130 B3
Wellington *capital of* New Zealand 133 D5
Wellington, Isla *island* Chile 47 B7
Wells, Lake *lake* Australia 129 C5
Wenden *see* Cēsis
Wenzhou China 111 D6
Werder *see* Virtsu
Werro *see* Võru
Wesenberg *see* Rakvere

Weser *river* Germany 76 B3
Wessel Islands *island group* Australia 130 B2
West Antarctica *region* Antarctica 134 B3
West Bank *disputed territory* SW Asia 101 A5
West Bengal *state* India 117 F4
Western Australia *state* Australia 128-129
Western Dvina *river* E Europe *Bel.* Dzvina, *Ger.* Düna, *Latv.* Daugava, *Rus.* Zapadnaya Dvina 88 C4
Western Ghats *mountain range* India 106 B3, 114 C1
Western Isles *see* Outer Hebrides
Western Sahara *region occupied by* Morocco N Africa 52 A3
Western Sierra Madre *see* Sierra Madre Occidental
Westerschelde *inlet* Netherlands 69 B5
West Falkland *island* Falkland Islands 47 D7
West Frisian Islands *see* Waddeneilanden
West Indies *island group* North America 48 A4
West Palm Beach Florida, USA 31 F4
Westport New Zealand 133 C5
West Siberian Plain *see* Zapadno-Sibirskaya Ravnina
West Virginia *state* USA 22-23
Wetar Strait *sea feature* Indonesia 121 F5
Wexford Ireland 71 B6
Whakatane New Zealand 132 E3
Whangarei New Zealand 132 D2
Wharton Basin *undersea feature* Indian Ocean 123 D5
Wheeling Ohio, USA 22 D4
Whitehorse Canada 18 D4
White Nile *river* Sudan / South Sudan 55 B5
White Sea *see* Beloye More

Yazd — Zwolle